Spintronics

Wiley Series in Materials for Electronic & Optoelectronic Applications

www.wiley.com/go/meoa

Published Titles

Bulk Crystal Growth of Electronic, Optical and Optoelectronic Materials, Edited by P. Capper
Properties of Group-IV, III—V and II—VI Semiconductors, S. Adachi
Charge Transport in Disordered Solids with Applications in Electronics, Edited by S. Baranovski
Optical Properties of Condensed Matter and Applications, Edited by J. Singh
Thin Film Solar Cells: Fabrication, Characterization, and Applications, Edited by J. Poortmans and V. Arkhipov
Dielectric Films for Advanced Microelectronics, Edited by M. R. Baklanov, M. Green, and K. Maex
Liquid Phase Epitaxy of Electronic, Optical and Optoelectronic Materials, Edited by P. Capper and M. Mauk
Molecular Electronics: From Principles to Practice, M. Petty
Luminescent Materials and Applications, A. Kitai
CVD Diamond for Electronic Devices and Sensors, Edited by R. S. Sussmann
Properties of Semiconductor Alloys: Group-IV, III—V and II—VI Semiconductors, S. Adachi
Mercury Cadmium Telluride, Edited by P. Capper and J. Garland
Zinc Oxide Materials for Electronic and Optoelectronic Device Applications, Edited by C. Litton, D. C. Reynolds, and T. C. Collins
Lead-Free Solders: Materials Reliability for Electronics, Edited by K. N. Subramunian
Silicon Photonics: Fundamentals and Devices, M. Jamal Deen and P. K. Basu
Nanostructured and Subwavelength Waveguides: Fundamentals and Applications, M. Skorobogatiy
Photovoltaic Materials: From Crystalline Silicon to Third-Generation Approaches, Edited by G. Conibeer and A. Willoughby
Glancing Angle Deposition of Thin Films: Engineering the Nanoscale, M.M. Hawkeye, M.T. Taschuk and M.J. Brett
Physical Properties of High-Temperature Superconductors, R. Wesche
Spintronics for Next Generation Innovative Devices, Edited by K. Sato and E. Saitoh
Inorganic Glasses for Photonics: Fundamentals, Engineering and Applications, A. Jha
Amorphous Semiconductors: Structural, Optical and Electronic Properties, K. Morigaki, S. Kugler and K. Shimakawa
Microwave Materials and Applications, Two-volume set, Edited by M. T. Sebastian, R. Ubic and H. Jantunen
Molecular Beam Epitaxy: Materials and Applications for Electronics and Optoelectronics, Edited by H. Asahi and Y. Korikoshi
Metalorganic Vapor Phase Epitaxy (MOVPE): Growth, Materials Properties, and Applications, Edited by S. Irvine and P. Capper
Optical Properties of Materials and Their Applications, Second Edition, Edited by J. Singh
Oxide Electronics, Edited by A. Ray
Organic Semiconductors for Optoelectronics, Edited by H. Naito
Phosphors for Radiation Detection, Edited by T. Yanagida and M. Koshimizu
Spintronics: Materials, Devices and Applications, Edited by Kaiyou Wang, Meiyin Yang and Jun Luo.

Spintronics

Materials, Devices and Applications

Edited by

Kaiyou Wang
State Key Laboratory of Superlattices and Microstructures
Institute of Semiconductors
Chinese Academy of Sciences
Beijing, China

Meiyin Yang and Jun Luo
Key Laboratory of Microelectronic Devices & Integrated Technology
Institute of Microelectronics
Chinese Academy of Sciences
Beijing, China

WILEY

This edition first published 2022

Registered Offices
John Wiley & Sons, Inc., 111 River Street, Hoboken, NJ 07030, USA
John Wiley & Sons Ltd, The Atrium, Southern Gate, Chichester, West Sussex, PO19 8SQ, UK

Editorial Office
The Atrium, Southern Gate, Chichester, West Sussex, PO19 8SQ, UK

For details of our global editorial offices, customer services, and more information about Wiley products visit us at www.wiley.com.

Wiley also publishes its books in a variety of electronic formats and by print-on-demand. Some content that appears in standard print versions of this book may not be available in other formats.

Library of Congress Cataloging-in-Publication Data
Names: Wang, Kaiyou (Physicist), editor. | Yang, Meiyin, editor. | Luo, Jun (Professor of microelectronics), editor. | John Wiley & Sons, publisher.
Title: Spintronics: materials, devices and applications / edited by Kaiyou Wang, Meiyin Yang and Jun Luo.
Other titles: Wiley series in materials for electronic & optoelectronic applications.
Description: Hoboken, NJ: John Wiley & Sons, 2022. | Series: Wiley series in materials for electronic & optoelectronic applications | Includes bibliographical references and index.
Identifiers: LCCN 2022011186 (print) | LCCN 2022011187 (ebook) | ISBN 9781119698975 (hardback) | ISBN 9781119698944 (pdf) | ISBN 9781119698951 (epub) | ISBN 9781119698968 (obook)
Subjects: LCSH: Spintronics.
Classification: LCC TK7874.887 .S665 2022 (print) | LCC TK7874.887 (ebook) | DDC 621.3--dc23/eng/20220511
LC record available at https://lccn.loc.gov/2022011186
LC ebook record available at https://lccn.loc.gov/2022011187

Cover image: © bubaone/Getty Images
Cover design by Wiley

Set in 10/12pt TimesLTStd-Roman by Integra Software Services Pvt. Ltd, Pondicherry, India

Contents

List of Contributors

Yan Cui Key Laboratory of Microelectronic Devices and Integrated Technology, Institute of Microelectronics, Chinese Academy of Sciences, Beijing, China

Haifeng Du The Anhui Key Laboratory of Condensed Matter Physics at Extreme Conditions, High Magnetic Field Laboratory and University of Science and Technology of China, Chinese Academy of Science, Hefei, Anhui Province, China

Kevin William Edmonds School of Physics and Astronomy, University of Nottingham, Nottingham, United Kingdom

Xiaobin Guo School of Physics&Optoelectric Engineering, Guangdong University of Technology, Guangzhou Higher Education Mega Centre, Guangzhou, China

Qunwen Leng School of Integrated Circuit Science and Engineering, Beihang University, Beijing, China
Beihang-Goertek Joint Microelectronics Institute, Qingdao Research Institute, Beihang University, Qingdao, China

Yucai Li State Key Laboratory of Superlattices and Microstructures, Institute of Semiconductors, Chinese Academy of Sciences, Beijing, China
College of Materials Science & Opto-Electronic Technology and Center for Excellence in Topological Quantum Computation, University of Chinese Academy of Science, Beijing, China

Xionghua Liu State Key Laboratory of Superlattices and Microstructures, Institute of Semiconductors, Chinese Academy of Sciences, Beijing, China
Center of Materials Science and Optoelectronics Engineering, University of Chinese Academy of Sciences, Beijing, China

Jun Luo Key Laboratory of Microelectronic Devices and Integrated Technology, Institute of Microelectronics, Chinese Academy of Sciences, Beijing, China

Kejing Ran School of Physical Science and Technology, ShanghaiTech University, Shanghai, China
ShanghaiTech Laboratory for Topological Physics, ShanghaiTech University, Shanghai, China

Renata Saha Department of Electrical and Computer Engineering, University of Minnesota, Minneapolis, MN, United States

Dongsheng Song Institutes of Physical Science and Information Technology, Anhui University, Hefei, China

Diqing Su Department of Chemical Engineering and Material Science, University of Minnesota, Minneapolis, MN, United States

Huayao Tu Suzhou Institute of Nano-tech and Nano-bionics, Chinese Academy of Sciences, Suzhou, China

Jian-Ping Wang Department of Electrical and Computer Engineering, University of Minnesota, Minneapolis, MN, United States
Department of Chemical Engineering and Materials Science, University of Minnesota, Minneapolis, MN, United States

Hailong Wang State Key Laboratory of Superlattices and Microstructures, Institute of Semiconductors, Chinese Academy of Sciences, Beijing, China
Center of Materials Science and Optoelectronics Engineering, University of Chinese Academy of Sciences, Beijing, China

Kaiyou Wang State Key Laboratory of Superlattices and Microstructures, Institute of Semiconductors, Chinese Academy of Sciences, Beijing, China
College of Materials Science & Opto-Electronic Technology and Center for Excellence in Topological Quantum Computation, University of Chinese Academy of Science, Beijing, China
Beijing Academy of Quantum Information Sciences, Beijing, China

Weiwei Wang Institutes of Physical Science and Information Technology, Anhui University, Hefei, China

Kai Wu Department of Electrical and Computer Engineering, University of Minnesota, Minneapolis, MN, United States

Li Xi Key Laboratory for Magnetism and Magnetic Materials of Ministry of Education, Lanzhou University, Lanzhou, China

Meiyin Yang Key Laboratory of Microelectronic Devices and Integrated Technology, Institute of Microelectronics, Chinese Academy of Sciences, Beijing, China

Tengzhi Yang Key Laboratory of Microelectronic Devices and Integrated Technology, Institute of Microelectronics, Chinese Academy of Sciences, Beijing, China

Zhongming Zeng Suzhou Institute of Nano-tech and Nano-bionics, Chinese Academy of Sciences, Suzhou, China

Kun Zhang School of Integrated Circuit Science and Engineering, Beihang University, Beijing, China

Shilei Zhang School of Physical Science and Technology, ShanghaiTech University, Shanghai, China
ShanghaiTech Laboratory for Topological Physics, ShanghaiTech University, Shanghai, China

Jianhua Zhao State Key Laboratory of Superlattices and Microstructures, Institute of Semiconductors, Chinese Academy of Sciences, Beijing, China
Center of Materials Science and Optoelectronics Engineering, University of Chinese Academy of Sciences, Beijing, China

Zitong Zhou School of Integrated Circuit Science and Engineering, Beihang University, Beijing, China

Series Preface

Wiley Series in Materials for Electronic & Optoelectronic Applications

This book series is devoted to the rapidly developing class of materials used for electronic and optoelectronic applications. It is designed to provide much-needed information on the fundamental scientific principles of these materials, together with how they are employed in technological applications. The books are aimed at (postgraduate) students, researchers, and technologists, engaged in research, development, and the study of materials in electronics and photonics, and industrial scientists developing new materials, devices, and circuits for the electronic, optoelectronic, and communications industries.

The development of new electronic and optoelectronic materials depends not only on materials engineering at a practical level, but also on a clear understanding of the properties of materials, and the fundamental science behind these properties. It is the properties of a material that eventually determine its usefulness in an application. The series therefore also includes such titles as electrical conduction in solids, optical properties, thermal properties, and so on, all with applications and examples of materials in electronics and optoelectronics. The characterization of materials is also covered within the series in as much as it is impossible to develop new materials without the proper characterization of their structure and properties. Structure–property relationships have always been fundamentally and intrinsically important to materials science and engineering.

Materials science is well known for being one of the most interdisciplinary sciences. It is the interdisciplinary aspect of materials science that has led to many exciting discoveries, new materials, and new applications. It is not unusual to find scientists with a chemical engineering background working on materials projects with applications in electronics. In selecting titles for the series, we have tried to maintain the interdisciplinary aspect of the field, and hence its excitement to researchers in this field.

Arthur Willoughby
Peter Capper
Safa Kasap

Preface

More than 100 years ago the spin properties of electrons have been theoretically predicted and experimentally found, thus opening the door for the study of electron spin. In recent decades, the concept of spintronics, and its application in information technology, has received tremendous attention from academia and industry. Spintronic materials and devices have been used in storage, sensing, and many other fields, but there is still room for further development. We expect an increasing number of researchers, especially young scientists, to make a difference in the field of spintronics. So, it is necessary for them to have a comprehensive understanding of the history, current situation, and future development of spintronics materials, devices, and applications. This book, Spintronics – Materials, Devices and Applications, introduces the progress of the fundamental research of spintronic materials and devices, the strategies and methods used to solve problems, and the future direction and opportunities from an application point of view. The editors and authors of this book have many years of research experience in the field.

Here is an outline of the book. At first, the authors briefly introduce the history of spintronics, with a survey of materials, devices, and their applications. Then the book starts with commercialized spintronic devices, including GMR/TMR sensors and STT-MRAM. Their existing problems, solutions, and future research and development trends are described. The authors go on to present the existing issues and solutions of spintronic materials and devices, which are likely to be commercialized in the near future, such as the spin-orbit torque (SOT) technology of the next-generation MRAM and the magnetic tunnel junction in artificial intelligence neural networks. Then they present spintronic materials and devices with the potential to have applications in the long term, such as electrically controlled antiferromagnetic, multiferroic, dilute magnetic semiconductor, skymion-generating materials, as well as the spin oscillator. They present their current research, its problems, and possible methods of solution in detail. Finally, they give an outlook to future developments of spintronic materials, devices, and their applications.

I think this book will be of great value to postgraduates, engineers, and researchers in spintronics and related fields. Readers can imbibe a comprehensive understanding of the current fundamental research and applications of spintronics materials and devices. I strongly recommend that scientists, engineers, and particularly postgraduates who are working in or are interested in the related fields read this book. I believe it can not only broaden and deepen their knowledge but also illuminate their current studies, foster progress and broaden applications for spintronics.

Beijing, China *Baogen Shen*

1

Introduction

Kaiyou Wang[1,2,3]

[1]*State Key Laboratory of Superlattices and Microstructures, Institute of Semiconductors, Chinese Academy of Sciences, Beijing, China*
[2]*College of Materials Science & Opto-Electronic Technology and Center for Excellence in Topological Quantum Computation, University of Chinese Academy of Science, Beijing, China*
[3]*Beijing Academy of Quantum Information Sciences, Beijing, China*

Research in spintronics exploits both charge and spin degrees of freedom of electrons to provide additional advantages to conventional charge-based microelectronic devices, such as non-volatility and reduced power consumption. The discovery of giant magnetoresistance (GMR) in 1988 kick-started the field of spintronics, a result for which Albert Fert and Peter Grünberg were awarded the Nobel Prize in Physics. After that, many phenomena have been studied, including the magnetotransport properties of nanostructured magnetic systems controlled by electrical current, the interplay of charge and spin transport, and spin dynamics in magnetic and nonmagnetic systems. For large-scale commercial applications, the GMR-based spin valves and magnetic tunnel junctions (MTJs) have been extensively used as magnetic field sensors in the read heads of hard disk drives, position or proximity sensors in automated industrial tools, and data storage like magnetic random access memory (MRAM). However, the magnetization manipulation in MRAM based on GMR/tunneling magnetoresistance (TMR) effects by applying an external magnetic field still restricts practical applications.

Since the proposal of spin-transfer torque (STT) by Slonczewski and Berger in 1996, which tunes the transfer of spin angular momentum between two magnetic layers with non-collinear magnetization, electrical approaches have become a direct and effective method to switch magnetization and act as the main stream for manipulating the bit states in MRAM (STT-MRAM). STT has been widely studied in magnetic sandwich structures, including spin valve and MTJ, where a spacer layer (nonmagnetic metal and oxide tunneling barriers, respectively) is sandwiched by two ferromagnetic electrodes. However, the requirement to write information by passing a high current density directly through a spacer layer impacts upon its stability. Moreover, only minority electrons at fermi levels are involved in the magnetization switching from a parallel state to an anti-parallel state, which also decreases the energy efficiency.

Alternatively, the current-induced spin-orbit torque (SOT), originating from the strong spin-orbit coupling (SOC), can transfer charge current into spin current in nonmagnetic semiconductors, heavy metals or topological materials, which results in sustained magnetic oscillations or switching of magnetic structures. Spintronic devices based on SOT have the advantages of non-volatility, fast speed, large endurance, and low power consumption, which are promising candidates used in energy efficient data storage and information processing,

Spintronics: Materials, Devices and Applications, First Edition. Edited by Kaiyou Wang, Meiyin Yang, and Jun Luo.

as well as unconventional computing schemes like neuromorphic and probabilistic computing. Current-assisted magnetization switching by SOT was first observed in dilute magnetic semiconductor (Ga,Mn)As in 2009. Thereafter, manipulation of magnetization via SOT has been extended to ferromagnetic metal multilayers. Recent developments contain the generation and interconversion of charge and spin based on non-equilibrium spin-orbit interaction effects, such as spin Hall and Rashba-Edestein effects. Importantly, SOT has been demonstrated to manipulate the magnetism of most magnetic materials, including ferromagnetic (ferrimagnetic or antiferromagnetic) metals, semiconductors and insulators. In particular, taking advantage of a lateral wedge oxide, a ferroelectric substrate, interlayer exchange coupling by a ferromagnetic layer, or antiferromagnetic layer, current-induced magnetization switching has been achieved without an external magnetic field. Furthermore, SOT can also be used to drive non-uniform magnetization states, including domain walls and skyrmions. However, it should be noted that numerous problems remain to be overcome in SOT spintronics, such as the three-terminal configuration, new materials and mechanisms for high efficient SOT switching.

In addition, using an electrical field instead of an electrical current to manipulate the electronic properties of magnetic materials is promising for realizing ultralow-energy-consuming memory devices due to the suppression of Joule heating, especially when the devices are scaled down to the nanoscale. Spintronic devices based on magnetoelectric and multiferroic materials (such as Cr_2O_3 and $BiFeO_3$) have become an active area and numerous emergent phenomena have been reported over the past decades. However, the practical application of voltage-controlled MRAM is still challenging. For example, the use of moderate voltages (of around 1 V) cannot obtain a sufficient voltage-induced change of magnetic anisotropy (VCMA) to produce magnetization switching, while guaranteeing the magnetic thermal stability of the devices in standby; moreover, the write time window in these devices is small and size-dependent, leading to reliability issues.

In this book, we describe recent progress on materials, devices, and applications of spintronics. Chapters 2 and 3 focus on GMR and TMR materials and devices for magnetic sensors. We introduce the underlying mechanisms of GMR and TMR effects and several industrial and biomedical applications of the GMR and TMR devices, as well as the GMR and TMR sensors' surface modifications, sensors combined with different auxiliary tools, such as microfluidic channels and magnetic flux concentrators (MFCs), and methods of improving the signal-to-noise ratio. Two recently crucial methods for electrical manipulation of magnetization switching (STT and SOT) are described in Chapters 4 and 5, which include the fundamental physics and the progress of STT and SOT on materials and devices, as well as the challenges and opportunities for future development of the related materials and devices. Then the STT- and SOT- based spintronic devices (like MTJ), used in spin oscillators and artificial neural networks, are discussed in detail in Chapters 6 and 7. More fundamental research about skyrmions is described in Chapter 8. In addition to conventional ferromagnetic materials, multiferroelectric materials, typical dilute magnetic semiconductor (Ga,Mn)As, and antiferromagnetic materials, which are viewed as the active materials for future spintronic devices, are also introduced. This book contains spintronic materials ranging from metals, semiconductors and insulators, with ferromagnetic, ferrimagnetic or antiferromagnetic behaviors. It also introduces representative spintronic devices, such as GMR spin valves, STT- and SOT-MTJs, which have great potential applications in information storage and processing. This book will provide a comprehensive and deep understanding of spintronics.

2

Giant Magnetoresistance (GMR) Materials and Devices for Biomedical and Industrial Applications

Kai Wu[1], Diqing Su[2], Renata Saha[1], and Jian-Ping Wang[1,2]

[1]*Department of Electrical and Computer Engineering, University of Minnesota, Minneapolis, MN, United States*
[2]*Department of Chemical Engineering and Materials Science, University of Minnesota, Minneapolis, MN, United States*

CHAPTER MENU

2.1 Introduction

Giant magnetoresistance (GMR) is the large change in electrical resistance of metallic layered structures when the magnetizations of ferromagnetic layers are reoriented relative to one another under the application of an external magnetic field. This effect was at first

Spintronics: Materials, Devices and Applications, First Edition. Edited by Kaiyou Wang, Meiyin Yang, and Jun Luo.

discovered in 1988 by a team led by Albert Fert on (001) Fe/(001) Cr multilayers [1] and, independently, by another team led by Peter Grünberg on multilayers of Fe and Cr on (110) GaAs substrate [2]. The discovery of GMR was to a large extent due to the significant progress in the thin film deposition technique: molecular beam epitaxy (MBE). By stacking such thin films with nearly a monolayer precision, one can fabricate multilayered structures with properties that are totally distinct from those of the constitutive bulk materials.

The publications reporting the discovery of GMR rapidly attracted attention for its fundamental interest as well as for many potential applications. Nowadays, it has been commercialized in many areas such as magnetic field sensors including biosensors and read/write heads in hard disk drives, as well as detectors of oscillations in microelectromechanical system (MEMS) [3–12]. These achievements wouldn't be possible without an in-depth understanding of the physics of the GMR effect, which requires a quantum-mechanical insight into the electronic spin-dependent transport in magnetic nanomaterials. In this chapter, we will introduce the origin of the GMR effect in Section 2.2 where spin-dependent scattering and two-channel model are explained as well as some basic GMR stack structures are given. In Section 2.3, different types of GMR sensors are introduced, including sensors with different geometries that are fabricated on both rigid and flexible substrates, printable GMR flakes, and granular GMR sensors. GMR sensor surface modifications for biomedical applications, magnetic flux concentrators (MFC) to focus and amplify weak magnetic fields, and microfluidic channels for quantitative and fully automatic on chip assays on sub-microliter volume of biological fluids are introduced in Section 2.4. In Sections 2.5 and 2.6, various biomedical and industrial applications based on GMR sensors are reviewed. The roles of GMR sensors in position sensing, current sensing, and material defect inspection are discussed.

2.2 Giant Magnetoresistance (GMR) Effect

GMR is a quantum mechanical magnetoresistive effect that exists in multilayers composed of alternating ferromagnetic (FM) and non-magnetic (NM) conductive layers. Pioneers in this field, Albert Fert and Peter Griinbergwere are awarded the 2007 Nobel Prize in Physics [1, 2]. The physical origin of the GMR effect is caused by a spin-dependent scattering of the conduction electrons with different spins relative to the magnetizations of FM layers. Briefly, in magnetically ordered materials, the electrical resistance is affected by the scattering of electrons on the magnetic sublattices of the crystal; namely, the non-zero magnetic moments. This scattering is weaker when the electron spins and magnetic moments are parallel and stronger when they are antiparallel. A good review of this mechanism is provided in reference [13].

To qualitatively understand GMR, a simplified Mott model is introduced herein [14]. Assuming that the scattering of conduction electrons at the interface between the FM and NM layers is small, and the electron spins persist long enough, the conduction of up-spin and down-spin electrons are viewed as two conducting channels lying in parallel configuration.

As shown in Figure 2.1 (A1), taking a FM/NM/FM multilayer structure as an example, when the magnetizations of FM layers are aligned parallel, the up-spin electrons pass through the structure almost without scattering since the spins are parallel to the magnetizations of FM layers. On the contrary, the down-spin electrons encounter strong scattering within both FM layers since their spins are antiparallel to the magnetizations of both layers. The conduction occurs in parallel for the two spin channels (also called the two-channel model): thus, the total resistance Rp of this multilayer structure is the sum of resistance in parallel of electrons with up-spin and down-spin, which is expressed as:

$$\frac{1}{R_p} = \left(\frac{1}{R\uparrow\uparrow + R\uparrow\uparrow} + \frac{1}{R\uparrow\downarrow + R\uparrow\downarrow} \right). \tag{2.1}$$

On the other hand, the R_{AP}, defined as the resistance of this multilayer structure when the magnetizations of FM layers are antiparallel as shown in Figure 2.1(A2), is expressed as:

$$\frac{1}{R_{AP}} = \left(\frac{1}{R\uparrow\uparrow + R\uparrow\downarrow} + \frac{1}{R\uparrow\downarrow + R\downarrow\downarrow} \right). \tag{2.2}$$

Assuming that for the NM layer the resistance is negligible for both electron spins (up and down). To further simplify equations (2.1) and (2.2), the resistances of up-spin and down-spin channels in Figure 2.1(A1) are expressed by $R^{\uparrow}$ and $R^{\downarrow}$:

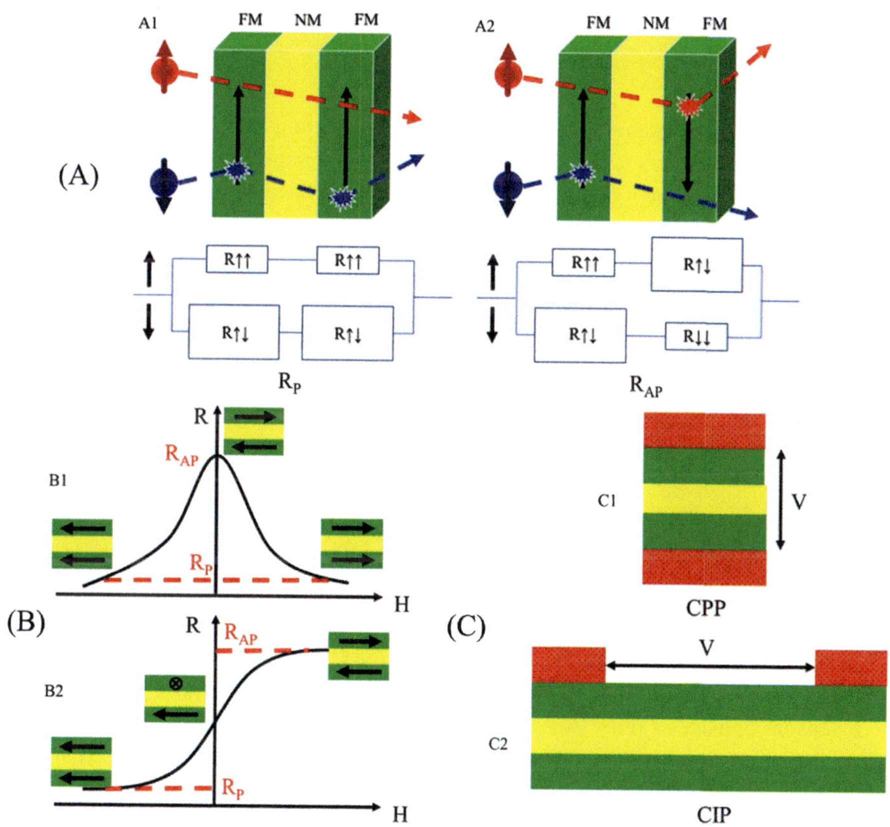

Figure 2.1 (A) The two-channel resistor model of a FM/NM/FM multilayer GMR structure when the magnetizations of FM layers are aligned parallel (A1) and antiparallel (A2). (B) RH curves of GMR multilayers with easy axis of the free layer (B1) parallel and (B2) perpendicular to the pinned layer. (C) The CPP (Cl) and CIP (C2) geometries.

$$R^{\uparrow} = R\uparrow\uparrow + R\uparrow\uparrow \tag{2.3}$$

$$R^{\downarrow} = R\uparrow\downarrow + R\uparrow\downarrow. \tag{2.4}$$

Thus, the simplified *R*P and *R*AP for a FM/NM/FM GMR multilayer structure with magnetizations of FM layers aligned parallel and antiparallel are expressed as:

$$\frac{1}{R_p} = \left(\frac{1}{R^{\uparrow}} + \frac{1}{R^{\downarrow}}\right) \tag{2.5}$$

$$\frac{1}{R_{AP}} = \left(\frac{4}{R^{\uparrow} + R^{\downarrow}}\right). \tag{2.6}$$

The GMR ratio (also referred as the MR ratio, unit: %) can be modeled by a simple resistor model expressed as:

$$MR = \frac{\Delta R}{R_P} = \frac{R_{AP} - R_P}{R_P} = \frac{(R^{\uparrow} - R^{\downarrow})^2}{4R^{\uparrow}R^{\downarrow}}. \tag{2.7}$$

Apart from the MR ratio, another important figure of merit that is used to evaluate the performance of the MR sensors is the detectivity *D*. This parameter is better known as the magnetic field resolution or the magnetic field equivalent noise level expressed by:

$$D = \frac{\sqrt{S_v}}{\Delta V / \Delta H}. \tag{2.8}$$

Where $\sqrt{S_v}$ is the noise voltage density (unit: $V/\sqrt{H_Z}$), ΔV is the total output voltage and ΔH is the range of magnetic field that the MR sensor can measure. For MR sensors, the ratio $\Delta V/\Delta H$ is expressed as

$$\Delta V / \Delta H = \frac{\Delta R}{R_P} \cdot \frac{V_b}{\Delta H} \tag{2.9}$$

where $\Delta R/R_P$ is the MR ratio and *V*b is the bias voltage applied to the sensor. Therefore, in order to obtain a high sensitivity GMR sensor, the detectivity *D* needs to be sufficiently small.

There are two most commonly used stack structures for GMR sensors. Multilayer GMR sensor consists of multiple alternating ferromagnetic layers and nonferromagnetic layers. Spin valve sensors consist of only two ferromagnetic layers separated by a nonmagnetic layer, where the magnetization of one ferromagnetic layer can rotate freely, while the magnetization of the other ferromagnetization layer is fixed to a certain direction.

Typical RH curves of GMR multilayers as well as the spin valves are shown in Figure 2.1(B). For multilayer sensors (Figure 2.1(B1), the magnetizations of the FM layers rotate with the external magnetic field, with the maximum resistance being achieved at zero field where the magnetizations of the two adjacent FM layers are antiparallel. For spin valve sensors where the easy axis of the free layer is perpendicular to the magnetization of pinned layer (Figure 2.1(B2), only the magnetization of the free layer can rotate with the magnetic field, leading to a maximum resistance when the magnetic field is antiparallel to the magnetization of the pinned layer. There are two variations of GMR sensors based on how electric current is passed through the multilayers, as shown in Figure 2.1(C). In the current perpendicular to plane (CPP) configuration (Figure 2.1(C1), the current is passed perpendicular to the layers, and the electrodes are located on top and bottom sides of the multilayer stacks. For the current in plane (CIP) geometry (Figure 2.1(C2), the current flows along the layers, and the electrodes are located on one side of the structure. The CPP geometry results in more than twice higher GMR but is difficult to realize in practice than the CIP configuration [15].

Initially, the observation of the GMR effect was restricted to multilayer stacks deposited by MBE only. However, this effect found manifold applications in the year 1990, when Parkin *et al.* demonstrated the room temperature GMR effect in Co/Cr and Co/Ru multilayer stacks by depositing them through magnetron sputtering techniques [16]. Ever since then, the industry has observed a boom of commercial applications for GMR-based devices which primarily includes biosensors [17–20], position and angular sensors [21, 22], as well as GMR-based magnetic read/write heads in hard disk drives [23–26] and magnetic random-access memory (MRAM) systems [27–29]. Of all, spin valve sensors for magnetic hard disk drives are probably the most celebrated applications of this sensor. The fact that the resistance of GMR device shows a steep change in an extremely small field range makes it so unique. As the magnetic domains on the hard drive pass under the read head, the orientation of the magnetization of the free layer in the spin valve sensors changes to a different direction which results in the resistance change. There are four main components in the sensing part of the read head: shields, a magnetic sensing region that contains GMR sensors, electrical contacts that pass the current through the sensing element, and a stabilization layer that biases the free layer in the single domain state to eliminate hysteresis and reduce noise [30]. One of the early reports of CPP GMR for future high density magnetic recording was made by Nagasaka *et al.* in 2005, which focused on the most important properties of a read sensor such as high output, high signal-to-noise ratio (SNR) and low resistance [24]. Eventually it led towards filing of several patents based on GMR-based read heads from several industrial companies [23, 31, 32].

2.3 Different Types of GMR Sensors

Since the first discovery of the GMR effect, different types of GMR device have been explored to meet the requirements for various applications such as high sensitivity, high flexibility, ease of fabrications, low cost, etc. In this section, different geometries of thin film-based GMR sensors on either rigid or flexible substrates are first introduced, followed by the printable GMR flakes and granular GMR sensors. The dimensions and MR values of these GMR sensors are summarized in Table 2.1. It is expected that the pros and cons of different sensing configurations can be explored by comparing the fabrication processes, mechanical properties as well as the magnetic properties of these GMR sensors.

2.3.1 Rigid GMR Sensors

2.3.1.1 Long-strip GMR Sensors

A linear response with minimum hysteresis to the external magnetic field is one of the key parameters to be considered in the design of magnetic field sensors. High linearity can be

Table 2.1 Comparison between different types of GMR Sensors.

GMR sensor types	Sensor dimensions	MR ratio	Ref.
Long-stripe GMR sensors	150 × 0.75 μm	4%	[33]
Large-area GMR sensors	200 × 150 μm	16%	[34]
Flexible GMR sensors	100 μm in width	9.9%	[35]
Printable GMR sensors	250 × 250 μm	37%	[36]
Granular GMR sensors (thin film-based)	NA	12%	[37]
Granular GMR sensors (solution-based)	1.2 cm × 0.5 cm	260%	[38]

achieved through different approaches such as optimizing magnetic stack designs [39], introducing shape anisotropy [40, 41], and applying external bias fields [42]. Fabricating sensing elements with high aspect ratios to introduce shape anisotropy in the free layer is the most straightforward and thus the most used approach to increase the linearity. In spin valve sensors, the magnetization direction in the pinned layer is predefined in the short axis of the sensor while the magnetization in the free layer at zero field is along the long axis of the sensor due to shape anisotropy. This orthogonal configuration leads to a linear transfer curve prior to free layer saturation [43]. Wang *et al.* investigated the influence of sensor dimension on the SNR [44]. It was found that for the detection of magnetic nanoparticles (MNPs) with spin valve sensors, a larger electrically active length is beneficial for the measurement of targets with low concentrations, while a small sensor width is beneficial for both low-concentration and high-concentration detections. To date, most GMR sensors for low-frequency and small-field detection are fabricated into stripes with large aspect ratio which are connected in series and in parallel to fulfill better linearity while maintaining optimal sensitivity (Figure 2.2A) [45]. As an example, spin valve sensors with a size of 150 × 100 μm were fabricated by Su *et al* [33]. Each sensor consists of 24 spin valve stripes, each 150 μm long and 0.75 μm wide. The RH transfer curve of the sensor exhibited high linearity from −30 Oe–30 Oe with a MR ratio of 2% in the linear regime.

2.3.1.2 Large-area GMR Sensors

Although stripe GMR sensors exhibit high linearity, the active sensing area is limited due to the requirement of large aspect ratio, which reduces the amount of MNPs that can be captured to the sensor surface in biological detections. Furthermore, it was found that the signals of the stripe GMR sensors vary with the position of the MNPs on the sensor surface [46]. Due to the interaction between the stray fields from the MNPs and the sensing elements, the MNPs that land on the edge of the stripe element contribute higher signal compared to those land in the middle of the stripe element. To address this issue, large-area GMR sensors with lower aspect ratios and larger active sensing areas are proposed with a different switching mechanism, where the magnetization reversal within the large-area sensing element is non-uniform. Su *et al.* explored the switching of a permalloy film under different concentrations of MNPs through micromagnetic simulations [20]. It was found that a localized magnetization reversal happened in the permalloy film in proximity to the MNPs rather than the uniform switching happening in the stripe sensors. Due to the unique magnetization reversal process, the change of MR values after the addition of MNPs turned out to be over 20 times larger than the stripe sensors. Feng *et al.* fabricated large-area GMR sensors with a dimension of 40 × 80 μm followed by the deposition of 15 nm FeCo MNPs (Figure 2.2B) [47]. The active sensors exhibited a significant signal with a mean value of 0.2 (normalized MR change) and a standard error of 0.0045 when the peak value difference was taken as the sensor signal, while a mean value of 4.2 (normalized MR change) and a standard error of 0.13 when the difference in the total integrated area was taken as the sensor signal. Due to their high sensitivity, several attempts have been made to utilize large-area sensors for biological detection. Yang *et al.* fabricated GMR sensors with a size of 200 × 150 um for the detection of carcinoembryonic antigen (CEA) protein, which resulted in a MR value of 16% under a field range of −1000 ~ 1000 Oe and a detection limit down to femtomole leve1 [34]. Another large-area GMR sensor was fabricated by Srinivasan *et al.* for the detection of human interleukin 6 (IL-6) with a detection limit of 2.08 × 10^6 molecules [48].

2.3.2 Flexible GMR Sensors

Flexible electronics have attracted great attention in the past decade due to the low cost, light weight, and potential applications in wearable devices. Flexible magnetic nanostructures also possess their unique advantages in the sensing of stress or strain direction based on the magnetoelastic effect (Figure 2.2C) [51]. Different fabrication technologies have been employed to obtain flexible GMR structures include directly depositing GMR stacks on the flexible substrates (direct deposition) [35, 52], transferring GMR stacks from rigid substrates to flexible substrates (substrate transfer) [49], or thinning of silicon substrates after the stack deposition (substrate thinning) [53, 54]. Direct deposition is the most straightforward way in

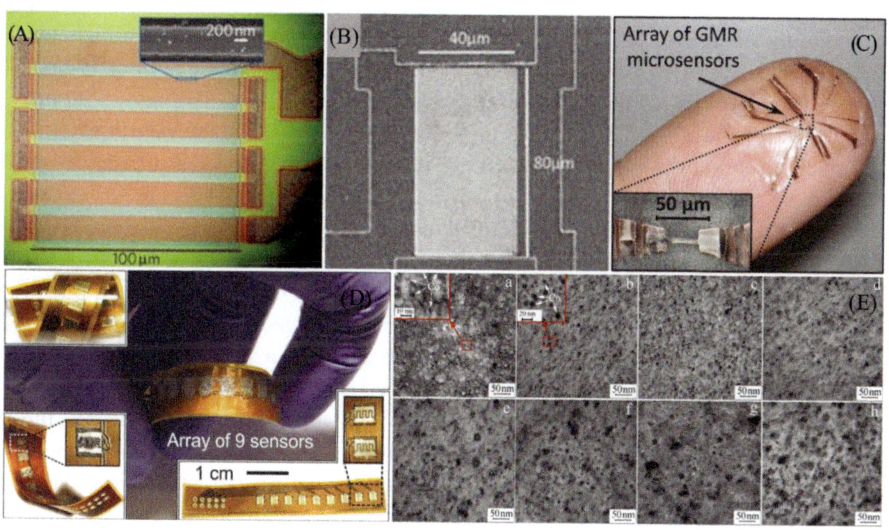

Figure 2.2 Different types of GMR sensors. (A) Long-stripe GMR sensors. (B) Large-area GMR sensors. (C) Flexible GMR sensors. (D) Printable GMR sensors. (E) Granular GMR sensors. Subfigures (a–h) are the TEM images of the granular alloy films with different Co concentrations of 21% (a, e), 28% (b, f), 40% (c, g), and 71% (d, h) in unannealed (a–d) and annealed states (e–h). *Source:* (A) Reprinted with permission from [45], copyright 2011 Springer Nature. (B) Reprinted with permission from [47], copyright 2017 AIP Publishing. (C) Reprinted from [49] under the terms of the Creative Commons Attribution – NonCommercial License. (D) Reprinted from [36] under the terms of the Creative Commons Attribution – NonCommercial License. (E) Reprinted with permission from [50], copyright 2018 Elsevier B.V.

fabricating the flexible structures. Guo *et al.* fabricated a top-pinned spin valve with a structure of $Co_{90}Fe_{10}$/Cu/$Co_{90}Fe_{10}$/IrMn on a polyethylene terephthalate substrate by magnetron sputtering and studied the influence of strain on the MR response of the devices [55]. It was found that at bending angles of 0°, 90°, and 180°, the MR was 0.71%, 1.21%, and 2.20%, which can be attributed to the strain-induced uniaxial anisotropy. When the magnetic field is parallel to the easy axis, the uniaxial anisotropy will increase after the application of a compressive strain, which leads the ferromagnetic moments to orient along the easy axis, and thus an increase in the MR value.

To minimize the influence of residual strain on GMR performance, Li *et al.* proposed a strain-relief structure by combining the strain-engineered periodic wrinkles and parallel ribbons [35]. Prior to deposition, a series of uniaxial tensile strains were applied to the poly (dimethylsiloxane) (PDMS) substrate. To obtain a ribbon-like structure, a shadow mask with a pattern of bands oriented in the pre-strain direction was employed during the deposition process. It was shown that a wrinkled GMR structure was formed after the strain relief, which was able to accommodate the biaxial strain induced by the stretching process. The resulting device exhibited a high tolerance to the external strains with an unchanged MR of 9.9% over a tensile strain ranging from 0% to 35%.

Contrary to Li's work, Ota *et al.* took advantage of the strain-sensitive MR response and demonstrated the possibility of utilizing flexible GMR devices for the sensing of strain direction [51]. The GMR stack contains a strain-sensitive Co layer and a strain-insensitive permalloy layer, separated by a non-magnetic Cu layer. Upon the application of a strain in the in-plane direction at different angles, the easy axis of the strain-sensitive layer is always parallel to the strain direction and rotates accordingly, while the easy axis of the strain-insensitive layer remains unchanged, leading to a change in the RH transfer curves. It was also shown that the change of RH curves only occurs beyond a threshold strain value since the magnetization in the Co layer was hampered by the stray field from the permalloy layer. Consequently, reducing the interlayer coupling field can increase the sensitivity of the sensor, which makes it possible to detect the strain direction at small strain levels (<0.1%).

Despite much effort on optimizing the direct deposition method, it suffers from several process limitations. For example, the flexible substrates like PDMS are not compatible with standard nanofabrication processes, which are required for the patterning of the devices. Furthermore, the stretching of the functional elements also induces cracks, which further inhibits the functionality and minimization of the devices. Melzer *et al.* reported a direct transfer method to transfer the magnetic stacks grown and patterned on a silicon substrate to a flexible substrate [49]. GMR stacks were at first deposited on a silicon substrate coated with a poly(acrylic acid) (PAA) sacrificial layer, followed by the patterning of the magnetic devices. In the meantime, a PDMS layer was spin coated on an anti-stick layer and was pre-stretched. Oxygen plasma treatment was then applied to both surfaces of the GMR sensors and the PDMS substrate. The activated surfaces were laminated to each other for the bonding process at 60°C for 30 min. Finally, the sacrificial layer was dissolved, releasing the pre-strained GMR structure on the PDMS substrate. The transferred GMR elements were capable of enduring strains of about 30% without the formation of cracks. GMR's characteristics also remained unchanged during the entire range of applied strains.

Perez *et al.* explored the possibility of fabricating flexible GMR devices on silicon membranes by thinning down the silicon wafer from backside to 100 μm and 50 μm after the stack deposition [54]. The minimum bending radius turned out to be 15.5 mm and 6.8 mm for the 100 μm and 50 μm Si membranes, respectively with the MR value remaining at 15.3% for all bending radius. Broadening of the GMR curve was observed during the increase of the curvature due to the strain-induced uniaxial anisotropy, which also altered the slope of the linear regime of the MR curve. The effective magnetostriction constant for the multilayer stacks was estimated to be 1.7×10^6 based on the change of the MR curves upon bending. Similarly, Yin *et al.* reported a two-step thinning method for the silicon substrate with grind thinning followed by inductively coupled plasma (ICP) etching [53]. The thickness of the resulting silicon membrane was 80 μm and the MR ratio of the device was 4.24%, which remained unchanged comparing to the MR value before the substrate thinning process.

2.3.3 Printable GMR Sensors

Conventional GMR sensors are fabricated by thin film fabrication technologies, which results in highly sensitive performance but also possesses the disadvantages such as high cost and relatively low fabrication efficiency. To overcome these issues, Karnaushenko *et al.* proposed for the first time to fabricate printable GMR sensors with a mixture of GMR flakes and polymer binders (Figure 2.2D) [36, 56]. The GMR multilayers were first deposited on a sacrificial layer on a rigid substrate, followed by the dissolving of the sacrificial layer and the release of GMR flakes. The collected GMR flakes were then milled and screened to obtain metallic pieces in regular shapes with an average flake size of 36 ± 5 μm. Finally, the GMR flakes were mixed with a polymeric binder to prepare the printable GMR paste. The influence of GMR flake sizes as well as the choice of polymeric binders on the MR value were investigated. It was shown that the GMR ratio increased from 14% to 21% when the flake size was reduced from 150 to 36 μm. Of all the polymer binders tested, polyepichlorohydrin (PCH) resulted in the highest MR ratio of 20% while methylenebis (phenyl isocyanate)-polyester/polyether polyurethane (Pu) resulted in the lowest MR ratio of 7%. The optimum MR ratio of 37% was achieved in the PCH-based binder solution with a flake size of 36 μm and a flake concentration of 99%. The thermal stability of the GMR sensors, when printed on flexible printed circuit (FPC) boards, was also evaluated. The performance of the printed sensors remains unchanged up to the transition temperature of the binder, which indicated that an elastomer with small transition temperature was desired and an operation temperature of up to 95°C was achieved. This printed GMR sensor technology not only reduces the complexity and cost in the magnetic sensor fabrication process but also provides an alternative approach to integrate magnetic devices with flexible electronics.

2.3.4 Granular GMR Sensors (Thin Film- and Solution-based)

Although the GMR effect was at first discovered in thin-film systems, it can also be found in granular systems where MNPs are embedded in the conductive, non-magnetic matrix (Figure 2.2E). When no magnetic field is applied, the magnetizations of the MNPs are randomly distributed, leading to a high-resistance state. Upon application of the magnetic field, the magnetizations of the MNPs align with the field direction, resulting in a low resistance state. It was shown that the origin of the GMR was the spin-dependent scattering taking place at the interfaces between the granules and the matrix [57]. As a result, the magnetoresistive properties depend on multiple factors that could influence the scattering process such as the MNP and matrix materials, the volume fraction of MNPs as well as the size distribution of MNPs [58]. The influence of the microstructures of the granular films on the transport properties have been studied by several groups. Shpetnyi *et al.* investigated the microstructures of the Cu–Co granular films during multiple heat treatments [50]. It was found that the annealing process increased the size of the Co granules due to the coalescence of the material, which facilitated the transition from superparamagnetic to ferromagnetic single-domain state. After annealed at 800 K, the samples with Co concentrations of 15% and 21% exhibited 5-fold and 2-fold increases in magnetoresistance compared to the unannealed samples, respectively. The magnetotransport properties have also been studied in other granular alloy systems such as NiFe-Ag [59], Fe-Ag [60], and Co-Ag [61].

The granular systems mentioned above utilized top-down methods to deposit granular films on the substrate by co-sputtering or co-evaporation. Bottom-up approaches were also proposed to form granular alloys [62]. Dupuis *et al.* synthesized Co and Fe nanoparticles with an intense cluster beam of selected size produced by a laser vaporization source, which were then deposited with Ag matrix material on a cold substrate at the same time [37]. With this additive method, both the size and the concentration of the granules can be adjusted independently by altering the growth parameters during particle synthesize without influencing the deposition of the matrix material. A magnetic percolation threshold of 25% and a MR ratio of 12% were achieved for Co clusters in an Ag matrix. Meyer *et al.* proposed a solution-based approach where carbon-coated Co MNPs were dispersed in different conductive gel matrix [38]. A MR ratio of 260% was achieved for a water-based conductive gel. The high MR ratio originates from the formation of chain fragments during the application of magnetic field. The uniaxial anisotropic particle density in the field direction results in increased spin-dependent scattering along the current path, leading to a high MR ratio. The magnetoresistive properties of the solution-based GMR sensors were shown to be dependent on the volume concentration of MNPs as well as the conductivity and viscosity of the matrix. It was also found that the MR ratio decreased dramatically upon drying of the matrix due to the limited movement of MNPs during measurement, which can be partially solved by the application of magnetic field during sample preparation process.

2.4 GMR Sensors: Surface Modification and Auxiliary Tools

2.4.1 GMR Sensor Surface Modification for Biomedical Applications

For most biomedical applications, GMR sensor surfaces are passivated with tens to hundreds of nanometers thick insulating thin films such as SiO_2, $A1_2O_3$, and Si_3N_4 [63–67]. There are three functions for these passivation layers: 1) to isolate magnetic materials from external biological environments since most magnetic materials are highly cytotoxic; 2) to prevent the leakage current flowing towards biofluid samples in contact; 3) to provide the basis for subsequent sensor surface biofunctionalization [11.] For biomedical applications, the deposition of passivation layer must be non-toxic thus, chemical vapor deposition (CVD) techniques are not allowed. Moretti *et al.* have evaluated the cytotoxicity of a multilayer thin film stack ($SiO_2/Si_3N_4/SiO_2$) deposited by the physical vapor deposition (PVD) system [68]. It is reported that the neuron cells can grow on the thin film surface for 31 days while maintaining their functionality and viability. The underlying magnetoresistive sensors' performances are unchanged for several weeks in contact with neurons and neuronal medium. GMR sensors passivated with these biocompatible thin films are ready for some biomedical applications (e.g., biomagnetic field recording introduced in Section 2.5.3) that requires directly contact with biological tissues.

However, most GMR-based biomedical applications (e.g., immunoassays and genotyping introduced in Sections 2.5.1 and 2.5.2) rely on detecting the magnetic signal from magnetic tags (i.e., MNPs and magnetic microbeads). Since the stray fields from the magnetic tags decay rapidly with distance, the passivation layer thickness requires optimization to achieve isolation purpose and minimal magnetic signal loss. Furthermore, the passivation layers must be of low energy states to favor the surface immobilization of biological molecules. Herein, we will introduce two popular methods for GMR sensor surface biofunctionalization on top of passivation layers: 1) the APTES and Glu method [33, 66, 67, 69] in Box 2.1 and 2) the NHS and EDC method [63–65] in Box 2.2.

After the above biomodification steps, the GMR sensor surface is exposed with aldehyde groups that can readily bind to the amino groups from capture probes (such as capture antibodies and amino-modified capture DNA). A robotic spotter is then used to deposit several hundred picoliters to several nanoliters of capture probe solution on each active GMR sensor. And 1% phosphate buffered saline (PBS) solutions of BSA (bovine serum albumin) is added to block the remaining aldehyde group binding sites in order to prevent non-specific binding during the assay step. GMR chips at this step are ready for biomedical applications such as immunoassays (sensor surface coated with capture antibody) and genotyping (sensor surface coated with capture DNA).

Box 2.1 APTES and Glu Method

1. Sequentially wash the GMR chip with acetone, methanol, isopropanol, and de-ionized water.
2. Further clean chip surface by exposing it to oxygen plasma (e.g., for 3 minutes) or UV ozone (e.g., for 10 min) to remove organic residues.
3. Immerse the chip in a 0.5–1% (3-Aminopropyl)triethoxysilane (APTES) in toluene mixture (e.g., for 15 min), then rinse with acetone and followed by de-ionized water.
4. Immerse the chip in a 5.0% glutaraldehyde (Glu) solution (e.g., for 5 h), then rinse with de-ionized water.
5. Store the chip in a desiccated environment before further step.

Box 2.2 NHS and EDC Method

1. Sequentially wash the GMR chip with acetone, methanol, isopropanol, and de-ionized water.
2. Further clean chip surface by exposing it to oxygen plasma (e.g., for 3 mi) or UV ozone (e.g., for 10 min) to remove organic residues.
3. Immerse the chip in a 1% aqueous solution of poly(allylamine hydrochloride) (e.g., for 5 min), then rinse with de-ionized water and followed by baking at 110–120°C for 1 h.
4. Immerse the chip in a 2% aqueous solution of poly(ethylene-alt-maleic anhydride), then rinse with de-ionized water.
5. Finally, the chip surface is activated by adding a mixture of l-ethyl-3-[3- dimethylaminopropyl]carbodiimide (EDC) and N-hydroxysuccinimide (NHS) in de-ionized water.
6. Store the chip in a desiccated environment before further step.

2.4.2 Integration of a Magnetic Flux Concentrator (MFC)

In pursuit of higher sensitivity GMR field sensors for detecting ultra-low quantity of magnetic tags in biomedical applications and weak magnetic fields such as the biomagnetic fields generated by the electrical activities of organs and cells, magnetic flux concentrators (MFC) with different materials and geometry designs are proposed to focus and amplify the magnetic fields in the sensor region. The effectiveness of a MFC is defined by the gain, G, which is the ratio of concentrated magnetic field in the sensor region and the external field [70–72]. There are two types of MFC that are frequently reported: the soft-ferromagnetic material- and superconducting material-based MFCs.

Although the GMR sensor, combined with MFC, can achieve highly sensitive magnetic field detection, it suffers from 1/f noise, limiting its applicability for DC and low-frequency field detection. For example, 1/f noise can greatly weaken the sensor's detectivity for slowly moving vehicles whose frequencies are below 1 Hz. To suppress this 1/f noise, some researchers have combined GMR sensors with MFC and MEMS [8, 73–76]. The MFC combined with MEMS, also known as a MEMS flux concentrator, modulates the low frequency magnetic fields to kilohertz (kHz) frequencies, where the 1/f noise in the GMR sensor can be effectively suppressed. Herein, we present some examples using MFC or MEMS flux concentrator to improve the sensitivities of all kinds of MR sensors (such as GMR, MTJ (magnetic tunnel junction), and AMR (anisotropic magnetoresistance)).

2.4.2.1 Superconducting MFC

The superconducting MFC is first proposed by Pannetier *et al.*, where they combined the GMR sensor with a superconducting loop made of niobium (Nb) or yttrium barium copper oxide (YBCO), which was electrically isolated from the GMR by a 400 nm thick Si_3N_4 layer and closed above the GMR by a constricted area [77]. The GMR is patterned in a yoke-type shape and titanium contacts which allows the measurement of the central, domain-free part. The superconducting film is coated on top of the GMR without covering the entire measurement part. This superconducting loop works as a flux-to-field transformer. When an external magnetic field is applied perpendicular to the loop, a supercurrent, I_S, is induced as a result of the Meissner effect. This current leads to a very high coplanar magnetic field which can be detected by the GMR sensor positioned underneath. The GMR sensor used here is designed to be sensitive to in-plane magnetic fields. Based on this design, the authors tested two superconducting materials, Nb and YBCO. Later in 2011, the same group reported the application of this GMR combined with superconducting MFC loop for magnetocardiography (MCG) recordings over the chest of healthy volunteers at 4 K and 77 K, in a magnetically shielded room [78].

2.4.2.2 Soft-ferromagnetic Material-based MFC

Just like the SQUID, the superconducting MFC, although very small in size, still requires very low temperature for high sensitivity magnetic field detection. This disadvantage gives the soft-ferromagnetic material-based MFC opportunities to develop room-temperature magnetic field sensors. By combining soft-magnetic material MFCs with MEMS cantilevers, these MEMS flux concentrators become popular alternative tools for ultrasensitive field detections even for low frequency fields. Herein, we give one example of MEMS flux concentrator.

As shown in Figure 2.3(A), the external low frequency magnetic field (f1) flows through the MFC on the MEMS cantilevers. Both cantilevers mechanically oscillate at resonant frequency f_0, which induces an oscillatory magnetic field at frequency of $2f_0$ (for an external DC magnetic field) or $2f_0 \pm f_l$ (for an external low frequency f_l field) that is sensed by the GMR sensor. This AC magnetic field induced by MEMS modulates the low frequency magnetic fields to kilohertz range where the 1/f noise in the GMR sensors can be neglected. The modulation efficiency, α, defined by the ratio of field modulated to high frequency over the original low frequency/DC field, is measured by sweeping the external field as shown in Figure 2.3(B1). For detecting low frequency magnetic fields, a 17 nT, 1 Hz field is applied to the system and the resulting GMR signal is monitored by recording the sidebands at $2f_0 \pm f_l$. As is plotted in Figure 2.3(B2), the sidebands' amplitudes are approximately half of the external field's amplitude. Figure 2.3(C) compares the noise spectra of a standalone GMR and the GMR-MEMS hybrid sensor. Since the GMR-MEMS hybrid sensor operates at the modulation frequency of $2f_0$, the 1/f noise is tremendously weakened. Thus, the only limiting factors contributing to low frequency region noise are the modulation efficiency and the GMR sensor thermal noise at

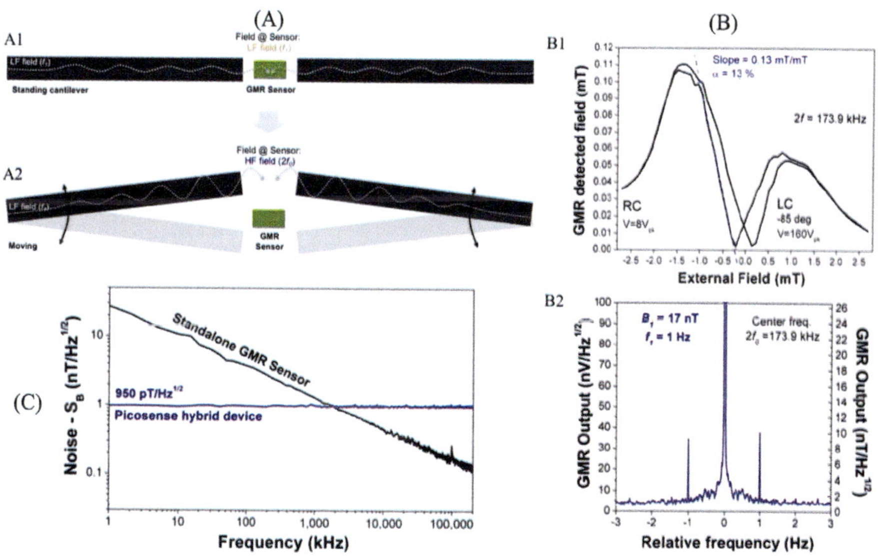

Figure 2.3 (A) GMR combined with MEMS MFC. (A1): Side view of a MEMS MFC. (A2): The mechanical movements of MEMS cantilevers induce AC magnetic field at the GMR sensor region. (B) MEMS MFC for detecting low frequency fields. (B1): Detected field vs applied external field. The calculated modulation efficiency of this MEMS MFC is α=13%. (B2): Detection of a 17 nT low frequency field (1 Hz). The sidebands' amplitudes are half of the external applied field amplitude. (C) Noise spectra of a standalone GMR and GMR-MEMS hybrid sensor. *Source:* Figure reprinted from [79] under the terms and conditions of the Creative Commons Attribution (CC BY) license.

$2f_0$. At 1 Hz, the GMR-MEMS hybrid sensor shows more than 10 times better sensitivity than a standalone GMR sensor, enabling a sensitivity of 950 $pT/\sqrt{H_z}$ at room temperature.

2.4.3 Integration of Microfluidic Channels

For GMR-based biomedical applications related to the detection of target analytes from sub-microliter volume of biological fluids, microfluidic channels are often reported as auxiliary tools for quantitative and fully automatic on chip assays [80–82]. These microfluidic channels can confine, manipulate, and mix very small amount of liquids on a single chip: thus, meeting the growing demand for lab-on-a-chip platforms. Polymers such as PDMS are among the mostly used materials for microfluidic channels due to their transparency, biocompatibility, flexibility, and cost-effectiveness. The manufacture process of a microfluidic channel starts with a negative photoresist mold patterned by standard photolithography processes, and upon which the PDMS is casted, cured, peeled off and placed on top of the GMR chip. In addition to the standard photolithography processes for preparing the molds, through the years, researchers have economized the process of mold fabrication by using 3D printers. Briefly, the master molds are fabricated by 3D printing of cured photopolymer and then transferred to PDMS for subsequent processing to construct the device. Fabricated devices are reported with channel dimensions in the order of 200–300 μm [83–86]. Once the PDMS microfluidic channels are made, they are integrated with the silicon-based GMR chips either by applying pressure over the channel or treating with oxygen plasma to form irreversible chemical bonds [87–89].

2.5 GMR-based Biomedical Applications

2.5.1 GMR-based Immunoassays

In the past decade, GMR biosensors have developed into a sensitive, rapid, and quantitative immunoassay tool. GMR-based immunoassays operate in a similar way to the enzyme-linked immunosorbent assay (ELISA) where capture antibodies are first immobilized on the GMR sensor surface then analyte of interest specifically bind to the capture antibody and followed by the binding of detection antibody labeled with tags. However, GMR biosensor relies on magnetic tags whereas ELISA relies on optical tags (fluorescent molecules). Biological samples such as blood, urine, serum, etc. exhibit virtually no magnetic background noise. Thus, biosensors utilizing magnetic tags can achieve high sensitivity measurements on minimally processed samples. To achieve similar performance characteristics as that of the current optical and electrochemical platforms, GMR biosensors are usually arranged in arrays and can perform multiplexed detection on a panel of target analytes in a single assay. Furthermore, benefiting from the advances in modern nanofabrication, GMR biosensing chips can be manufactured at an extremely low cost in mass quantities and deployed in a single time use-and-throw format, in addition, they are compatible with the current VLSI (very large scale integration) technology [12]. In this section, we will introduce different types of GMR-based immunoassays and review some representative works reporting the protein biomarker quantification and disease diagnosis.

2.5.1.1 Wash-free and Non-wash-free Immunoassays

Immunoassay is a biochemical test that measures the presence of target biomolecules from biofluidic samples using antibodies or antigens. Generally, GMR-based biosensors utilize a "capture antibody – target antigen – detection antibody" sandwich immunoassay method [90–92]. The general steps for running a GMR-based sandwich immunoassay are listed in Table 2.2 and schematically drawn in Figure 2.4. There are two types of sandwich immunoassay modes regarding whether it requires washing off unbound reagents; namely, non-wash-free and wash-free immunoassays. For the wash-free immunoassay, the biological sample containing target analytes is incubated with detection antibody–magnetic tag complexes (incubation time varies from 20 minutes to 1 hour) before adding to GMR sensor arrays (also called reaction well). Wash-free is a trending immunoassay mode that allows one-step, rapid, on-site disease diagnostics by non-technicians with minimal training requirements. This mode is more likely to be used in future point-of-care (POC) devices handled by patients outside the laboratory.

2.5.1.2 Different Immunoassay Methods

Sandwich immunoassay is so far the most frequently used detection method however, based on different application scenarios, there are several immunoassay methods reported for GMR biosensors. Herein, we highlight four immunoassay methods as shown in Figure 2.5: direct assay, competitive assay, sandwich assay, and indirect assay. In a direct assay (Figure 2.5(A)), the GMR sensor surface is directly coated with target analytes (for example, antigen in Figure 2.5(A2)), then magnetic tag conjugated with antibodies are added for incubation and followed by a wash step to remove unbound antibody–magnetic tag complexes from the medium. The measured magnetic signal is directly proportional to the concentration/quantity

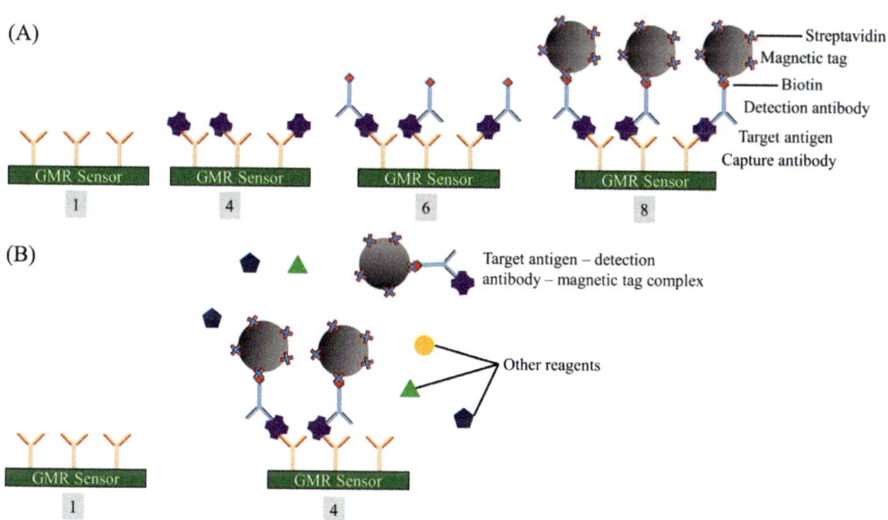

Figure 2.4 Schematic views of GMR-based (A) non-wash-free and (B) wash-free sandwich immunoassay steps. The numbers highlighted in gray represent the step number in Table 2.2.

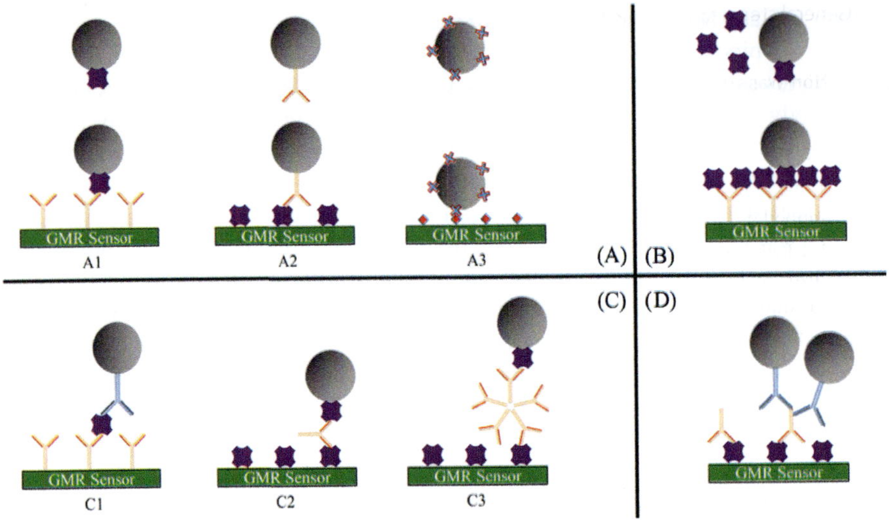

Figure 2.5 Schematic views of different detection methods based on GMR biosensors. (A) Direct assay. Target antibodies (A1), or antigens (A2), or biotin (A3) immobilized on a GMR sensor surface, are detected and quantified by using the (A1) antigen, or the (A2) antibody, or the (A3) streptavidin, conjugated with a magnetic tag. (B) Competitive assay. (C) Sandwich assay. (D) Indirect assay.

of antigens. An example of this direct assay is the positive control group frequently used in GMR-based immunoassays where the biotins (or biotin BSA) are coated on GMR sensors and streptavidin conjugated magnetic tags are added (Figure 2.5(A3)). Due to the high biotin-streptavidin binding affinity, magnetic tags will occupy all the possible binding sites and yield the highest achievable MR signal. This control group serves as indicator of whether GMR biosensors are working and shows the upper limit of signal level.

In a competitive assay (Figure 2.5(B)), a limited number of antibodies are immobilized on GMR sensor surface, target free antigens (or "unlabeled antigens") from the sample will compete for limited antibody binding sites with the antigens that conjugated with magnetic tags. During the incubation, samples with larger antigen concentration/quantity result in more unlabeled antigens bound and less magnetic tags captured. Thus, this detection method produces an inverse relationship between magnetic signal and antigen concentration/quantity. Competitive immunoassays are often used to measure small biomolecules with molecular weight generally less than 10,000 Daltons.

Sandwich assay, also called immunometric assay, is the most frequently used immunoassay method due to its high detection specificity (two antibodies are used for capture and detection). This type of assay method can be further divided into double-antibody (Figure 2.5(C1), mostly used and referred as sandwich assay) and double-antigen (Figure 2.5(C2) & (C3) sandwich assays). This kind of assay is already introduced in Table 2.2 and Figure 2.4. However, it's worth mentioning that double-antigen sandwich assays are capable of detecting IgG (Figure 2.5(C2)) and IgM (Figure 2.5(C3)) antibodies.

In the indirect assay, primary antibodies are first incubated in antigen coated GMR sensors, then secondary antibody conjugated magnetic tags are added for a second step incubation (Figure 2.5(D)). This secondary antibody specifically binds to primary antibody to assist the detection and is usually polyclonal. Thus, more than one secondary

Table 2.2 General steps for running GMR-based sandwich immunoassays: wash-free and non-wash-free.

Modes	Non-Wash-free	Wash-free
Steps	1. Immobilize capture antibodies to GMR sensor surface.	1. Immobilize capture antibodies to GMR sensor surface.
	2. Wash off unbound capture antibodies.	2. Wash off unbound capture antibodies.
	3. Block non-specific binding sites on GMR sensor surface to reduce background noise.	3. Block non-specific binding sites on GMR sensor surface to reduce background noise.
	4. Add biological sample containing target background noise. antigens and allow specific binding (incubation time varies from 20 minutes to 1 hour)	4. Add target antigen – detection antibody – magnetic tag complexes to GMR reaction well to generate signal.
	5. Wash off unbound antigens and other reagents.	5. Calibration curve fitting, data analysis and quantification of target antigens.
	6. Add detection antibodies and allow specific binding (incubation time varies from 20 minutes to 1 hour).	
	7. Wash off unbound detection antibodies.	
	8. Add magnetic tags to GMR reaction well to generate signal.	
	9. Calibration curve fitting, data analysis and quantification of target antigens.	
Pros & Cons	More sensitive by washing off the unbound reagents. Time consuming and needs to be handled by technicians in lab.	Allows rapid detection and can be handled by non-technician with minimal training requirements. Less sensitive since unbound reagents in the mixture can interfere the signal.

antibody–magnetic tag complexes can bind to each primary antibody. This detection method inherently becomes highly sensitive but cross-reactivity and non-specific signal may occur due to the secondary antibody.

2.5.1.3 GMR for Disease Diagnosis

There is a growing panel of protein biomarkers that are being investigated for minute expression changes in order to diagnose a wide variety of diseases. Gani *et al.* applied the GMR biosensors for the multiplexed quantitative detection of a panel of Hepatitis B virus (HBV) serology biomarkers: HBV "e" antigen (HBeAg), HBV surface antigen (HBsAg), and the antibody against HBsAg (anti-HBs) [93]. The calibration curves in Figure 2.6(A) show that this platform achieved three orders of magnitude in dynamic range for each biomarker. The GMR sensor signal, $\Delta MR/MR_0$, is the change of MR (due to the binding of magnetic tags) normalized by sensor's initial MR, expressed in parts per million (ppm). The dashed lines are cut-off points representing negative control sensor signals and indicating the limit of detection (LOD) of the assay platform. Figure 2.6(B) gives the schematic drawings of different

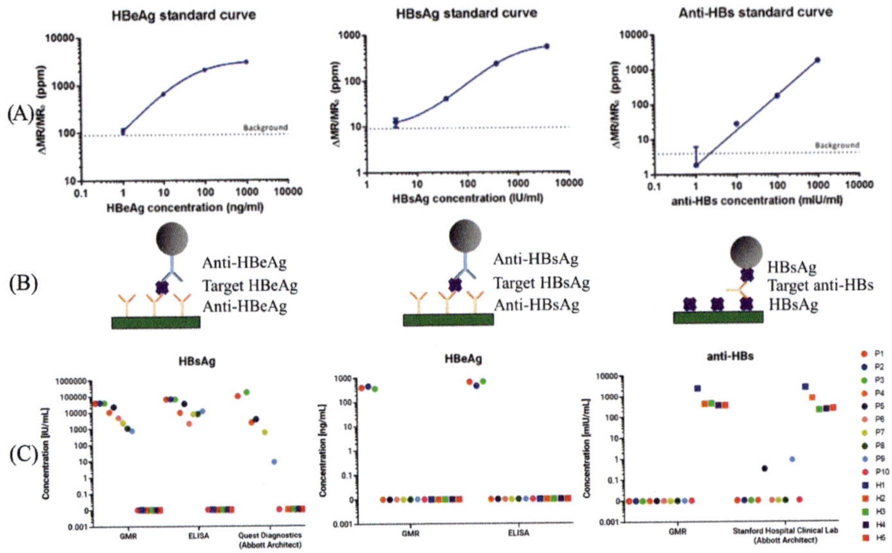

Figure 2.6 (A) HBeAg, HBsAg, and anti-HBs assay standard curves obtained with the automated GMR biosensor system. (B) Schematic view of the corresponding sandwich immunoassay method used. (C) Quantitative comparisons between the serum measurement results obtained with the GMR and other diagnostic methods (ELISA and/or Abbott Architect). Each colored dot represents a single patient's analyte concentration obtained across the different platforms. Healthy sera (squares) do not have HBeAg or HBsAg but may have anti-HBs (e.g., from vaccination). HBV sera (circles) have either HBeAg or HBsAg but do not have free anti-HBs. The measurement results, done with the GMR, ELISA, and Abbott Architect, are consistent and show strong correlation. *Source:* (A) & (C) are reprinted from [93] under the terms of the Creative Commons Attribution 4.0 International License.

detection methods used here: for detecting HBeAg and HBsAg the double-antibody sandwich assay method is used, and for detecting anti-HBs the double-antigen sandwich assay method is used. Blinded analysis of 10 HBV-infected patients (P1-P10) and 5 individuals with no history of HBV infection (H1-H5) as controls are tested on GMR platform. As shown in Figure 2.6(C), all 5 healthy serum samples show negative results for HBeAg and HBsAg as well as adequate anti-HBs levels for protection against HBV infection. While all 10 HBV patient serum samples show low levels of anti-HBs. The multiplexed GMR immunoassay results show excellent agreements with the single-plexed commercial immunoassays – ELISA and Abbott Architect platforms.

GMR-based sandwich immunoassay method has also been reported for the multiplexed detection of ovarian cancer [94] biomarkers (cancer antigen 125 (CA125 II), human epididymis protein 4 (HE4), and interleukin 6 (IL6)), cardiovascular disease [69] biomarkers (pregnancy-associated plasma protein A (PAPP-A), proprotein convertase subtilisin/kexin type 9 (PCSK9), and suppression of tumorigenicity 2 (ST2)), myocardial injury [95] (NT-proBNP and cTnI), and tumor [96] biomarkers (AFP, CEA, CYFRA21-1, NSE, SCC, PG I, PG II, CA19-9, total PSA, free PSA, free-β-hCG, Tg) for the screening of patients with lung cancer, digestive tract cancer, prostatic cancer, etc. Different GMR-based immunoassay works are summarized and compared in Table 2.3.

Table 2.3 Summary of different GMR-based immunoassays.

Diseases	Target biomarkers	LODs	Evaluated matrices	Detection methods	Wash-free?	Assay time	Ref.
HBV	HBeAg	1 ng/mL (~1 IU/mL)	Serum	Sandwich assay	No	NA	[93]
	HBsAg	3 IU/mL					
	anti-HBs	1 mIU/mL					
SLE	Autoantibodies	NA	Serum	Indirect assay	No	NA	[104]
IAV	Nucleoprotein	0.3 nM	Buffer	Sandwich assay	Yes	4 min	[33]
	H1N1 CA/09, H1N1 IA/73, H3N2 VI/11	250 TCID50/mL	Nasal swab				
Tumor [a]	AFP [b]	0.52 ng/mL	NA	Sandwich assay	No	15 min	[96]
	CEA [b]	0.27 ng/mL					
	CYFRA21-1 [b]	0.25 ng/mL					
	NSE [b]	0.5 ng/mL					
	SCC [b]	0.3 ng/mL					
	PG I [b]	1 ng/mL					
	PG II [b]	0.5 ng/mL					
	CA19-9 [b]	2 U/mL					
	total PSA [b]	0.02 ng/mL					
	free PSA [b]	0.07 ng/mL					
	free-β-hCG [b]	0.3 ng/mL					
	Tg [b]	1 ng/mL					
Ovarian cancer	CA125 II	3.7 U/mL	Buffer	Sandwich assay	No	20–30 min	[94]
	HE4	7.4 pg/mL					
	IL6	7.4 pg/mL					
Cardiovascular disease	PAPP-A	1 ng/mL	Buffer and serum	Sandwich assay	No	40 min	[69]
	PCSK9	433.4 pg/mL					
	ST2	40 pg/mL					
Myocardial injury	NT-proBNP	0.33 pg/mL	Buffer	Sandwich assay	No	20 min	[95]
	cTnI	1.43 pg/mL					

a) Tumor including lung cancer, liver cancer, digestive tract cancer, prostatic cancer, etc.
b) Alpha-fetoprotein (AFP, 70 kDa), carcinoembryonic antigen (CEA, 180 kDa), cytokeratin 19 fragment (CYFRA21-1, 40 kDa), neuron specific enolase (NSE, 47 kDa), squamous cell carcinoma (SCC, 45 kDa), pepsinogen I (PG I, 42 kDa), pepsinogen II (PG II, 42 kDa), carbohydrate antigen 19-9 (CA19-9, > 400 kDa), total prostate specific antigen (total PSA, including PSA-ACT (90 kDa) and free PSA), free prostate specific antigen (free PSA, 34 kDa), free β-subunit of human chorionic gonadotropin (free-β-hCG, 23.5 kDa), thyroglobulin (Tg, 660 kDa).

Another example of GMR-based multiplexed detection of systemic lupus erythematosus (SLE) was reported by Lee *et al.* where an indirect assay method was used. SLE is a chronic and autoimmune disease that the immune system of the body mistakenly attacks healthy tissues. Individuals with severe SLE show dysregulated interferon pathways and, the chemokine score, based on levels of three interferon-regulated chemokines (MCP1 (Monocyte chemotactic protein 1, CCL2), IP 10 (Interferon gamma-induced protein 10, CXCL10) and MIP3β (Macrophage inflammatory protein-3-beta, CCL19)), was reported to be positively correlated with the disease activity [97–99]. In this work, the authors used GMR sensor arrays to analyze serum samples from SLE patients and found that individuals with reportedly high chemokine scores show significantly greater reactivity to 13 auto antigens than individuals with low chemokine scores. Herein, GMR sensor array chips each with 72 effective sensors are used for multiplexed detection. The schematic view of assaying antibody reactivity to autoantigens is given in Figure 2.7(A). Autoantigens are first coated on individual GMR sensor's surface (Figure 2.7(A1)), then serum sample is added to the reaction well allowing antibodies to bind to the corresponding autoantigens (Figure 2.7(A2). After a wash step to remove unbound reagents, the biotinylated anti-IgG antibodies are added as the secondary antibody (Figure 2.7(A3). After another wash step to remove unbound antibodies, streptavidin-coated magnetic tags are added to bind with the biotinylated antibodies (Figure 2.7(A4)), the respective GMR sensor detects the stray field generated by the captured magnetic tags. Serum samples from 15 SLE patients with high chemokine scores and 15 SLE patients with low chemokine scores are drawn (15 healthy samples as control group are also measured but not shown here), as shown in Figure 2.7(B). The Significance Analysis of

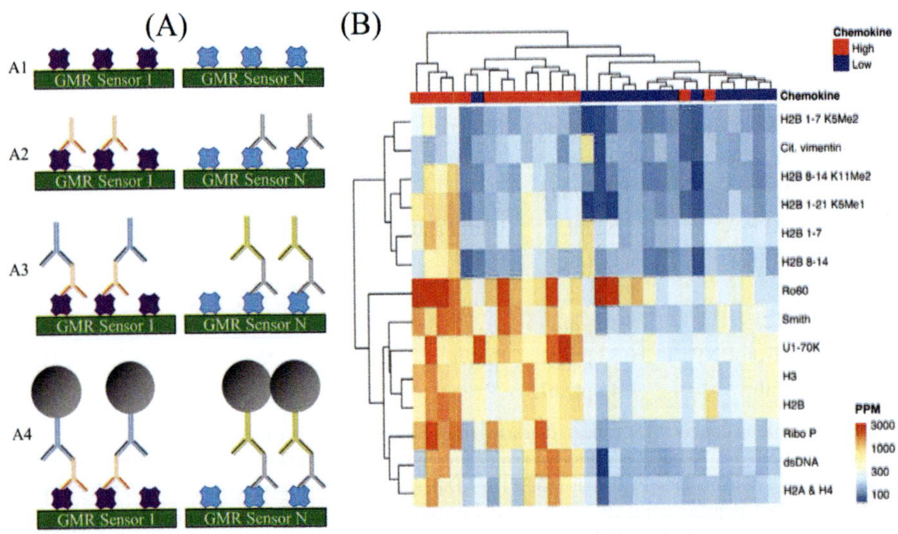

Figure 2.7 (A) Schematic view of GMR-based indirect assay method for assaying antibody reactivity to autoantigens. (B) Significance analysis of microarrays (SAM) was used to identify autoantigens with significantly different antibody reactivity between the groups ($q < 0.001$, Wilcoxon). An unsupervised hierarchically clustered heatmap (Euclidean distance, complete linkage) shows the candidate autoantigens identified by SAM. *Source:* (B) Reprinted from [104] under the terms of the Creative Commons Attribution 4.0 International License.

Microarrays (SAM) algorithm is used to identify statistically significant differences in autoantigen reactivity between the high and low chemokine score groups. SAM identified 13 autoantigens and double-stranded DNA (dsDNA) show greater reactivity in SLE patients with high chemokine scores than low chemokine scores. Although only anti-dsDNA [98, 99] has been confirmed to be associated with chemokine score in SLE, this work further confirmed other autoantibody associations reported previously, including smith [100–102], Ribo P [100] and specific histone proteins (H1, H2A, H2B, and H3/H4) [103]. Identifying autoantibodies associated with the dysregulated interferon pathways in SLE can provide useful information for understanding SLE pathogenesis and monitoring the treatment of those dysregulated interferon pathways. This work reveals the feasibility of applying GMR sensor arrays for multiplexed detection of autoantibodies in autoimmune diseases.

Wash-free GMR-based immunoassay method has emerged in recent years and is attracting great interest for future on-site health diagnostics. This one-step, rapid, and easy-to-use immunoassay method allows the disease diagnosis in a variety of locations such as outdoor, at home, in the physician office, in the urgent care clinic and other point-of-care locations. Su *et al.* reported a one-step, wash-free, sandwich assay method-based GMR platform for detecting influenza A virus (IAV) [33]. To perform wash-free GMR immunoassay, they premixed 5 μL of 150 μg/mL biotinylated IAV detection antibody, 25 μL of magnetic tags, and 20 μL of biofluid sample. This mixture with a total volume of 50 μL is placed on a shaker incubator for 1 hour for the specific binding of target analytes, then transferred to GMR reaction well for magnetic signal measurement. The capture antibodies from GMR sensor surface specifically capture the target analyte – detection antibody – magnetic tag complexes, forming a sandwich structure as shown in Figure 2.8(A). The real-time $\Delta MR/MR_0$ signals are monitored after the adding of sample mixtures in a period of 4 min, as shown in Figure 2.8(B1). Due to the capture of magnetic tags, the magnetic signal increases linearly (except for 17.7 nM sample). In order to quantify the concentrations of target analytes from unknown samples, calibration curves are plotted. One calibration curve is shown in Figure 2.8(B2) using the $\Delta MR/MR_0$ values collected at t=4 min. The other calibration curve, given in Figure 2.8(B3), is based on signal slope values obtained by linear regression of real-time $\Delta MR/MR_0$ curves from t=0 to t=4 min.

To validate this wash-free assay in detecting IAV from complex matrix, nasal swab samples from IAV negative pig are spiked with different concentrations of human pandemic influenza A/California/04/2009 (H1N1 CA/09), human influenza A/Victoria/2011 (H3N2 VI/11) and swine influenza virus A/Swine/Iowa/73 (H1N1 IA173). As shown in Figure 2.8(C), different concentrations of H1N1 CA/09 (Figure 2.8(C1)) and H3N2 (Figure 2.8(C2)) virus are spiked into nasal swab samples. To provide a reference, ELISA testing is carried out in parallel and compared with GMR testing results. Different unpurified IAV isolates, H1N1 IA/73, along with H1N1 CA/09 and H3N2, with unknown concentrations, are randomly selected from and tested on GMR and ELISA platforms. All these three IAV strains show positive signals and similar trends on both platforms. Proving the feasibility of applying wash-free GMR immunoassay for detecting diseases from complex biological matrix. Although compared to other GMR-based, non-wash-free methods, this wash-free approach shows slightly lower sensitivity, it dramatically simplifies the assay preparation steps and which is crucial for a device designed for non-technicians to perform rapid and onsite diagnostics.

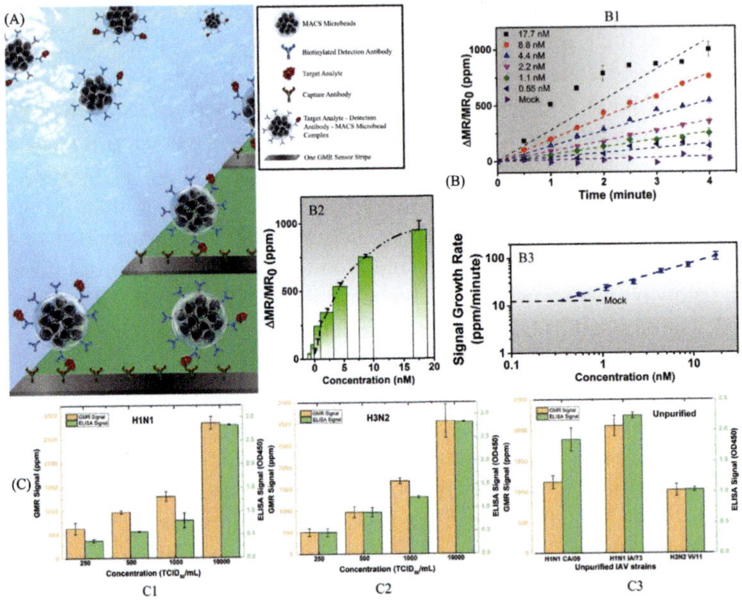

Figure 2.8 (A) Schematic view of one-step, wash-free, sandwich immunoassay method-based GMR immunoassay. 50 μL of premixed target analyte – detection antibody – magnetic tag complexes are added to GMR reaction well. (B) The real-time binding curves and calibrations curves. (B1): real-time binding curves of different concentrations of IAV nucleoprotein samples over a 4 minute testing window. (B2): Calibration curve of IAV nucleoprotein bioassays, Y-axes is $\Delta MR/MR_0$. The mock level is plotted in dashed line. (B3): Calibration curve of IAV nucleoprotein bioassays, Y-axes is magnetic signal slope obtained by linear regression from t=0 to t=4 min. (C) GMR and ELISA results for nasal swab samples spiked with 250, 500, 1000, and 10000 $TCID_{50}$/mL of H1N1 CA/09 (CI) and H3N2 (C2) virus as well as other IAV strains. *Source:* Figure reprinted from [33] under the terms of the Creative Commons Attribution License (CC BY).

2.5.1.4 GMR-based Point-of-Care (POC) Devices

Considering a large number of analytes that require routine monitoring and the increasing demands for more accessible and inexpensive routine diagnostics from patients, point-of-care (POC) diagnostic kits is one strategy to address these challenges, to improve the disease surveillance and to mitigate the financial burden on healthcare systems. In this section, we introduce three GMR-based POC devices independently developed by different teams. Figure 2.9(A) shows a GMR-based POC device, Z-Lab, reported by a University of Minnesota team. This device is capable of real-time data collection and transmission through wireless connection via Bluetooth to a smartphone, a tablet and a laptop, or through a USB connection to a computer. This Z-Lab platform consists of three parts: a plastic cartridge, an electrical interface connecting the electrodes from GMR chip to the circuit board, and a handheld device. Figure 2.9(B) shows another GMR-based POC device reported by a Stanford University team, the Eigen Diagnosis Platform (EDP). It consists of three parts: a disposable cartridge, a reader station, and a smartphone running the Eigen Diagnosis App. The reader acquires signals from the GMR biosensors in the cartridge and transfers the data to the smartphone for signal processing. Figure 2.9(C) shows a commercialized GMR-based POC product from Dongguan Bosh Biotechnologies, Ltd. This platform mainly consists of 4 parts: a temperature control system, a microfluidic actuating system, a signal processing system, and a data processing system.

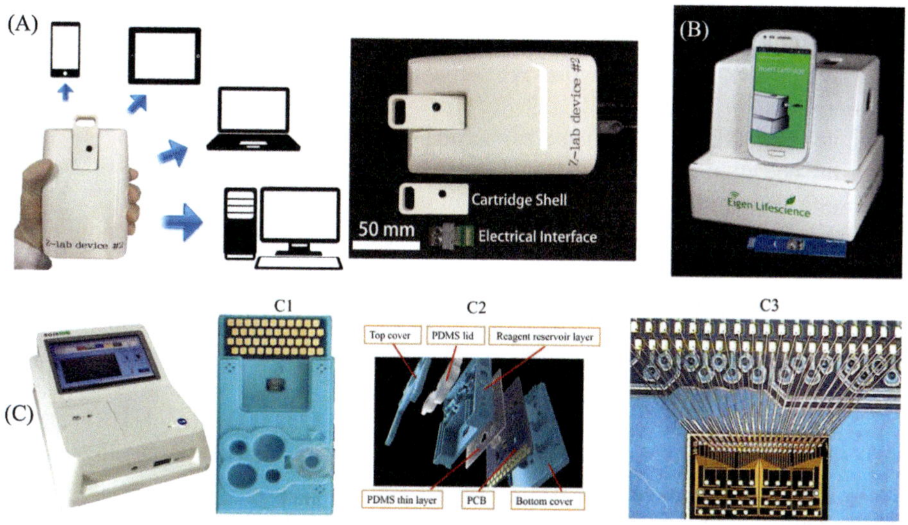

Figure 2.9 Images of different GMR POC devices reported. (A) The Z-Lab diagnosis platform reported by a University of Minnesota group. (B) The Eigen Diagnosis Platform (EDP) reported by a Stanford University group. (C) GMR immunoassay analyzer product from Dongguan Bosh Biotechnologies, Ltd. (C1): the test card; (C2): the multilayer structure of the test card; (C3): one GMR chip and its connections to the circuit board. *Source:* (A) Reprinted with permission from [67], copyright 2017 American Chemical Society. (B) Reprinted with permission from [17], copyright 2016 Elsevier B.V. (C) Reprinted with permission from [96], copyright 2018 Elsevier B.V.

2.5.2 GMR-based Genotyping

Genotyping is a technology that identifies small variations in genetic sequence within populations. This kind of genetic variant may lead to physical differences that make us unique and pathological changes underlying disease. This technology has a broad range of applications including the personalized healthcare by providing disease prognosis and designing targeted therapies specific to the genetic basis of diseases [105–107]. Genotyping is important in the research of genes and gene variants associated with diseases. It also applies to a broader range of organisms including microorganisms such as viruses and bacteria. Nowadays, many genotyping methods are developing in parallel including polymerase chain reaction (PCR), DNA sequencing, and hybridization to DNA microarrays or beads. Coupling these techniques with standard nucleic acid amplification technologies, GMR sensor arrays have been reported for detecting ultra-low-concentration/quantity of targe nucleic acids for disease prognosis [108–111]. In this section, we will introduce several GMR-based genotyping combined with different nucleic acid amplification technologies.

GMR-based genotyping relies on the mature nucleic acid amplification methods to amplify target analytes. Generally, capture probes complementary to target nucleic acid strands are coated on GMR sensors followed by the addition of the generated amplicons to allow sequence-specific binding. Usually the amplicons are carrying biotin for magnetic tags to conjugate. Thus, a capture probe – target analyte – magnetic tag structure is formed on GMR sensors and magnetic signals proportional to concentration/quantity of target analytes are monitored. It is worth mentioning that the nucleic acid amplification products (i.e., amplicons) are initially dsDNA: thus, actions need to be taken to split these amplicons into singe-stranded DNA (ssDNA) so that they can hybridize with the capture probes, which is not the focus of this book [19, 110]. Herein, we give some examples to explain the mechanisms underlying GMR-based genotyping. Different types of GMR-based genotyping works are also summarized in Table 2.4.

Zhi *et al.* combined the loop-mediated isothermal amplification (LAMP) with GMR biosensors for the detection of HBV DNA molecules, they achieved a LOD of 10 copies/mL within a testing period of 1 h [109]. As shown in Figure 2.10(A), three conserved LAMP primers B3, F3, BIP (B1c–B2) and biotinylated primer FIP (Flc–F2) are used to amplify the conserved domain of HBV genome. Each LAMP reaction contains 1.611M each FIP and BIP, 0.2 μM each F3 and B3, 1.4 μM dNTP, 1 M Betaine, 20 mM Tris-HC1 (pH 8.8), 10 mM KC1, 10 mM $(NH4)_2SO4$, 8 mM $MgSO_4$, 0.1% Triton X-100, 8U Bst DNA polymerase large fragment and 1 μL HBV DNA template. This mixture is incubated at 63°C for 1 h. The capture probes are coated on the GMR sensor surface: probe T can hybrid with all kinds of HBV DNA genotypes, while probes B and C can only specifically hybrid with HBV DNA genotypes B and C, respectively. During the 1 h LAMP incubation period, these probes will specifically hybridize with targe HBV DNA fragments produced by LAMP. After the hybridization step, streptavidin conjugated magnetic tags are added and captured by the biotins locate at the terminals of target DNA strands, this incubation step takes 15 min at room temperature and the unbound reagents are washed off for 3 min.

Genotypes B and C HBV plasmids as target DNA are prepared for LAMP amplification. As shown in Figure 2.10(B), electrical resistances of different GMR sensors coated with probes T, B, and C are collected. There are 4 GMR sensors, there of them are coated with probe T, probe B, and probe C, respectively and one is blank control group (NTC). The results show strong specificity in detecting different HBV genotypes. In Figure 2.10(B1), GMR sensors coated with T and B probes show positive signal changes while NTC and C

Table 2.4 Summary of different GMR-based genotyping.

Diseases	Target analytes	Amplification technologies	LODs	Evaluated matrices	Ref.
HBV	HBV genotypes B and C	LAMP	10 copies/mL	NA	[109]
HPV	HPV16, 18, 33, 45 strains	PCR	10 pM	NA	[111]
Cancer (a)	cfDNA (a) fragment ALU115	PCR	30 pM	Plasma	[110]
	cfDNA (a) fragment ALU247		20 pM		
DNA mutation and (hyper-)methylation	Muted and methylated DNA	PCR	NA	NA	[65]
Melanoma cancer	Methylated DNA	PCR	0.1% (b)	Buffer	[113]

a) cfDNA fragments ALU115 and ALU247 are reported for the diagnosis of Gliomas, lobular breast cancer and colorectal cancer [114–116].
b) The LOD is defined by differentiating the melting temperatures of samples mixed with down to 0.1% methylated DNA.

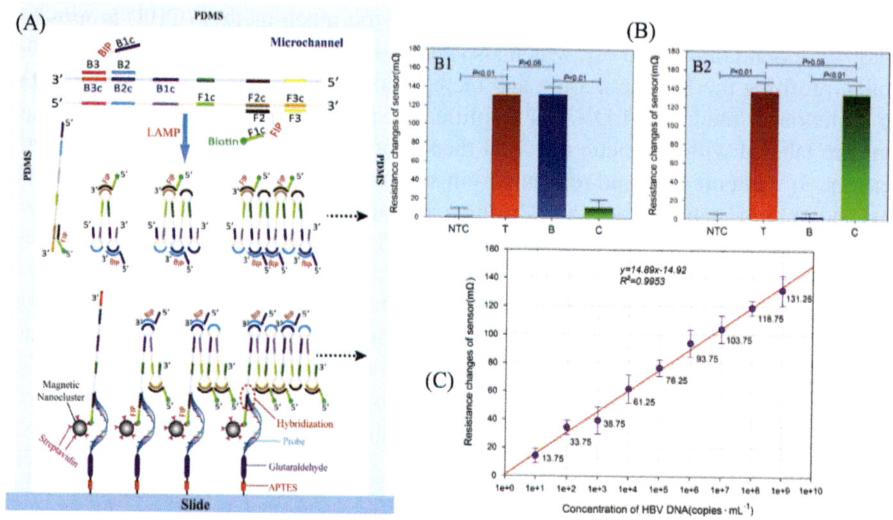

Figure 2.10 (A) Graphic of LAMP and principle of genotyping on a GMR sensor array. (B) Electrical resistances of GMR sensors monitored for detecting genotype B (B1) and genotype C (B2) HBV DNA. (C) GMR-based genotyping calibration curve for detecting genotype C plasmids of HBV. *Source:* Figure reprinted with permission from [109], copyright 2013 Elsevier B.V.

sensors show negligible changes, which means the sample is genotype B HBV DNA. Similarly, in Figure 2.10(B2), only GMR sensors coated with T and C probes show positive signal for sample genotype C HBV DNA. This result indicates the high specificity of capture probes in detecting different genotypes of HBV. A calibration curve based on detecting 10-fold dilutions of genotype C plasmids of I-IBV is given in Figure 2.10(C). Linear magnetic responses of GMR sensors are observed and a LOD of 10 copies/mL target HBV DNA is achieved.

Xu *et al.* reported the GMR sensors for HPV genotyping combined with PCR technology [111]. HPV plasmids for HPV16, 18, 33, and 45 are first amplified by PCR then added to GMR sensor arrays that coated with corresponding probes, followed by the addition of magnetic tags. Dias *et al.* applied GMR sensors to the detection of circulating cell-free DNA (cfDNA), which is seen as a promising source for liquid biopsies in cancer diagnostics [110]. The analysis of total cfDNA levels in blood along with its integrity (size of cfDNA fragments) give insights on the assessment of cancer, tumor staging and metastatic potential [112]. Herein, DNA capture probes complementary to target cfDNA fragments ALU115 and ALU246 are coated on GMR sensors and followed by a washing step to remove unbound probes. The authors used PCR to amplify two cfDNA fragments: ALU115 (115 bp) and ALU247 (247 bp). The generated dsDNA amplicons are purified from contaminants, then "digested" by Lambda exonuclease into ssDNA. Then these PCR products are conjugated to magnetic tags through streptavidin-biotin interactions before adding to GMR reaction well.

In addition to the direct detection of target nucleic acids for disease diagnosis, GMR sensor arrays have also been reported for the detection of DNA mutation and (hyper-)methylation that caused by virus/pathogen infections [65, 113]. The viruses target DNA promoter regions of the genome, which are sections of DNA that initiate the transcription of a particular gene. These regions are abundant in CpG nucleotides and act as switches for gene expression. When they are hypermethylated (by adding too much methyl (CH3) groups to DNA), they are essentially turned off, which also silences that gene [117].

Generally, profiling the DNA mutation and methylation includes the following steps: 1) genomic mutation or methylated DNA is amplified by nondiscriminatory primers; 2) the amplicons are labeled with magnetic tags and then hybridized to the capture probes from GMR sensors; 3) wash off unbound reagents from sensors; 4) carry out melting curve measurements and monitor magnetic signals. During a melting curve measurement, the temperature of GMR sensors is ramped at a constant rate from 20°C to 65°C (or even higher) where all the DNA hybrids are melted. The GMR sensor signal, ΔMR, is corrected for its temperature dependency. The melting temperature is defined as the temperature at which the signals, ΔMR, dropped to the half of its initial signal (at 20°C). Due to the mutation and methylation in DNA, the base mismatch between amplicons and capture probes results low-stringency DNA hybridization thus, shows lower melting temperature. With the addition of multiplexing capabilities, GMR sensor offers the advantage of analyzing multiple DNA templates' hybridization stringencies in a single platform.

2.5.3 GMR-based Bio-magnetic Field Recording

Generally, GMR-based immunoassays and genotyping rely on magnetic tags (e.g., magnetic nanoparticles or magnetic microbeads) to alter the local magnetic fields in the presence of external excitation fields. Magnetic tags convert the concentration/quantity of biological analytes to detectable magnetic fields for GMR biosensors and then, GMR biosensors convert these magnetic fields to readable electric signals (e.g., sensor's electrical resistance, ΔMR, $\Delta MR/MR_0$).

In this section, we will introduce the application of GMR sensors for the direct detection of magnetic fields produced by the electrical activities of living cells; namely, bio-magnetic fields. The human body produces electrical activities in several different types of cells such as neurons and muscle cells, which creates a magnetic field. Figure 2.11 gives a straightforward view of various environmental and biological magnetic fields and the currently available magnetic sensing technologies. Nowadays, superconducting quantum interference

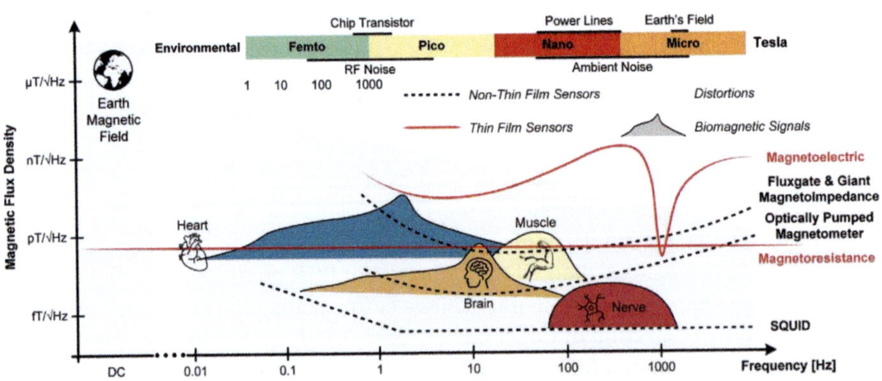

Figure 2.11 A summary of the biomagnetic field strengths of various examples in comparison to the existing magnetic sensing technologies. *Source:* Figure reprinted from [124] under the terms of the Creative Commons Attribution License.

device (SQUID) is the most sensitive device that can measure femto-Tesla (10^{-15} T) magnetic fields, and possibly achieve attoTesla (10^{-18} T) with a few days of averaged measurements. Such high sensitivity level puts SQUIDs for the applications of *in vivo* magnetoencephalography (MEG) and magnetocardiography (MCG) on animals [118–123]. However, running SQUID requires a magnetically shielded room with a cooling system for operation at liquid-helium temperature of 4.2K, which makes it expensive and bulky.

Another fast-developing magnetic sensing technology is the magnetoresistive (MR, including GMR, AMR and TMR) platform, which is seen as an effective alternative pico-Tesla (10^{-12} T) biosensing approach at room temperature. MR sensors hold many advantages over SQUIDs such as: 1) low cost, sensors can be mass manufactured inexpensively; 2) can measure the field at room temperature without cooling; 3) compatible with the current VLSI; 4) have wide dynamic range of detectable fields; 5) can operate without magnetic shield room (basic shielding preferred); 6) have excellent temperature characteristics, a Wheatstone Bridge circuit to remove the thermal fluctuation caused noise.

Herein, we will give some examples of GMR-based bio-magnetic fields detection. Caruso *et al.* reported GMR sensors for *in vivo* recording of neuronal activity from visual cortex of cats [125]. In their experiment, a visual stimulus is directly applied to the contralateral eye of cats (Figure 2.12(A)). The GMR sensors (Figure 2.12(A) inset) positioned within visual cortex collect event-related fields (ERF) and a tungsten electrode positioned 1 mm from the GMR simultaneously obtain an independent event-related potentials (ERP). The noise

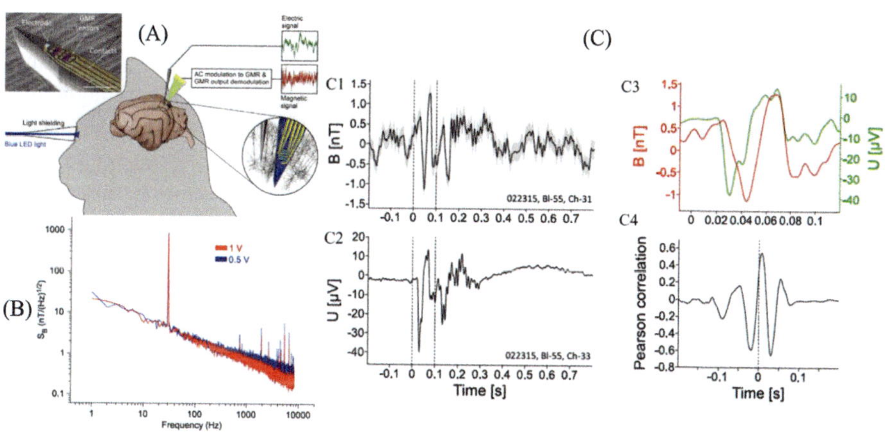

Figure 2.12 (A) Experimental setup for in vivo recording of neuronal activity from visual cortex of cats. The zoomed-in inset illustrates the expected configuration of the GMR sensor and tungsten electrode in the neuropil. The inset photograph shows the scanning electron microscopy (SEM) of sensing probe containing 2 GMR sensors. Scale bar, 100 μm. (B) Equivalent-field noise spectral density SB of GMR sensors from 1 Hz to 10 kHz for DC input voltages corresonding to AC input voltages of 0.5 V (blue) and 1 V (red). (C) Comparison of ERF and ERP obtained from cat. (C1): ERF signal obtained by averaging the GMR in-phase output over 1000 stimulus repetitions. (C2): ERP signal obtained by averaging the tungsten electrode over the same 1000 stimulus repetitions. The dashed vertical lines indicate onset and offset of 100 ms long visual stimulus. (C3): Direct comparison of the waveforms of the ERF (red) and the ERP (green). (C4): Pearson correlation coefficient between the ERF and the ERP as a function of time lag, with positive lag values indicating that the ERF lagged the ERP. Figure reprinted with permission from [125], copyright 2017 Elsevier Inc.

spectrum of GMR sensors at a typical input voltage of 0.5 V leads to sensitivities of $7\,nT/\sqrt{H_z}$ at 10 Hz, $2\,nT/\sqrt{H_z}$ at 100 Hz and $370\,pT/\sqrt{H_z}$ in the thermal noise regime above 1 kHz (Figure 2.12(B)). Figure 2.12(C) shows the comparison between simultaneously recorded ERF (by GMR sensors) and ERP (by tungsten electrode) from one cat. The ERF in Figure 2.12(C1) shows magnetic responses starting at 20 ms after the onset of stimulus which corresponds to the conduction delay between retina and primary visual cortex. One strong negative peak and one strong positive peak are observed at 36 ms and 61 ms, respectively. The peak to peak amplitude is 2.5 nT. The ERP in Figure 2.12(C2) shows a similar trend. The real-time (within the 100 ms visual stimulus duration) magnetic field data (ERF) and electric potential data (ERP) are scaled and superimposed in Figure 2.12(C3) to facilitate comparison. Figure 2.12(C4) shows the Pearson correlation coefficient between ERF and ERP as function of time lag, with positive values indicating ERF lagged the ERP. The maximum correlation value reaches to 0.55 for a lag of approximately 2 ms. Other works such as *in vitro* magnetic response recording of mice hippocampus slice upon stimulus [126–128] are summarized in Table 2.5.

Pannetier-Lecoeur *et al.* applied the GMR sensors for recording bio-magnetic fields generated by the electric activities of human heart from healthy volunteers [78, 129]. As shown in Figure 2.13(A), one sensing unit consists of 4 GMR sensors forming a Wheatstone Bridge and mounted on a superconducting flux-to-field transformer (i.e., Niobium loop). Upon the application of external magnetic fields, this Niobium (Nb) loop generates a supercurrent to counteract with the field. The specially designed structure of loop could amplify local

Table 2.5 Summary of different GMR-based biomagnetic field recording.

Function	Sensitivity of GMR sensor	Test objects	Detection mode	Magnetic shielding?	Ref.
MEG	7 nT/√*Hz* at 10 Hz [(a)]	Visual cortex of cats	*In vivo*	No	[125]
	0.15%/Oe [(b)]	Mice hippocampus slice	*In vitro*	No	[126]
	54 nT/√*Hz* at 5 Hz	Mice hippocampus slice	*In vitro*	No	[127]
	33 nT/√*Hz* at 1 kHz	Mice hippocampus slice	*In vitro*	No	[128]
	3 pT/√*Hz* at 1 Hz, T = 4 K	Human heart (chest)	*In vivo*	Yes	[129]
MCG	3 pT/√*Hz* at 1 Hz T = 4 K	Human heart (chest)	*In vivo*	Yes	[129]
	2 pT/√*Hz* at 1 Hz, 100 fT/√*Hz* above 300 Hz, T = 4 K	Human heart (chest)	*In vivo*	Yes	[78]
Heart rate recording	950 pT/√*Hz* at 1 Hz	Human heart (chest)	*In vivo*	No	[130]
	NA	Human wrist	*In vivo*	No	[130]
Muscle action potential	100 pT/√*Hz* at 1 Hz	Mouse muscle	*In vitro*	No	[131]

a) GMR sensor sensitivity is represented by the equivalent-field noise spectral density in the unit format of nT/Hz.
b) GMR sensor sensitivity is represented by the transfer curve slope in the unit format of %/Oe.

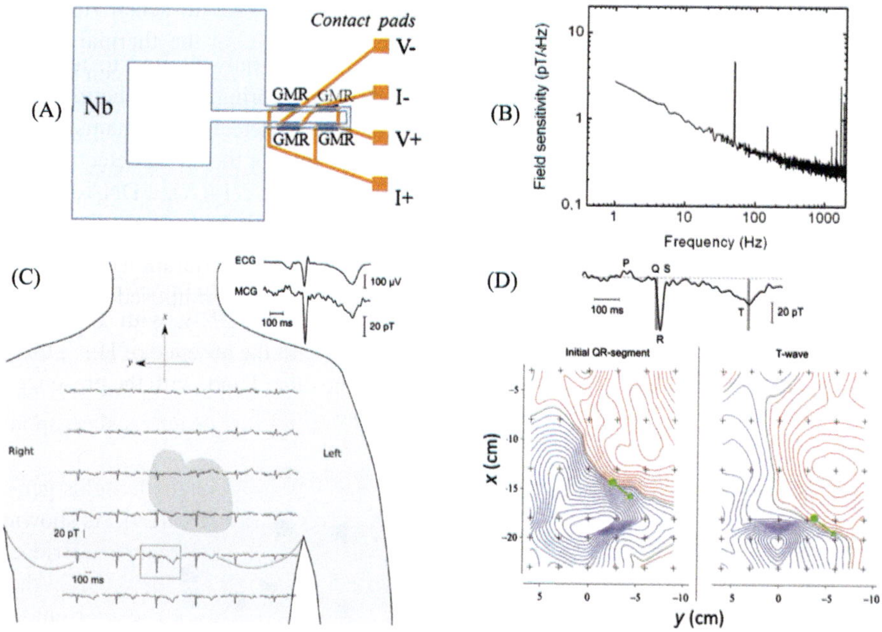

Figure 2.13 (A) Schematic view of 4 GMR sensors wired to form a Wheatstone Bridge and mounted on a Niobium loop. (B) Equivalent-field noise spectral density of GMR sensors measured at 4 K. (C) MCG signals measured at multiple locations on a single subject. The enclosed MCG trace is shown enlarged with the ECG signal. Passband 0.1–40 Hz. The measurement locations are separated by 3 cm horizontally and by 5 cm vertical (in the foot–head direction) except that the lowest rows are separated only by 2 and 3 cm due to space constraints. (D) Iso-contours of the normal component of the magnetic field during the initial QR-segment and T-wave; blue and red indicate field in and out, respectively, of the chest of the subject. The green arrows denote the fitted equivalent current dipoles. *Source:* Figure reprinted with permission from [129], copyright 2011 American Institute of Physics.

magnetic fields which are sensed by the GMR sensors [77, 132]. The equivalent-field noise spectral density of these GMR sensors are given in Figure 2.13(B). Since the MCG signals are on the order of several tens of pT, external magnetic field may interfere the results, so the authors carried out the MCG recording in magnetically shielded room. Volunteers lay down on his/her back with GMR sensors positioned 25–30 mm above from the skin. As shown in Figure 2.13(C), a 6 × 6 point grid of measurement locations is marked and MCG signals are recorded for 1 min (and averaged over 36 continuous recordings) at each location simultaneously with the ECG recording. By comparing the ECG and MCG waves (inset in Figure 2.13(C), the characteristic components (P-wave, QRS complex and T-wave as labeled P, Q, R, S, T in Figure 2.13(D) are all present in magnetic signals. In addition, the magnetic field maps are reconstructed in Figure 2.13(D) and equivalent current dipoles are modeled (green arrows).

Furthermore, GMR sensors have also been reported for the recording of heart rate (HR) from human and action potential (AP) from mouse skeletal soleus muscle [79, 130, 131]. Which will not be covered here but summarized in Table 2.5.

2.5.4 GMR-based Food and Drug Safety Supervision

In the area of biomedical applications, GMR biosensors are not only limited to testing biological samples but also been used in food and drug safety monitoring. In this section, we will briefly introduce these applications along with the underlying detection mechanisms.

Wang *et al.* have combined GMR biosensor and DNA chemistry for the direct detection of mercuric ion (Hg^{2+}) from drinking water [133]. As shown in Figure 2.14(A), a DNA-based sandwich assay method is used. Where complementary DNA pairs with deliberately designed thymine (T)- thymine (T) mismatches are used. These T-T mismatches can selectively detect Hg^{2+} by forming T- Hg^{2+}-T pairs. The capture DNA is at first immobilized on GMR biosensors then a mixture of biotinylated DNA strands (labeled as biotin-DNA, with T-T mismatches to capture DNA) and sample containing Hg^{2+} is added. In the absence of Hg^{2+}, this biotin-DNA can hardly hybridize to capture DNA while on the other hand, with the presence

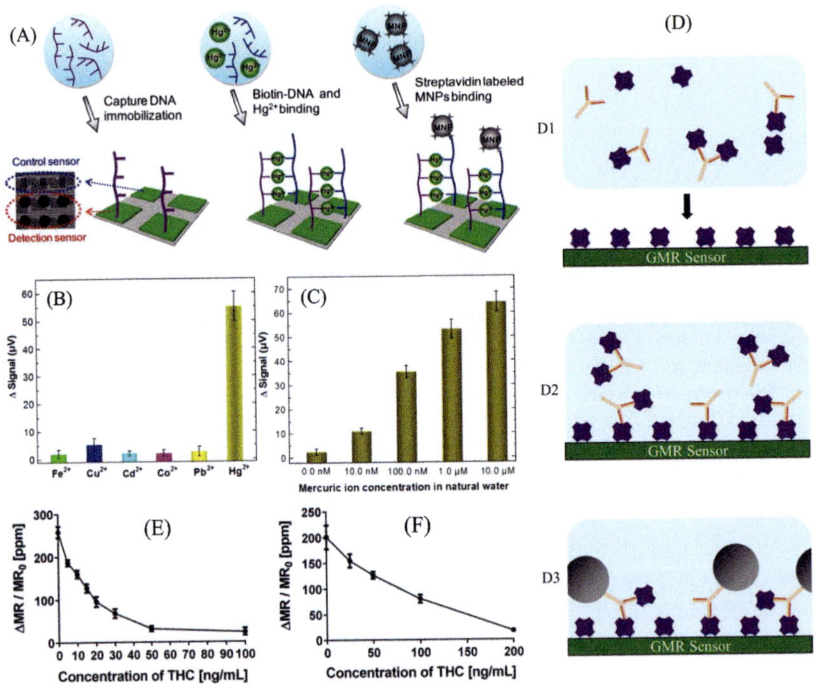

Figure 2.14 (A) Schematic view of DNA-based sandwich assay method for the detection of Hg^{2+}. (B) Average GMR signal changes for detecting various metal ions ([M^2] = 1.0 μM). The result shows high specificity of this DNA-DNA structure used for detecting Hg^{2+}. (C) Average GMR signal changes for detecting different Hg^{2+} concentrations from natural water. (D) Schematic view of competitive assay method for the detection of THC. (D1): biotinylated anti-THC antibodies mixed with THC-containing sample and preincubated to bind to THC. (D2): the mixture is added to THC coated GMR biosensors for unoccupied anti-THC antibodies to bind with. (D3) Wash off unbound reagents and add magnetic tags for GMR signal readout. (E) Calibration curve of detecting THC from saliva samples. THC concentrations are varied from 0 to 100 ng/mL and it requires 15 min of preincubation and 15 min of on-chip incubation. (F) Calibration curve of detecting THC from saliva samples. A 3 min rapid assay method is used. Error bars represent standard deviations. *Source:* (A-C) Reprinted with permission from [133], copyright 2014 American Chemical Society. (E) & (F) Reprinted with permission from [137], copyright 2016 American Chemical Society.

of Hg^{2+}, this biotin-DNA can hybridize to capture DNA due to the T- Hg^{2+}-T complex. The last step is to remove unbound reagents and add magnetic tags (coated with streptavidin to bind with biotin from biotin-DNA) for GMR signal collection. The detection specificity of this DNA-based assay is verified by testing other metal ions such as Cu^{2+}, Co^{2+}, Pb^{2+}, etc. as shown in Figure 2.14(B). Negligible GMR signal changes are observed from samples containing other metal ions, proving the high detection specificity of this assay method. As shown in Figure 2.14(C), this GMR and DNA-based assay shows a LOD of 10 nM for detecting Hg^{2+} from both buffer and natural water, which is the maximum mercury level in drinking water regulated by the US's Environmental Protection Agency (EPA).

This work extends the versatile GMR platform to the monitoring of heavy metal pollution from drinking water and it can be foreseen that this assay method could become a robust contender in the areas of environmental and food safety monitoring. There are other works that reported the GMR biosensors for food safety supervision including foodborne pathogen *Escherichia coli* (E. coli) [134], food allergens such as peanut allergens (Ara h1 and Ara h 2) and wheat allergen Gliadin [135], food toxin such as mycotoxins (aflatoxins B1, zearalenone and HT-2) [136]. These works are summarized in Table 2.6 for readers' information.

Another example of applying GMR biosensing platform to drug safety supervision is reported by Lee *et al.* where the tetrahydrocannabinol (THC), psychoactive substance from marijuana, has been successfully detected from saliva samples [137]. The schematic view of GMR-based competitive assay for THC detection is shown in Figure 2.14(D). Competitive assay method is used here for detecting small molecules such as THC (molecular weight, 314.5 Daltons). The biotinylated antiTHC antibodies are first preincubated with sample for 15 min to allow the binding to THC (Figure 2.14(D1)). Then the mixture is added to GMR biosensors that coated with THC thus, the unoccupied anti-THC antibodies will bind to THC from GMR biosensors (Figure 2.14(D2)). After a washing step to remove the unbound reagents, streptavidin coated magnetic tags are added and captured by anti-THC antibodies (Figure 2.14(D3). The measured GMR magnetic signals, $\Delta MR/MR_0$, are proportional to the bound magnetic tags and inversely proportional to the concentration/quantity of THC in the

Table 2.6 Summary of different GMR biosensors for food and drug safety supervision.

Applications	Target analytes	LOD	Evaluated matrices	Detection methods	Wash-free?	Assay time	Ref.
Food safety	Hg^{2+}	10 nM	Natural water	Sandwich assay	No	40 min	[133]
Foodborne pathogen	E. coli	4 Pm	Buffer	Sandwich	Yes	3 min	[134]
Food allergen	Ara h 1	7 ng/mL	Buffer	Sandwich	No	15–30 min	[135]
	Ara h 2	0.2 ng/mL					
	Gliadin	1.5 ng/mL					
Food toxin	Aflatoxin B1	50 pg/mL	Buffer	Sandwich	No	10 min	[136]
	Zearalenone	50 pg/mL					
	HT-2	33 ng/mL					
Drug, marijuana	THC	NA	Saliva	Competitive assay	No	30 min	[137]
		NA			Yes	3 min	

testing sample. The calibration curve for detecting THC from saliva samples is given in Figure 2.14(E) where it requires 15 min preincubation time and 15 min on-chip incubation time. In addition, a 3 min rapid assay is reported where the mixture of testing sample and anti-THC antibodies is immediately added to GMR chip and incubated for 2 min followed by 1 min of data collection. As shown in Figure 2.14(F), this one-step, wash-free assay method greatly cut down the assay time. Without any preincubation, the calibration curve shows a wider dynamic range for THC from 0 to 200 ng/mL. This shows a promising road-side testing.

2.6 GMR-based Industrial Applications

2.6.1 GMR for Position Sensing

Due to their nature of robustness, contact-free sensing scheme, and the tolerance of contamination, GMR sensors have been widely used in the field of position sensing [138, 139]. Rieger *et al.* developed a Co/Cu multilayer GMR sensor and explored its potential in several different sensing applications including distance, speed, angle and annular speed detection (Figure 2.15) [140]. A magnet can be positioned either above (Figure 2.15(A)) or beside (Figure 2.15(B)) the sensor for the application of a potentiometer or absolute angle measurement,

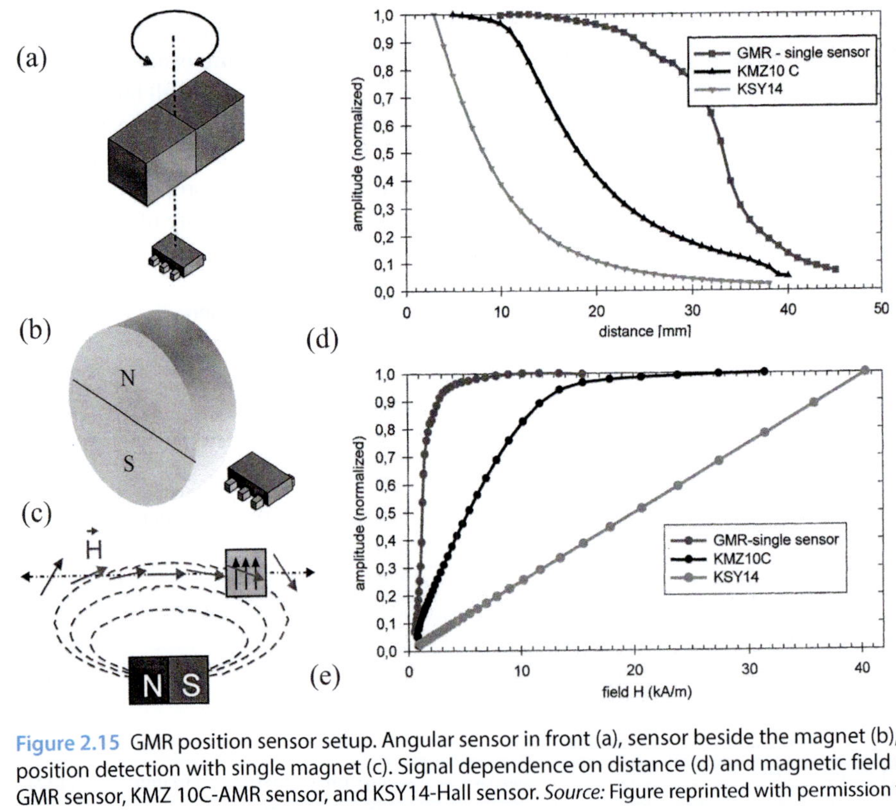

Figure 2.15 GMR position sensor setup. Angular sensor in front (a), sensor beside the magnet (b), linear position detection with single magnet (c). Signal dependence on distance (d) and magnetic field (e) for GMR sensor, KMZ 10C-AMR sensor, and KSY14-Hall sensor. *Source:* Figure reprinted with permission from [140], copyright 2001 Elsevier Science B.V.

respectively. The measurement of linear distance can be realized by measuring the angle variation of the dipolar field from a single rod magnet (Figure 2.15(C)). It was shown that by using a recalibration circuit, a maximum measurable distance of 10 mm with a resolution of 20 μm can be achieved. A comparison of the GMR signal with the signal from AMR and Hall sensors during distance and magnetic field measurement are shown in Figure 2.15 (D) and (E). Compared to Hall sensors and AMR sensors which are sensitive to the distance variation between the magnet and the sensor surface, GMR sensors developed for position sensing measure only the direction of the applied field, thus eliminating the tremendous mounting effort and the complex signal conditioning process. Furthermore, GMR sensors also exhibited more uniform signal in the rotation speed detection compared to Hall sensors.

Besides rotation speed sensing, GMR sensors have also been employed to monitor the traffic speed [141]. The developed GMR sensing system is highly sensitive with the capability of detecting the earth's magnetic field (30 μT to 60 μT). When a car passes by or stands still near the sensing system, the high ferrous content mainly under car parts could provoke local disturbances at around 1 μT, which leads to a change in the sensor signal. After signal amplification and filtration, a spectrum of the sensor signal can be obtained. By comparing the signal spectra at different sampling times, the speed and even the length of the car can be calculated, which is useful to carry out vehicle classification of different types.

Huang *et al.* developed a car positioning system to guide the driver to park into the appropriate position for wireless power charge/discharge [142]. In this setup, a magnetic tape array made with ferrite powders is attached to the ground. The magnetic tapes consist of multiple N- and S -pole sectors so that a detectable change of the output voltage can be generated by the GMR sensors when passing through the tapes. The GMR sensing system is installed at the car chassis. It consists of four sensors with different functions such as speed and position measurement and off-track detection from both sides of the magnetic tapes. The precision of the system is shown to be less than 10 mm, which ensures a wireless power transfer efficiency of more than 85%.

2.6.2 GMR for Current Sensing

Apart from the well-celebrated roles of GMR as a biosensor, in position and angular motion sensing, researchers have found yet another application of this magnetic sensor array in the field of current measurement. Musuroi *et al.* demonstrated the proof-of-concept of a differential GMR-based sensor for non-contact DC/AC current measurement (see Figure 2.16(A)–(C)) [42]. The novelty of their system lies in the double differential measurement system based on commercial GMR sensors, but with an easily adjustable biasing system that is used to linearize the field response of the system. The broad current measurement of the developed system reported was 75 mA for DC and 150 mA for AC. Other fascinating features of the differential system set-up includes a sensitivity between 0.0272 to 0.0307 V/A (i.e., for sensing a current of 1 A, the GMR sensor reports a readout voltage as 0.0272 V), a high immunity to external magnetic fields and low hysteresis effects of 40 mA and an extremely low temperature drift offset of about -2.59×10^{-4} A/°C. Overall, the design has been made such that the system suits the application of detecting fields of extremely low amplitudes. Marcellis *et al.* developed an application specific integrated circuit (ASIC) (see Figure 2.16(D)–(F)) that performs as an oscillator (i.e. can perform voltage-to-frequency conversion) [143]. Experimental results demonstrate a detection sensitivity of 100 Hz/mA with a resolution of ~5 μA (i.e. the lowest current which their GMR sensor can sense is 5 μA).

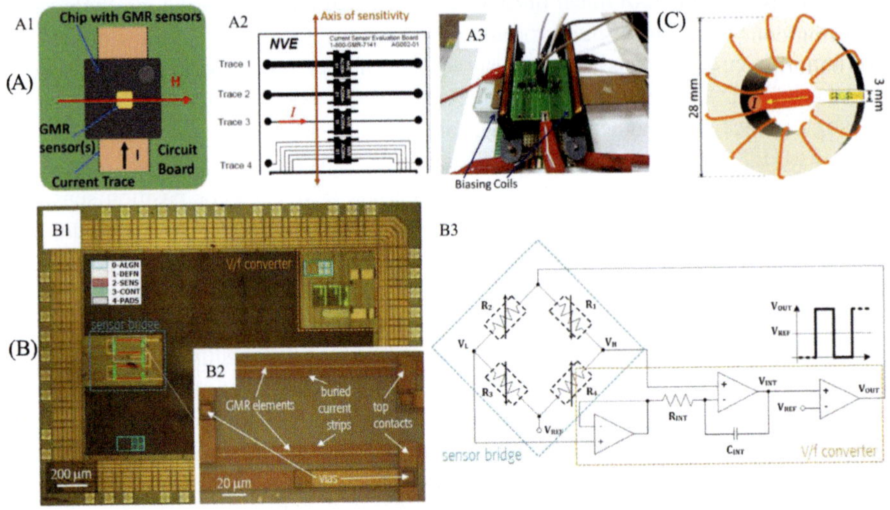

Figure 2.16 (A1) Top view of a GMR-based non-contacting current measurement on a PCB set-up. (A2) Schematic overview of the evaluation board built with commercially available GMR sensors. The red-colored arrow showing the current direction at trace 3 showed the current trace which the authors used for the tests. (A3) The complete image of the GMR-based DC/AC current sensing set-up as reported by Musuroi *et al.* (B1) The custom-built ASIC model for full-bridge, spin-valve GMR-based current sensing as reported by Marcellis *et al.* Inset (B2) Zoomed in image of the final GMR bridge. (B3) The complete schematic of the sensor-ASIC interface. (C) Schematic image of the GMR-based current sensor situated inside a C-shaped ferrite ring core. Source: (A) Reprinted from [42]under the terms of the Creative Commons Attribution License (CC BY). (B) Reprinted with permission from [143], copyright 2017 Elsevier Ltd. (C) Reprinted from [145] under the terms of the Creative Commons Attribution License (CC BY).

The authors calibrated the voltage readout of the GMR sensor in terms of frequency. For instance, for the aforementioned GMR-based oscillator circuit, for a current of 1 mA, the voltage readout of the GMR sensor corresponded to a frequency of 100 Hz. The Verilog-A model for the custom build ASIC circuit for the GMR-based current sensing has been reported [144]. However, this current sensitivity can be significantly improved with additional planarization steps. Jeng *et al.* reported a quasi-static current using field modulation of a full-bridge, GMR sensors surrounded by a ferrite core (see Figure 2.16 (G)) [145]. They applied an AC modulation current to modulate the output of the GMR sensor which resulted in reduction of hysteresis and therefore a linear operation range of –0.5 A–0.5 A. The sensitivity was reported at 0.1 mA at a 10 Hz bandwidth and the output was obtained by demodulating the AC signal or by filtering the output of inherently non-linear GMR sensor. According to the authors, this proposed current sensing scheme can measure DC up to 100 Hz bandwidth, unlike the set-up by Musuroi *et al.* which has decent sensitivity for both AC as well as DC measurements [42]. More breakthrough research in this respect was done by Schuina *et al.* when they integrated GMR sensors with neural networks for enhanced signal analysis and applied them for contactless measurement of direct electrical currents [146]. Although their data analysis seems promising, for better accuracy they need to isolate the sensor magnetically in addition to improving the conditioning circuit. This could reduce offset variations and interferences, generating a more coherent data.

From the analysis of the GMR-based current sensing schemes, it can be concluded that GMR-based non-contact current sensing schemes have a promising future for wirelessly charging of batteries, non-interrupting circuit probes and long-term current monitoring for power grids.

2.6.3 GMR for Material Defect Inspection

Nowadays, the non-destructive testing (NDT) plays an important role in guaranteeing structural integrity of engineering components by detecting and sizing flaws. Among all NDT techniques, eddy current testing (ECT) is the dominating tool to detect, locate, and geometrically characterize the surface and sub-surface defects (i.e., cracks) in electrically conducting non-ferromagnetic materials. Briefly, when passing an alternating electrical current through a conductive coil, an alternating magnetic field of the same frequency is produced around it. As this coil approaches a conductive material (i.e., the test object), an eddy current that opposites to the current in the coil is induced in the material. Variations in the electrical conductivity, magnetic permeability of the test object, as well as the presence of defects can cause a detectable change in impedance of the coil. Due to the nature of the eddy current phenomenon, this ECT technique is in general limited to detecting flaws within a depth of 5.0 mm due to the exponential decay of the eddy current [147].

In order to extend the application of ECT technique for detecting deep seated material flaws, providing higher sensitivity and spatial resolution testing for different NDT applications. GMR sensors combined with ECT coils are reported [148–151]. As shown in Figure 2.17(A), a ECT-GMR system consists of a coil for generation of eddy currents in the material and a GMR sensor receiver (or an array of GMR sensors) for measuring the induced magnetic field. Herein, GMR sensor as a receiver (or sensing probe) is attractive due to its high sensitivity for weak magnetic fields, superior spatial resolution and wide frequency range capability. For example, Jeng *et al.* have reported the ECT-GMR system for detecting subsurface flaws in multilayer aluminum sheets of 1.0 mm thickness [152]. Later in 2011, Cacciola *et al.* have proposed a real-time application of ECT–GMR system for inspection of printed circuit boards (PCBs). This combined system allows for crack inspection over flat surface, especially suitable for micro-defect detection on high density bare PCB. The GMR sensor herein can detect the magnetic field from the interaction between a planar coil exciter and PCB. The eddy current detected by the GMR sensor is at first acquired by a high speed analog-to-digital (A/D) converter, followed by a subsequent application of signal processing based on digital techniques [153]. Postolache *et al.* further modified this ECT-GMR system by adding a neural network processing architecture consisting of multilayer perceptron and a competitive neural network [154]. In this design, the GMR sensor detects a magnetic field component parallel to a plate surface, when an excitation field perpendicular to the plate is imposed. Then the neural network processing architecture classifies defects using the output GMR signal and its operation frequency. Other applications using ECT-GMR systems for detecting material flaws in thin steel sheets [155], riveted structures [156], overhead crane bridge [157], steel bridge [158], etc. A more comprehensive review on applying GMR sensors in non-destructive ECT can be found in Ref. [150].

Herein, we give one example of applying ECT-GMR system for deep seated flaw inspections. Sasi *et al.* carried out Finite Element Model (FEM) and experimental studies on 8.0 mm thick stainless steel plate with surface and subsurface notches of varying depth locations [148]. The ECT-GMR system setup is given in Figure 2.17(B). It consists of (i) an excitation coil driven by sinusoidal current of 100 mA at frequency of 1 kHz, (ii) two GMR sensors (NVE AA002-04) mounted adjacent to the excitation coil and (iii) DC biasing coil for fixing the operating point of the GMR sensors. In order to enable simultaneous measurement of B_x and B_Z, components of induced magnetic field, sensitive axis of the two GMR sensors are oriented parallel and perpendicular to the scanning surface, respectively. The integrated ECT-GMR sensor is mounted on the arm of XY scanner for automated scanning. For separation of phase angle and amplitude information from the acquired signal, I/Q detection technique-based digital lock-in amplifier is implemented in LabVIEW environment. The schematic view of the stainless-steel plate with machined notches of varying depths from 2.0 mm to 7.0 mm, on one side, is shown in Figure 2.17(C). For detecting surface flaws, the amplitude and phase angles of B_x and B, components at 0.5 mm lift-off are measured using voltage responses from horizontally and vertically placed GMR sensors. For detecting subsurface flaws, the stainless-steel plate is scanned from the far side with respect to the flaw location with a lift-off of 0.5 mm – 6.0 mm. The influence on detection sensitivity of the integrated ECT-GMR sensor under varying lift-off conditions has been evaluated. The depth below the surface varies from 1.0 mm to 6.0 mm for flaws D1–D6, as shown in the inset of Figure 2.17(D). The measured amplitude of B_x component from the subsurface flaws in the presence of lift-off varying as 0.5 mm, 1.0 mm, 3.0 mm, and 6.0 mm are shown in Figure 2.17(D). In this work, the horizontal and vertical components of induced magnetic field across deep seated flaws in 8.0 mm thick stainless-steel plate is reliably measured using the integrated ECT-GMR system. The enhanced detection sensitivity of integrated ECT-GMR sensor promises reliable detection of flaws at the early stage in the installed thick components.

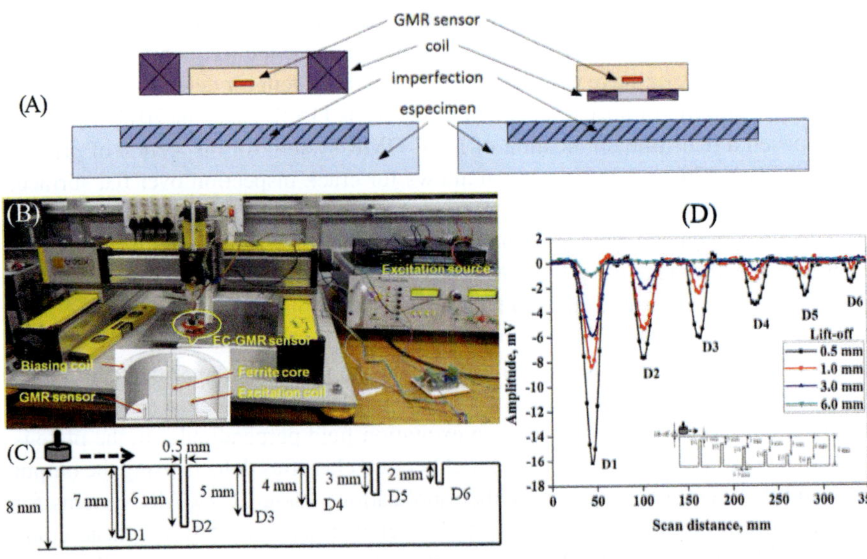

Figure 2.17 (A) Possible geometric arrangements of ECT excitation coils with GMR sensors in an ECT-GMR system. (B) ECT-GMR measurement system. (C) Schematic view of 8.0 mm thick stainless steel plate having surface flaws. (D) Change in amplitude of B_x component across subsurface flaws under varying lift-off conditions. *Source:* (A) Reprinted from [9] under the terms of the Creative Commons Attribution License (CC BY). (B-D) Reprinted with permission from [148], copyright 2018 Elsevier B.V.

2.7 Conclusions and Outlook

In this chapter, we intend to provide a comprehensive review of GMR sensors and its related applications. Based on the spin-dependent scattering of conduction electrons in alternating ferromagnetic and nonmagnetic metallic layers, the GMR effect converts the external magnetic field signal to a measurable electrical signal, which facilitates its application in biosensors, hard disk drives as well as MEMS technologies. Different GMR sensor configurations have been developed to optimize the device performance in these applications. In biosensing applications, thin-film-based GMR sensors with the shape of long stripes possess high aspect ratio, leading to high linearity in response to the external magnetic field, whereas large-area sensors with low aspect ratio increase the active sensing area and enables the capture of a large amount of target analytes.

GMR sensors with granular structures with either top-down and bottom-up approaches have also attracted great attention with comprehensive studies on their magnetotransport properties. Compared to the sensors deposited on rigid substrate, flexible GMR sensors either based on thin films or printable GMR inks exhibit the advantages of light weight, low cost, and the ability to perform strain measurement.

Besides the optimization of the sensor structure itself, various auxiliary technologies such as surface modification, MFCs, and microfluidic channels have been integrated with the GMR sensors for different applications. Surface modification technologies provide the approaches to functionalize magnetic immunoassays, DNA strains, or RNA strains on the sensor surface for biological detection. MFCs concentrate the magnetic field in a restrained region, which increases the sensitivity of the GMR sensors. Microfluidic channel not only reduces the amount of sample needed for bio-detection but also increases the efficiency and sensitivity of the detection process.

In the area of biomedical applications, GMR biosensors have been actively used for disease diagnosis and monitoring. For GMR-based immunoassays, four immunoassay methods are introduced: direct assay, competitive assay, sandwich assay, and indirect assay. These assay methods are explained theoretically and along with the real-life applications in disease (target biomarker) detections. Although the GMR-based immunoassay is still developing, POC devices based on GMR have been reported by many teams, these portable devices make diagnosis more accessible and inexpensive for patients from rural areas and developing countries where the shortage of medical equipment is a big challenge. As a result, POC devices improve the disease surveillance and mitigate the financial burden on local healthcare systems. GMR biosensors combined with standard nucleic acid amplification technologies for genotyping applications is a trending topic in recent years. It is important in research of genes and gene variants associated with diseases as well as infections by virus and bacteria. Due to the high sensitivity and linear magnetic response of GMR sensors, they have also been explored as field sensors for detecting ultra-low biological magnetic fields such as the *in vivo* MCG and MEG monitoring. Other GMR-based applications such as food and drug safety supervisions are also reported. Several examples including the detection of foodborne pathogens, heavy metal pollution from drinking water, and THC from marijuana are given. In the industry, in addition to the read/write heads in hard disk drives, GMR sensors are also used for position and current sensing and material defect inspection.

The inherit merit of GMR, combined with different technologies, make integrated GMR systems good candidates for weak magnetic field detection. Although GMR has been replaced by TMR in the hard disk drive, enormous interest has been focused on its biomedical and industrial applications due to the robustness, linear magnetic response, compatibility with the current VLSI, and its inexpensive ease of fabrication. We can foresee that GMR sensors will shine in these areas and transform today's technologies with higher sensitivity and performance.

References

1 Baibich, M.N., Broto, J.M., Fert, A., Van Dau, F.N., Petroff, F., Etienne, P., Creuzet, G., Friederich, A., and Chazelas, J. Giant Magnetoresistance of (001)Fe/(001)Cr Magnetic Superlattices. *Phys. Rev. Lett.* 11, 61 (21): 2472–2475.

2 Binasch, G., Griinberg, P., Saurenbach, F., and Zinn, W. Enhanced magnetoresistance in layered magnetic structures with antiferromagnetic interlayer exchange. *Phys. Rev. B* January 3, 9 (7): 4828–4830.

3 Lamberton, R., Seigler, M., Pelhos, K., Zhou, H., McCurry, M., Ormston, M., Yi, G., McClean, G., McLaughlin, T., Kolbo, P. et al. (2007). Current-in-plane GMR trilayer head design for hard-disk drives: characterization and extendibility. *IEEE Trans. Magn.* 43 (2): 645–650.

4. Dedrick, J. and Kraemer, K.L. (2015). Who captures value from science-based innovation? the distribution of benefits from GMR in the hard disk drive industry. *Res. Policy* 44 (8): 1615–1628.

5 Thompson, S.M. (2008). The discovery, development and future of GMR: The Nobel Prize 2007. *J. Phys. Appl. Phys.* 41 (9): 093001.

6 Miura, Y. (2002). Hard disk drive technology: past, present and future. *Digest of the Asia-Pacific Magnetic Recording Conference*, IEEE, AK1–AK1.

7 McFadyen, I.R., Fullerton, E.E., and Carey, M.J. (2006). State-of-the-art magnetic hard disk drives. *Mrs Bull* 31 (5): 379–383.

8 Guedes, A., Jaramillo, G., Buffa, C., Vigevani, G., Cardoso, S., Leitao, D.C., Freitas, P.P., and Horsley, D.A. (2012). Towards PicoTesla magnetic field detection using a GMR-MEMS hybrid device. *IEEE Trans. Magn.* 48 (11): 4115–4118.

9 Cubells-Beltran, M.-D., Reig, C., Madrenas, J., De Marcellis, A., Santos, J., Cardoso, S., and Freitas, P.P. (2016). Integration of GMR sensors with different technologies. *Sensors* 16 (6): 939.

10 Reig, C., Cardoso, S., and Mukhopadhyay, S.C. (2013). Giant Magnetoresistance (GMR) sensors. *Ssmi6* 1: 157–180.

11 Su, D., Wu, K., Saha, R., Peng, C., and Wang, J.-P. (2020). Advances in magnetoresistive biosensors. *Micromachines* 11 (1): 34.

12 Hall, D., Gaster, R., Lin, T., Osterfeld, S., Han, S., Murmann, B., and Wang, S. (2010). GMR biosensor arrays: a system perspective. *Biosens. Bioelectron.* 25 (9): 2051–2057.

13 Gregg, J.F., Petej, I., Jouguelet, E., and Dennis, C. (2002). Spin electronics–a review. *J. Phys. Appl. Phys.* 35 (18): R121.

14 Mott, N.F. (1936). The electrical conductivity of transition metals. *The Royal Society* 153: 699–717.

15 Bass, J. and Pratt, W.P., Jr. (1999). Current-Perpendicular (CPP) magnetoresistance in magnetic metallic multilayers. *J. Magn. Magn. Mater* 200 (1–3): 274–289.

16 Parkin, S.S.P., More, N., and Roche, K.P. (1990). Oscillations in exchange coupling and magnetoresistance in metallic superlattice structures: Co/Ru, Co/Cr,and Fe/Cr. *Phys. Rev. Lett.* 64 (19): 2304.

17 Choi, J., Gani, A.W., Bechstein, D.J., Lee, J.-R., Utz, P.J., and Wang, S.X. (2016). Portable, one-step, and rapid GMR biosensor platform with smariphone interface. *Biosens. Bioelectron.* 85: 1–7.

18 Gupta, S. and Kakkar, V. (2019). DARPin based GMR biosensor for the detection of ESAT-6 tuberculosis protein. *Tuberculosis* 118: 101852.

19 Rizzi, G., Lee, J.-R., Guldberg, P., Dufva, M., Wang, S.X., and Hansen, M.F. (2017). Denaturation strategies for detection of double stranded PCR products on GMR magnetic biosensor array. *Biosens. Bioelectron.* 93: 155–160.

20 Su, D., Wu, K., and Wang, J.-P. (2019). Large-area GMR bio-sensors based on reverse nucleation switching mechanism. *J. Magn. Magn. Mater.* 473: 484–489.

21 Xiao, M. (2016). A high-accuracy steering wheel angle sensor based on GMR. *2016 Sixth International Conference on Instrumentation & Measurement, Computer, Communication and Control (IMCCC);* IEEE, 1–4.

22 Sahoo, P.K., Sarkar, S., and Joseph, J. (2017). High sensitivity guided-mode-resonance optical sensor employing phase detection. *Sci. Rep.* 7 (1): 1–7.

23 Gill, H.S. (2008). CPP GMR with Hard Magnet in Stack Bias Layer. Google Patents.

24 Jogo, V.K.N.V.A., Ibusuki, V.T., Shimizu, V.H.o.V.Y., and Uzumaki, V.T. (2006). CPP-GMR technology for future high-density magnetic recording. *Fujitsu Sci. Tech. J.* 42 (1): 149–157.

25 Deac, A., Lee, K.J., Liu, Y., Redon, o., Li, M., Wang, P., Nozieres, J.P., and Dieny, B. (2005). Spin transfer effects in exchange-biased spin-valves for current-perpendicular-to-plane magnetoresistive heads. *J. Magn. Magn. Mater.* 290: 42–47.

26 Daughton, J.M. and Chen, Y.J. (1993). GMR materials for low field applications. *IEEE Trans. Magn.* 29 (6): 2705–2710.

27 Bae, S.E., Egelhoff, W.F., Jr., Chen, P.J., Zurn, S., Sheppard, L., Torok, E.J., and Judy, J.H. (2016). Application Of-Fe203 bottom GMR spin-valve for Magnetoresistive Random Access Memory (MRAM). In *Proceedings of the International Conference on Ferrites.*

28 Tannous, C. and Comstock, R.L. (2017). Magnetic information-storage materials. In: Springer Handbook of Electronic and Photonic Materials, 1–1. Springer.

29. Bhatti, S., Sbiaa, R., Hirohata, A., Ohno, H., Fukami, S., and Piramanayagam, S.N. (2017). Spintronics based random access memory: a review. *Mater Today* 20 (9): 530–548.

30 Childress, J.R. and Fontana, R.E., Jr. (2005). Magnetic recording read head sensor technology. *Comptes. Rendus. Phys.* 6 (9): 997–1012.

31 Ohinata, Y., Shimomura, K., Kimura, A., and Horiguchi, Y.(2001). Magnetic Recording Control System for Disk Storage Drive. Google Patents.

32 Katada, H., Shiimoto, M., and Hoshiya, H. (2012). Magnetic Read Head. Google Patents.

33 Su, D., Wu, K., Krishna, V., Klein, T., Liu, J., Feng, Y., Perez, A.M., Cheeran, M.C., and Wang, J.-P. (2019). Detection of influenza a virus in swine nasal swab samples with a wash-free magnetic bioassay and a handheld giant magnetoresistance sensing system. *Front. Microbiol.* 10: 1077.

34 Yang, N., Li, T., Zhang, P.P., Chen, X., and Hu, X. (2016). An early cancer diagnosis platform based on micro-magnetic sensor array demonstrates ultra-high sensitivity. *J. Nanomed. Nanotechnol.* 7 (1): 344.

35 Li, H., Zhan, Q., Liu, Y., Liu, L., Yang, H., Zuo, Z., Shang, T., Wang, B., and Li, R.-W. (2016). Stretchable spin valve with stable magnetic field sensitivity by ribbon-patterned periodic wrinkles. *AO Nano* 10 (4): 4403–4409.

36 Karnaushenko, D., Makarov, D., Stober, M., Karnaushenko, D.D., Baunack, S., and Schmidt, O.G. (2015). High-performance magnetic sensorics for printable and flexible electronics. *Adv. Mater* 27 (5): 880–885.

37 Dupuis, V., Tuaillon, J., Prevel, B., Perez, A., Melinon, P., Guiraud, G., Parent, F., Steren, L.B., Morel, R., and Barthelemy, A. (1997). From the superparamagnetic to the magnetically ordered state in systems of transition metal clusters embedded in matrices. *J. Magn. Magn. Mater.* 165 (1–3): 42–45.

38 Meyer, J., Rempel, T., Schafers, M., Wittbracht, F., Muller, C., Patel, A.V., and Flatten, A. (2013). Giant magnetoresistance effects in gel-like matrices. *Smart Mater Struct.* 22 (2): 025032.

39 Chung, K.H., Kim, S.N., and Lim, S.H. (2018). Magnetic parameters in giant magnetoresistance spin valve and their roles in magnetoresistance sensitivity. *Thin Solid Films* 650: 44–50.

40 Nguyen, A.T., Hoang, Q.K., Nguyen, T.N., Nguyen, A.T., and Nguyen, T.A. (2018). Magnetoresistive performances in exchange-biased spin valves and their roles in low-field magnetic sensing applications. *J. Sci. Adv. Mater. Devices* 3 (4): 399–405.

41 RiedmUller, B., Huber, F., and Herr, U. (2014). Towards Sub-200 Nm nano-structuring of linear giant magneto-resistive spin valves by a direct focused ion beam milling process. *J. Appl. Phys.* 115 (6): 064307.

42 Musuroi, C., Oproiu, M., Volmer, M., and Firastrau, I. (2020). High sensitivity differential Giant Magnetoresistance (GMR) based sensor for non-contacting DC/AC current measurement. *Sensors* 20 (1): 323.

43 Heim, D.E., Fontana, R.E., Tsang, C., Speriosu, V.S., Gurney, B.A., and Williams, M.L. (1994). Design and operation of spin-valve sensors. *IEEE Trans. Magn.* 30 (2): 316–321. doi: 10.1109/20.312279.

44 Wang, S.X. and Li, G. (2008). Advances in giant magnetoresistance bio sensors with magnetic nanoparticle tags: review and outlook. *IEEE Trans. Magn.* 44 (7): 1687–1702.

45 Gaster, R.S., Xu, L., Han, S.-J., Wilson, R.J., Hall, D.A., Osterfeld, S.J., Yu, H., and Wang, S.X. (2011). Quantification of protein interactions and solution transport using high-density GMR sensor arrays. *Nat. Nanotechnol.* 6 (5): 314–320.

46 Klein, T., Wang, Y., Tu, L., Yu, L., Feng, Y., Wang, W., and Wang, J.-P. (2014). Comparative analysis of several GMR strip sensor configurations for biological applications. *Sens. Actuators Phys.* 216: 349–354.

47 Feng, Y., Liu, J., Klein, T., Wu, K., and Wang, J.-P. (2017). Localized detection of reversal nucleation generated by high moment magnetic nanoparticles using a large-area magnetic sensor. *Appl. Phys.* 122 (12): 123901.

48 Srinivasan, B., Li, Y., Jing, Y., Xu, Y., Yao, X., Xing, C., and Wang, J. (2009). A detection system based on giant magnetoresistive sensors and high-moment magnetic nanoparticles demonstrates zeptomole sensitivity: potential for personalized medicine. *Angew. Chem. Int. Ed.* 48 (15): 2764–2767.

49 Melzer, M., Karnaushenko, D., Lin, G., Baunack, S., Makarov, D., and Schmidt, O.G. (2015). Direct transfer of magnetic sensor devices to elastomeric supports for stretchable electronics. *Adv. Mater.* 27 (8): 1333–1338.

50 Shpetnyi, I.O., Kondrakhova, D.M., Vorobiov, S.I., Scheibe, B., Grebinaha, V.I., Derecha, D.O., Gorobets, Y.I., and Protsenko, I.Y. (2019). The structural-phase state and magnetoresistive properties of thin film alloys obtained by co-evaporated Cu and Co. *J. Magn. Magn. Mater.* 474: 624–631.

51 Ota, S., Ando, A., and Chiba, D. (2018). A flexible giant Magnetoresistive device for sensing strain direction. *Nat. Electron.* 1 (2): 124–129.

52 Chen, Y., Mei, Y., Kaltofen, R., Monch, J.I., Schumann, J., Freudenberger, J., KlauB, H.-J., and Schmidt, O.G. (2008). Towards flexible magnetoelectronics: buffer-enhanced and mechanically tunable GMR of Co/Cu multilayers on plastic substrates. *Adv. Mater.* 20 (17): 3224–3228.

53 Yin, C., Xie, D., Xu, J.-L., and Ren, T.-L. (2014). Two-step thinning fabrication of giant magnetoresistance sensors for flexible applications. *Mod. Phys. Lett. B* 28 (10): 1450081.

54 Perez, N., Melzer, M., Makarov, D., Ueberschar, O., Ecke, R., Schulz, S.E., and Schmidt, O.G. (2015). High-performance giant magnetoresistive sensorics on flexible Si membranes. *Appl. Phys. Lett.* 106 (15): 153501.

55 Guo, Q., Xu, X.G., Zhang, Q.Q., Liu, Q., Wu, Y.J., Zhou, Z.Q., Zhu, W.M., Wu, Y., Miao, J., and Jiang, Y. (2016). Strain-controlled giant magnetoresistance of a spin valve grown on a flexible substrate. *RSV Adv.* 6 (91): 88090–88095.

56 Karnaushenko, D., Makarov, D., Yan, C., Streubel, R., and Schmidt, O.G. (2012). Printable giant magnetoresistive devices. *Adv. Mater* 24 (33): 4518–4522.

57 Xiao, J.Q., Jiang, J.S., and Chien, C.L. (1992). Giant magnetoresistance in nonmultilayer magnetic systems. *Phys. Rev. Lett.* 68 (25): 3749.

58 Shpetnyi, I.O., Vorobiov, S.I., Kondrakhova, D.M., Shevchenko, M.S., Duplik, L.V., Panina, L.V., Grebinaha, V.I., Gorobets, Y.I., Satrapinskyy, L., and Lucitiski, T. (2020). Correlation between the structural state and magnetoresistive properties of granular CoxAg100-x alloy thin films. *Vacuum* 109329.

59 Badia, F., Batlle, X., Labarta, A., Watson, M.L., Johnston, A.B., and Chapman, J.N. (1997). Magnetotransport properties of NiFe–Ag granular alloys: origin of the thermal behavior. *Appl. Phys.* 82 (2): 677–687.

60 Spizzo, F., Angeli, E., Bisero, D., Ronconi, F., Vavassori, P., Allia, P., Selvaggini, V., Coisson, M., Tiberto, P., and Vinai, F. (2003). GMR as a function of temperature in FeAg granular samples: the effect of magnetic interactions. *J. Magn. Magn. Mater.* 262 (1): 88–91.

61 Xiong, P., Xiao, G., Wang, J.Q., Xiao, J.Q., Jiang, J.S., and Chien, C.L. (1992). Extraordinary hall effect and giant magnetoresistance in the granular Co-Ag system. *Phys. Rev. Lett.* 69 (22): 3220.

62 Ennen, I., Kappe, D., Rempel, T., Glenske, C., and Hiitten, A. (2016). Giant magnetoresistance: basic concepts, microstructure, magnetic interactions and applications. *Sensors* 16 (6): 904.

63 Osterfeld, S.J., Yu, H., Gaster, R.S., Caramuta, S., Xu, L., Han, S.-J., Hall, D.A., Wilson, R.J., Sun, S., and White, R.L. (2008). Multiplex protein assays based on real-time magnetic nanotag sensing. *Proc. Natl. Acad. Sci.* 105 (52): 20637–20640.

64 Kim, D., Marchetti, F., Chen, Z., Zaric, S., Wilson, R.J., Hall, D.A., Gaster, R.S., Lee, J.-R., Wang, J., and Osterfeld, S.J. (2013). Nanosensor dosimetry of mouse blood proteins after exposure to ionizing radiation. *Sci. Rep.* 3 (1): 1–8.

65 Rizzi, G., Lee, J.R., Dahl, C., Guldberg, P., Dufva, M., Wang, S.X., and Hansen, M.F. (2017). Simultaneous profiling of DNA mutation and methylation by melting analysis using

magnetoresistive biosensor array. *Acs Nano* 11 (9): 8864–8870. doi: 10.1021/acsnano.7b03053.

66 Wang, W., Wang, Y., Tu, L., Klein, T., Feng, Y.L., and Wang, J.P. (2013). Surface modification for protein and DNA immobilization onto GMR biosensor. *IEEE Trans. Magn.* 49 (1): 296–299. doi: 10.1109/tmag.2012.2224327.

67 Wu, K., Klein, T., Krishna, V.D., Su, D., Perez, A.M., and Wang, J.-P. (2017). Portable GMR handheld platform for the detection of Influenza A Virus. *ACS Sens.* 2: 1594–1601.

68 Moretti, D., DiFrancesco, M.L., Sharma, P.P., Dante, S., Albisetti, E., Monticelli, M., Bertacco, R., Petti, D., Baldelli, P., and Benfenati, F. (2018). Biocompatibility of a magnetic tunnel junction sensor array for the detection of neuronal signals in culture. *Front. Neurosci.* 12: 909.

69 Wang, Y., Wang, W., Yu, L.N., Tu, L., Feng, Y.L., Klein, T., and Wang, J.P. (2015). Giant magnetoresistive-based biosensing probe station system for multiplex protein assays. *Biosens. Bioelectron.* 70: 61–68.

70 Zheng, Y.K., You, D., and Wu, Y.H. (2002). Flux-enhanced giant magnetoresistive head design and simulation. *IEEE Trans. Magn.* 38 (5): 2268–2270.

71 Coelho, P., Leitao, D.C., Antunes, J., Cardoso, S., and Freitas, P.P. (2014). Spin valve devices with synthetic-ferrimagnet free-layer displaying enhanced sensitivity for nanometric sensors. *IEEE Trans. Magn.* 50 (11): 1–4.

72 Shen, H.-M., Hu, L., and Fu, X. (2018). Integrated giant magnetoresistance technology for approachable weak biomagnetic signal detections. *Sensors* 18 (1): 148.

73 Tian, W., Hu, J., Pan, M., Chen, D., and Zhao, J. (2013). Magnetic flux vertical motion modulation for 11$ {f} $ noise suppression in magnetoresistance field sensors using MEMS device. *IEEE Trans. Magn.* 52 (2): 1–6.

74 Ozbay, A., Nowak, E.R., Edelstein, A.S., Fischer, G.A., Nordman, C.A., and Cheng, S.F. (2006). Magnetic-field dependence of the noise in a magnetoresistive sensor having MEMS flux concentrators. *IEEE Trans. Magn.* 42 (10): 3306–3308.

75 Edelstein, A.S., Fischer, G., Pulskamp, J., Pedersen, M., Bernard, W., and Cheng, S.F. (2004). Minimizing the effect of 1/f noise with a MEMS flux concentrator. *SENSORS, 2004 IEEE*, IEEE, 1562–1565.

76 Edelstein, A.S. and Fischer, G.A. (2002). Minimizing 1/f noise in magnetic sensors using a microelectromechanical system flux concentrator. *J. Appl. Phys.* 91 (10): 7795–7797.

77 Pannetier, M., Fermon, C., Le Goff, G., Simola, J., and Kerr, E. (2004). Femtotesla magnetic field measurement with magnetoresistive sensors. *Science* 304 (5677): 1648–1650.

78 Pannetier-Lecoeur, M., Polovy, H., Sergeeva-Chollet, N., Cannies, G., Fermon, C., and Parkkonen, L. (2011). Magnetocardiography with GMR-based sensors. *Journal of Physics: Conference Series*, Vol. 303, 012054. TOP Publishing.

79 Guedes, A., Macedo, R., Jaramillo, G., Cardoso, S., Freitas, P.P., and Horsley, D.A. (2018). Hybrid GMR sensor detecting 950 PT/Sqrt (Hz) at 1 Hz and room temperature. *Sensors* 18 (3): 790.

80 Cardoso, S., Leitao, D.C., Dias, T.M., Valadeiro, J., Silva, M.D., Chicharo, A., Silverio, V., Gaspar, J., and Freitas, P.P. (2017). Challenges and trends in magnetic sensor integration with microfluidics for biomedical applications. *J. Phys. Appl. Phys.* 50 (21): 21. doi: 10.1088/1361-6463/aa66ec.

81 Giouroudi, I. and Keplinger, F. (2013). Microfluidic biosensing systems using magnetic nanoparticles. *Int. J. Mol. Sci.* 14 (9): 18535–18556.

82 Giouroudi, I. and Kokkinis, G. (2017). Recent advances in magnetic microfluidic biosensors. *Nanomaterials* 7 (7): 171.

83 Choudhury, S., Dutta, S., and Chatterjee, S. (2019). Cost-effective template development for the microfluidic device. *Micro Nano Lett.* 14 (8): 860–864.

84 Rogers, C.I., Qaderi, K., Woolley, A.T., and Nordin, G.P. (2015). 3D printed microfluidic devices with integrated valves. *Biomicrofluidics* 9 (1): 016501.

85 Shallan, A.I., Smejkal, P., Corban, M., Guijt, R.M., and Breadmore, M.C. (2014). Cost-effective three-dimensional printing of visibly transparent microchips within minutes. *Anal. Chem.* 86 (6): 3124–3130.

86 Anderson, J.R., Chiu, D.T., Jackman, R.J., Cherniavskaya, O., McDonald, J.C., Wu, H., Whitesides, S.H., and Whitesides, G.M. (2000). Fabrication of topologically complex three-dimensional microfluidic systems in PDMS by rapid prototyping. *Anal. Chem.* 72 (14): 3158–3164.

87 Gao, R., Ko, J., Cha, K., Jeon, J.H., Rhie, G., Choi, J., DeMello, A.J., and Choo, J. (2015). Fast and sensitive detection of an anthrax biomarker using SERS-based solenoid microfluidic sensor. *Biosens. Bioelectron.* 72: 230–236.

88 Pekas, N., Porter, M.D., Tondra, M., Popple, A., and Jander, A. (2004). Giant magnetoresistance monitoring of magnetic picodroplets in an integrated microfluidic system. *Appl. Phys. Lett.* 85 (20): 4783–4785.

89 Devkota, J., Kokkinis, G., Berris, T., Jamalieh, M., Cardoso, S., Cardoso, F., Srikanth, H., Phan, M.-H., and Giouroudi, I. (2015). A novel approach for detection and quantification of magnetic nanomarkers using a spin valve GMR-integrated microfluidic sensor. *RSC Adv.* 5 (63): 51169–51175.

90 Sun, X., Lei, C., Guo, L., and Zhou, Y. (2016). Giant magneto-resistance based immunoassay for the tumor marker carcinoembryonic antigen. *Microchim. Acta* 183 (3): 1107–1114.

91 Krishna, V.D., Wu, K., Perez, A.M., and Wang, J.P. (2016). Giant magnetoresistance-based biosensor for detection of Influenza A Virus. *Front. Microbiol.* 7: 8.

92 Wang, W., Wang, Y., Tu, L., Feng, Y., Klein, T., and Wang, J.-P. (2014). Magnetoresistive performance and comparison of supermagnetic nanoparticles on giant magnetoresistive sensor-based detection system. *Sci. Rep.* 4.

93 Gani, A.W., Wei, W., Shi, R.-Z., Ng, E., Nguyen, M., Chua, M.-S., So, S., and Wang, S.X. (2019). An automated, quantitative, and multiplexed assay suitable for point-of-care hepatitis B virus diagnostics. *Sci. Rep.* 9 (1): 1–11.

94 Klein, T., Wang, W., Yu, L., Wu, K., Boylan, K.L., Vogel, R.I., Skubitz, A.P., and Wang, J.-P. (2019). Development of a multiplexed giant magnetoresistive biosensor array prototype to quantify ovarian cancer biomarkers. *Biosens. Bioelectron.* 126: 301–307.

95 Huo, W., Gao, Y., Zhang, L., Shi, S., and Gao, Y. (2015). A novel high-sensitivity cardiac multibiomarker detection system based on microfluidic chip and GMR sensors. *IEEE Trans. Magn.* 51 (11): 1–4.

96 Gao, Y., Huo, W., Zhang, L., Lian, J., Tao, W., Song, C., Tang, J., Shi, S., and Gao, Y. (2019). Multiplex measurement of twelve tumor markers using a GMR multi-biomarker immunoassay biosensor. *Biosens. Bioelectron.* 123: 204–210.

97 Baechler, E.C., Batliwalla, F.M., Karypis, G., Gaffney, P.M., Ortmann, W.A., Espe, K.J., Shark, K.B., Grande, W.J., Hughes, K.M., and Kapur, V. (2003). Interferon-inducible gene expression signature in peripheral blood cells of patients with severe lupus. *Proc. Natl. Acad. Sci.* 100 (5): 2610–2615.

98 Bauer, J.W., Baechler, E.C., Petri, M., Batliwalla, F.M., Crawford, D., Ortmann, W.A., Espe, K.J., Li, W., Patel, D.D., and Gregersen, P.K. (2006). Elevated serum levels of interferon-regulated chemokines are biomarkers for active human systemic lupus erythematosus. *PLoS Med* 3 (12): e491.

99 Bauer, J.W., Petri, M., Batliwalla, F.M., Koeuth, T., Wilson, J., Slattery, C., PanoskaltsisMortari, A., Gregersen, P.K., Behrens, T.W., and Baechler, E.C. (2009). Interferon-regulated chemokines as biomarkers of systemic lupus erythematosus disease activity: a validation study. *Arthritis Rheum. Off J. Am. Coll. Rheumatol.* 60 (10): 3098–3107.

100 Li, Q.-Z., Zhou, J., Lian, Y., Zhang, B., Branch, V.K., Carr-Johnson, F., Karp, D.R., Mohan, C., Wakeland, E.K., and Olsen, N.J. (2010). Interferon signature gene expression is correlated with autoantibody profiles in patients with incomplete lupus syndromes. *Clin. Exp. Immunol.* 159 (3): 281–291.

101 Petri, M., Singh, S., Tesfasyone, H., Dedrick, R., Fry, K., Lal, P.G., Williams, G., Bauer, J.W., Gregersen, P.K., and Behrens, T.W. (2009). Longitudinal expression of type I interferon responsive genes in systemic lupus erythematosus. *Lupus* 18 (11): 980–989.

102 Kirou, K.A., Lee, C., George, S., Louca, K., Peterson, M.G., and Crow, M.K. (2005). Activation of the interferon-a pathway identifies a subgroup of systemic lupus erythematosus patients with distinct serologic features and active disease. *Arthritis Rheum* 52 (5): 1491–1503.

103 Price, J.V., Tangsombatvisit, S., Xu, G., Yu, J., Levy, D., Baechler, E.C., Gozani, O., Varma, M., Utz, P.J., and Liu, C.L. (2012). On silico peptide microarrays for high-resolution mapping of antibody epitopes and diverse protein-protein interactions. *Nat. Med.* 18 (9): 1434.

104 Lee, J.-R., Haddon, D.J., Wand, H.E., Price, J.V., Diep, V.K., Hall, D.A., Petri, M., Baechler, E.C., Balboni, I.M., and Utz, P.J. (2016). Multiplex giant magnetoresistive biosensor microarrays identify interferon-associated autoantibodies in systemic lupus erythematosus. *Sci. Rep.* 6: 27623.

105 van der Schoot, C.E., de Haas, M., and Clausen, F.B. (2017). Genotyping to prevent Rh disease: has the time come? *Curr. Op/n. Hematol.* 24 (6): 544–550.

106 Bik, E.M., Bird, S.W., Bustamante, J.P., Leon, L.E., Nieto, P.A., Addae, K., Alegria-Mera, V., Bravo, C., Bravo, D., and Cardenas, J.P. (2019). A novel sequencing-based vaginal health assay combining self-sampling, HPV detection and genotyping, STI detection, and vaginal microbiome analysis. *PloS One* 14 (5): e0215945.

107 Kawaguchi, R., Tamura, H., and Rentschler, E. (2017). Direct-to-consumer genotyping in pre-disease field in Japan and the United States. *J. Hum. Bio. Health Edu.* 1: 002.

108 Ravi, N., Rizzi, G., Chang, S.E., Cheung, P., Utz, P.J., and Wang, S.X. (2019). Quantification of CDNA on GMR biosensor array towards point-of-care gene expression analysis. *Biosens. Bioelectron.* 130: 338–343.

109 Zhi, X., Deng, M., Yang, H., Gao, G., Wang, K., Fu, H., Zhang, Y., Chen, D., and Cui, D. (2014). A novel HBV genotypes detecting system combined with microfluidic chip, loop-mediated isothermal amplification and GMR sensors. *Biosens. Bioelectron.* 54: 372–377.

110 Dias, T.M., Cardoso, F.A., Martins, S.A.M., Martins, V.C., Cardoso, S., Gaspar, J.F., Monteiro, G., and Freitas, P.P. (2016). Implementing a strategy for on-chip detection of cell-free DNA fragments using GMR sensors: a translational application in cancer

diagnostics using ALU elements. *Anal. Methods* 8 (1): 119–128. doi: 10.1039/c5ay01587a.

111 Xu, L., Yu, H., Akhras, M.S., Han, S.-J., Osterfeld, S., White, R.L., Pourmand, N., and Wang, S.X. (2008). Giant magnetoresistive biochip for DNA detection and HPV genotyping. *Biosens. Bioelectron.* 24 (1): 99–103.

112 Crowley, E., Di Nicolantonio, F., Loupakis, F., and Bardelli, A. (2013). Liquid biopsy: monitoring cancer-genetics in the blood. *Nat. Rev. Clin. Oncol.* 10 (8): 472.

113 Nesvet, J., Rizzi, G., and Wang, S.X. (2019). Highly sensitive detection of DNA hypermethylation in melanoma cancer cells. *Biosens. Bioelectron.* 124: 136–142.

114 Shi, W., Lv, C., Qi, J., Zhao, W., Wu, X., Jing, R., Wu, X., Ju, S., and Chen, J. (2012). Prognostic value of free DNA quantification in serum and cerebrospinal fluid in glioma patients. *J. Mol. Neurosci.* 46 (3): 470–475.

115 Stotzer, O.J., Lehner, J., Fersching-Gierlich, D., Nagel, D., and Holdenrieder, S. (2014). Diagnostic relevance of plasma DNA and DNA integrity for breast cancer. *Tumor Biol.* 35 (2): 1183–1191.

116 da Silva Filho, B.F., Gurgel, A.P.A.D., Neto, M.A.de F.L., de Azevedo, D.A., de Freitas, A.C., Neto, J.da C.S., and Silva, L.A.F. (2013). Circulating cell-free DNA in serum as a biomarker of colorectal cancer. *J. Clin. Pathol.* 66 (9): 775–778.

117 Kuss-Duerkop, S.K., Westrich, J.A., and Pyeon, D. (2018). DNA tumor virus regulation of host DNA methylation and its implications for immune evasion and oncogenesis. *Viruses* 10 (2): 82.

118 Li, H., Zhang, S., Zhang, C., and Xie, X. (2016). SQUID-based MCG measurement using a full-tensor compensation technique in an urban hospital environment. *IEEE Trans. Appl. Supercond.* 26 (6): 1–5.

119 Korber, R., Storm, J.-H., Seton, H., Makela, J.P., Paetau, R., Parkkonen, L., Pfeiffer, C., Riaz, B., Schneiderman, J.F., and Dong, H. (2016). SQUIDs in biomagnetism: a roadmap towards improved healthcare. *Supercond. Sci. Technol.* 29 (11): 113001.

120 Wang, C., Sun, L., Lichtenwalter, B., Zerkle, B., and Okada, Y. (2016). Compact, ultra-low vibration, closed-cycle helium recycler for uninterrupted operation of MEG with SQUID magnetometers. *Cryogenics* 76: 16–22.

121 Iivanainen, J., Zetter, R., Gron, M., Hakkarainen, K., and Parkkonen, L. (2019). On-scalp MEG system utilizing an actively shielded array of optically pumped magnetometers. *Neuroimage* 194: 244–258.

122 Pfeiffer, C. (2019). On-Scalp MEG Using High-Tc SQUIDs: Measuring Brain Activity with Superconducting Magnetometers. Sweden: Chalmers Tekniska Hogskola.

123 Inaba, T., Nakazawa, Y., Yoshida, K., Kato, Y., Hattori, A., Kimura, T., Hoshi, T., Ishizu, T., Seo, Y., and Sato, A. (2017). Routine clinical heart examinations using SQUID magnetocardiography at University of Tsukuba Hospital. *Supercond. Sci. Technol.* 30 (11): 114003.

124 Zuo, S., Heidari, H., Farina, D., and Nazarpour, K. (2020). Miniaturized magnetic sensors for implantable magnetomyography. *Adv. Mater. Technol.* 2000185.

125 Caruso, L., Wunderle, T., Lewis, C.M., Valadeiro, J., Trauchessec, V., Rosillo, J.T., Amaral, J.P., Ni, J., Jendritza, P., and Fermon, C. (2017). In vivo magnetic recording of neuronal activity. *Neuron* 95 (6): 1283–1291.

126 Amaral, J., Cardoso, S., Freitas, P., and Sebastido, A. (2011). Toward a system to measure action potential on mice brain slices with local magnetoresistive probes. *J. Appl. Phys.* 109 (7): 07B308.

127 Valadeiro, J., Silva, M., Cardoso, S., Martins, M., Gaspar, J., Freitas, P.P., and Sebastiao, A.M. (2017). Microneedles with integrated magnetoresistive sensors: a precision tool in biomedical instrumentation. *2017 IEEE Sensors Applications Symposium (SAS)*, 1–6. IEEE.

128 Amaral, J., Gaspar, J., Pinto, V., Costa, T., Sousa, N., Cardoso, S., and Freitas, P. (2013). Measuring brain activity with magnetoresistive sensors integrated in micromachined probe needles. *Appl. Phys. A* 111 (2): 407–412.

129 Pannetier-Lecoeur, M., Parkkonen, L., Sergeeva-Chollet, N., Polovy, H., Fermon, C., and Fowley, C. (2011). Magnetocardiography with sensors based on giant magnetoresistance. *Appl. Phys. Lett.* 98 (15): 153705.

130 Kalyan, K., Chugh, V.K., and Anoop, C. (2016). Non-invasive heart rate monitoring system using giant magneto resistance sensor. IEEE, 4873–4876.

131 Barbieri, F., Trauchessec, V., Caruso, L., Trejo-Rosillo, J., Telenczuk, B., Paul, E., Bal, T., Destexhe, A., Fermon, C., and Pannetier-Lecoeur, M. (2016). Local recording of biological magnetic fields using giant magneto resistance-based micro-probes. *Sci. Rep.* 6 (1): 1–10.

132 Campiglio, P., Caruso, L., Paul, E., Demonti, A., Azizi-Rogeau, L., Parkkonen, L., Fermon, C., and Pannetier-Lecoeur, M. (2012). GMR-based sensors arrays for biomagnetic source imaging applications. *IEEE Trans. Magn.* 48 (11): 3501–3504.

133 Wang, W., Wang, Y., Tu, L., Klein, T., Feng, Y., Li, Q., and Wang, J.-P. (2014). Magnetic detection of mercuric ion using giant magnetoresistance-based biosensing system. *Anal. Chem.* 86 (8): 3712–3716.

134 Koets, M., Van der Wijk, T., Van Eemeren, J., Van Amerongen, A., and Prins, M.W.J. (2009). Rapid DNA multi-analyte immunoassay on a magneto-resistance biosensor. *Biosens. Bioelectron.* 24 (7): 1893–1898.

135 Ng, E., Nadeau, K.C., and Wang, S.X. (2016). Giant magnetoresistive sensor array for sensitive and specific multiplexed food allergen detection. *Biosens. Bioelectron.* 80: 359–365. doi: 10.1016/j.bios.2016.02.002.

136 Mak, A.C., Osterfeld, S.J., Yu, H., Wang, S.X., Davis, R.W., Jejelowo, O.A., and Pourmand, N. (2010). Sensitive giant magnetoresistive-based immunoassay for multiplex mycotoxin detection. *Biosens. Bioelectron.* 25 (7): 1635–1639. doi: 10.1016/j.bios.2009.11.028.

137 Lee, J.-R., Choi, J., Shultz, T.O., and Wang, S.X. (2016). Small molecule detection in saliva facilitates portable tests of marijuana abuse. *Anal. Chem.* 88 (15): 7457–7461.

138 Giebeler, C., Adelerhof, D.J., Kuiper, A.E.T., Van Zon, J.B.A., Oelgeschlager, D., and Schulz, G. (2001). Robust GMR sensors for angle detection and rotation speed sensing. *Sens. Actuators Phys.* 91 (1–2): 16–20.

139 Treutler, C.P.O. (2001). Magnetic sensors for automotive applications. *Sens. Actuators' Phys.* 91 (1–2): 2–6.

140 Rieger, G., Ludwig, K., Hauch, J., and Clemens, W. (2001). GMR sensors for contactless position detection. *Sens. Actuators Phys.* 91 (1–2): 7–11.

141 Sebastia, J.P., Lluch, J.A., and Vizcaino, J.R.L. (2007). Signal conditioning for GMR magnetic sensors: applied to traffic speed monitoring GMR sensors. *Sens. Actuators Phys.* 137 (2): 230–235.

142 Huang, -C.-C., Lin, C.-L., Kao, -J.-J., Chang, -J.-J., and Sheu, G.-J. (2018). Vehicle parking guidance for wireless charge using gmr sensors. *IEEE Transactions on Vehicular Technology.*

143 De Marcellis, A., Reig, C., Cubells-Beltran, M.-D., Madrenas, J., Santos, J.D., Cardoso, S., and Freitas, P.P. (2017). Monolithic integration of GMR sensors for standard CMOS-IC current sensing. *Solid-State Electron* 135: 100–104.

144 Cubells-Beltran, M.D., Reig, C., DeMarcellis, A., Cardoso, S., and Freitas, P.P. (2018). Electrical modeling of monolithically integrated GMR based current sensors. *2018 Spanish Conference on Electron Devices (CDE)*, 1–3. IEEE.

145 Jeng, J.-T., Trinh, X.-T., Hung, C.-H., and Lu, -C.-C. (2019). Quasi-static current measurement with field-modulated spin-valve GMR sensors. *Sensors* 19 (8): 1882.

146 Schuina, C., Magalhaes, D.P., and Barbosa, C.H. (2018). GMR sensors and neural networks applied to the contactless measurement of direct electrical currents. *JPhCS* 975 (1): 012070.

147 Libby, H.L. (1971). Introduction to Electromagnetic Nondestructive Test Methods. Krieger.

148 Sasi, B., Arjun, V., Mukhopadhyay, C.K., and Rao, B.P.C. (2018). Enhanced detection of deep seated flaws in 316 stainless steel plates using integrated EC-GMR sensor. *Sens. Actuators Phys.* 275: 44–50.

149 Pelkner, M., Stegemann, R., Sonntag, N., Pohl, R., and Kreutzbruck, M. (2018). Benefits of GMR sensors for high spatial resolution NDT applications. *AIP Conference Proceedings*, Vol. 1949, 040001. AIP Publishing LLC.

150 Rifai, D., Abdalla, A.N., Ali, K., and Razali, R. (2016). Giant magnetoresistance sensors: a review on structures and non-destructive Eddy current testing applications. *Sensors* 16 (3): 298.

151 Dogaru-, T. and Smith, S.T. (2000). Edge crack detection using a giant magnetoresistance based Eddy current sensor. *Nondestruct. Test. Eval.* 16 (1): 31–53.

152 Jeng, J.-T., Lee, G.-S., Liao, W.-C., and Shu, C.-L. (2006). Depth-resolved Eddy-current detection with GMR magnetometer. *J. Magn. Magn. Mater.* 304 (1): e470–e473.

153 Cacciola, M., Megali, G., Pellicano, D., and Morabito, F.C. (2011). A GMR–ECT based embedded solution for applications on PCB inspections. *Sens. Actuators Phys.* 167 (1): 25–33.

154 Postolache, O., Ramos, H.G., and Ribeiro, A.L. (2011). Detection and characterization of defects using GMR probes and artificial neural networks. *Comput. Stand. Interfaces* 33 (2): 191–200.

155 Pelkner, M., Pohl, R., Kreutzbruck, M., and Commandeur, C. (2016). Development of adapted GMR probes for automated detection of hidden defects in thin steel sheets. *AIP Conference Proceedings*, Vol. 1706, 020018. AIP Publishing LLC.

156 Ye, C., Udpa, L., and Udpa, S. (2016). Optimization and validation of rotating current excitation with GMR array sensors for riveted structures inspection. *Sensors* 16 (9): 1512.

157 Yadav, A.K. and Szpytko, J. (2019). Development of portable wireless non-destructive crack identification method by using GMR sensor array for overhead crane bridges.

158 Wang, R. and Kawamura, Y. (2016). An automated sensing system for steel bridge inspection using GMR sensor array and magnetic wheels of climbing robot. *J. Sens.* 2016.

3

Tunneling Magnetoresistance (TMR) Materials and Devices for Magnetic Sensors

Zitong Zhou[1], *Kun Zhang*[1], *and Qunwen Leng*[1,2]

[1]*School of Integrated Circuit Science and Engineering, Beihang University, Beijing, China*
[2]*Beihang-Goertek Joint Microelectronics Institute, Qingdao Research Institute, Beihang University, Qingdao, China*

CHAPTER MENU

Spintronics: Materials, Devices and Applications, First Edition. Edited by Kaiyou Wang, Meiyin Yang, and Jun Luo.

3.1 Principle of Tunneling Magnetoresistance Effect

The principle and theoretical model of the tunneling magnetoresistance (TMR) effect are explained in detail, and the typical structure of the sensing unit is introduced in this section.

3.1.1 Tunneling Process

Tunneling phenomenon results from the volatility of particles, which means that particles trapped in one region can tunnel through an energy barrier into another region. These following two conditions are essential for the particle tunneling: a potential difference at both ends of the barrier and a thin enough barrier. The barrier is the middle layer between two electrodes and different tunneling junctions can be constructed by changing the electrode material. When applying a bias voltage to the tunneling junctions, the induced Fermi levels difference at both ends of the barrier would cause electrons tunneling through the barrier. According to previous studies, Simmons [1] derived a formula for describing the electrical tunneling process through an arbitrarily shaped barrier. The tunneling probability $D(\varepsilon_x)$ of the electron can be calculated by using the Wentzel-Kramers-Brillouin (WKB) approximation

$$D(\varepsilon_x) = \exp\left\{-(4\pi / h)\int_{x2}^{x1}\left[2m\left(V(x)-\varepsilon_x\right)\right]^{\frac{1}{2}} dx\right\} \tag{3.1}$$

where $\varepsilon_x = mv_x^2/2$, v_x is the x component of the electron velocity, and ε_x is the energy component of the incident electron in the x direction. $V(x)$ is the barrier potential of the insulating film and "h" is Planck's constant. As shown in Figure 3.1, an energy difference of "eV" is formed between the two sides of the tunnel junction when a voltage of "V" is applied between the two electrodes. The number of electrons (N_1) tunneling from electrode 1 to electrode 2 is expressed as

$$N_1 = \int_0^{v_m} v_x n(v_x) D(\varepsilon_x) dv_x = (1/m)\int_0^{\varepsilon_m} n(v_x) D(\varepsilon_x) d\varepsilon_x \tag{3.2}$$

where $v_x n(v_x)$ is the number of electrons per unit volume with velocity between v_x and $v_x + dv_x$, and m is the mass of electron. The number of electrons per unit area as it approaches an infinitesimal can be represented as

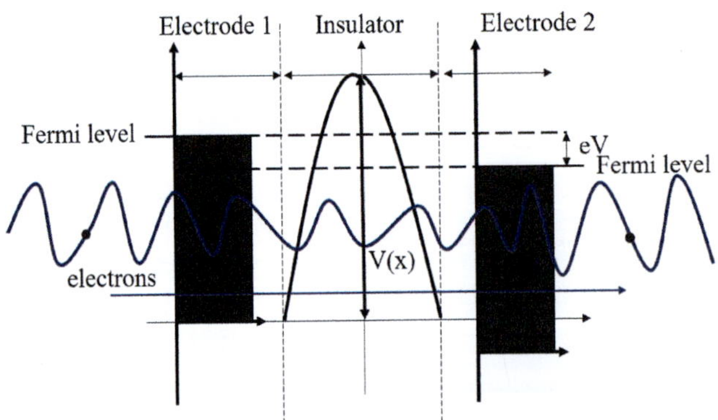

Figure 3.1 Schematic of the tunneling process of electrons in an electrode/insulator/electrode sandwich structure.

$$n(v)dv_x dv_y dv_z = \left(2m^4 / h^3\right) f(\varepsilon) dv_x dv_y dv_z \tag{3.3}$$

and $f(\varepsilon)$ is the Fermi-Dirac distribution function. According to formula (3.3), the density of electrons with energy ε_x is:

$$n(v_x) = \left(2m^4 / h^3\right) \int \int_{-\infty}^{\infty} f(\varepsilon) dv_y dv_z = \left(4\pi m^4 / h^3\right) \int_0^{\infty} f(\varepsilon) d\varepsilon_r. \tag{3.4}$$

Obviously, the formula (3.4) uses spherical coordinates $v_r^2 = v_y^2 + v_z^2$, so $\varepsilon_r = mv_r^2 / 2$. By combining the above all mentioned equations, the number of electrons tunneling from electrode 1 to electrode 2 is finally obtained as follows:

$$\begin{aligned} N_1 &= \left(4\pi m^4 / h^3\right) \int_0^{\varepsilon_m} D(\varepsilon_x) d\varepsilon_x \int_0^{\infty} f(\varepsilon) d\varepsilon_r \\ &= \left(4\pi m^4 / h^3\right) \int_0^{\varepsilon_m} D(\varepsilon_x) d\varepsilon_x \int_0^{\infty} f(\varepsilon_r + \varepsilon_x) d\varepsilon_r. \end{aligned} \tag{3.5}$$

Here, ε_m is the maximum energy of the electrons. Similarly, the number of electrons tunneling from electrode 2 to electrode 1 can be expressed as:

$$N_2 = \left(4\pi m^4 / h^3\right) \int_0^{\varepsilon_m} D(\varepsilon_x) d\varepsilon_x \int_0^{\infty} f(\varepsilon_r + \varepsilon_x + eV) d\varepsilon_r. \tag{3.6}$$

Therefore, the net number of electrons tunneling through the insulation layer is:

$$\begin{aligned} N &= N_1 - N_2 \\ &= \left(4\pi m^4 / h^3\right) \int_0^{\varepsilon_m} D(\varepsilon_x) d\varepsilon_x D(\varepsilon_x) d\varepsilon_x \int_0^{\infty} [f(\varepsilon_r + \varepsilon_x) - f(\varepsilon_r + \varepsilon_x + eV)] d\varepsilon_r. \end{aligned} \tag{3.7}$$

The tunneling current J is calculated as:

$$\begin{aligned} J &= -eN \\ &= \left(-4\pi m^4 e / h^3\right) \int_0^{\varepsilon_m} D(\varepsilon_x) d\varepsilon_x \int_0^{\infty} [f(\varepsilon_r + \varepsilon_x) - f(\varepsilon_r + \varepsilon_x + eV)] d\varepsilon_r. \end{aligned} \tag{3.8}$$

3.1.2 Spin-dependent Tunneling Process

The TMR effect refers to a kind of magnetoresistance (MR) effect found in the membrane structure, i.e., a thin insulating layer (IL) sandwiched between two ferromagnetic (FM) metal electrodes. In 1971, R. Meservey et al. observed spin-dependent tunneling process in superconductor/IL/FM tunnel joints, which is used as an effective method to measure spin polarization of FM [2]. In 1975, M. Julliére observed TMR value of 14% in Fe/Ge/Co three-layer structure at 4.2K, and established a theoretical model to describe the spin-dependent tunneling process [3].

The sandwich structure i.e., FM/IL/FM becomes the core structure of the magnetic tunnel junction (MTJ). Except for the shape, height, and width of insulation barrier, the tunneling probability of electrons also depends on the relative magnetization direction of the two FM layers when they pass through the MTJ. Figure 3.2 illustrates the schematic diagram of spin-dependent tunneling process in MTJ. At first, the density of spin up and spin down electrons in FM material is different due to exchange interaction, which is obviously different from the condition in normal metal. Meanwhile, two-current model [4] assumes that the tunneling current is transported independently by spin up and spin down electrons. If magnetization directions of two FM electrodes are parallel (P), majority-spin (minority-spin) electrons in electrode 1 will tunnel into majority-spin (minority-spin) subband in electrode 2 as shown in Figure 3.2a. In this case, there are enough empty states to accommodate the tunneling electrons, resulting in low resistance states of MTJ. If the magnetization directions of two FM electrodes are antiparallel (AP), majority-spin (minority-spin) electrons in electrode 1 will tunnel into minority-spin (majority-spin) subband in electrode 2 as shown in Figure 3.2b. In this case, the tunneling process becomes difficult because that large amounts majority-spin electrons have to enter into a small number of empty states provided by minority-spin subband, leading to high resistance states of MTJ.

3.1.3 The Julliére Model

As mentioned above, the spin-dependent tunneling process in amorphous-barrier MTJs could be well described by the Julliére model. This model assumes that tunneling junction conductance is proportional to the product of density of state (DOS) near Fermi level of the two FM electrode. At zero bias voltage, when the magnetization vectors of two FM layers are in P state, the tunneling junction conductance can be expressed as:

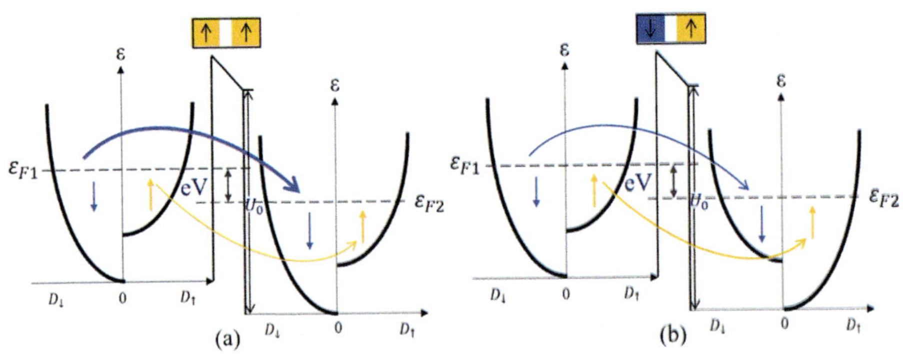

Figure 3.2 Schematic diagram of electron tunneling process in a MTJ when the magnetization directions of the two electrodes are parallel (a) and antiparallel (b). U_0 is the height of the tunneling barrier.

$$G_P \propto \alpha_1\alpha_2 + (1-\alpha_1)(1-\alpha_2) \tag{3.9}$$

When the magnetization vectors of two FM layers are in AP state, the tunneling junction conductance can be expressed as:

$$G_{AP} \propto \alpha_1(1-\alpha_2) + \alpha_2(1-\alpha_1) \tag{3.10}$$

G_P and G_{AP} are junction conductance under parallel and antiparallel states of the two FM layers. α_1 and α_2 are the denity of states of majority-spin electrons in two FM electrodes, respectively.

The TMR ratio could be defined as relative resistance variation,

$$TMR = \frac{\Delta R}{R_{AP}} = \frac{R_{AP} - R_P}{R_{AP}} = \frac{\Delta G}{G_P} = \frac{2P_1P_2}{1+P_1P_2} \tag{3.11}$$

or

$$TMR = \frac{\Delta R}{R_P} = \frac{R_{AP} - R_P}{R_P} = \frac{\Delta G}{G_{AP}} = \frac{2P_1P_2}{1-P_1P_2} \tag{3.12}$$

R_P and R_{AP} are the tunneling resistances when the magnetization directions of two FM layers are parallel and antiparallel, respectively. P_i is the spin polarization of the tunneling electron in the corresponding FM layer, and $P_i = 2\alpha_i - 1$. Obviously, nonzero spin polarization would induce MR effect and the higher the spin polarization is, the higher the TMR value is. If the tunneling electrons of the two FM electrodes are completely polarized ($P_1 = P_2 = 1$), the TMR value will tend to be infinite. This model is verified by spin polarization value obtained by Tedrow and Meservey's [5, 6] tunneling experiments.

However, there are also some limitations in the Julliére model. On the one hand, this model neglects the influence of other physical mechanisms on the TMR value. On the other hand, this model could not explain the negative TMR effect and Teresa derived the negative spin polarization coefficient experimentally [7], which is proved to be closely related to the barrier material. On the basis of the Julliére model, Slonczewski [8] proposed a new three-layer theoretical model to describe the carrier transport between two electrodes in 1989. This model adopted the spin polarized near-free electron model with different wave vectors, and considered the influence factors of potential barrier and interface. Assuming that P_{eff1} and P_{eff2} are the effective spin polarization of the two FM electrodes respectively, which can be expressed by the wave vector as follows [9]:

$$P_{eff1} = \frac{(k_1^{\uparrow} - k_1^{\downarrow})\left(k_1^2 - k_1^{\uparrow}k_1^{\downarrow}\right)}{(k_1^{\uparrow} + k_1^{\downarrow})\left(k_1^2 + k_1^{\uparrow}k_1^{\downarrow}\right)} \tag{3.13}$$

and

$$P_{eff2} = \frac{(k_2^{\uparrow} - k_2^{\downarrow})\left(k_2^2 - k_2^{\uparrow}k_2^{\downarrow}\right)}{(k_2^{\uparrow} + k_2^{\downarrow})\left(k_2^2 + k_2^{\uparrow}k_2^{\downarrow}\right)} \tag{3.14}$$

Here, $k=\sqrt{(2m/h^2)(U_0-\varepsilon_F)}$ and $k_i^{\uparrow(\downarrow)}=\sqrt{(2m/h^2)\left(\varepsilon_F-V_i^{\uparrow(\downarrow)}\right)}$, where k is the wave vector in the barrier, and $k_i^{\uparrow}$ and $k_i^{\downarrow}$ are the wave vectors of the spin up and spin down electrons of the two electrodes respectively. $V_i^{\uparrow(\downarrow)}$ is the spin-dependent electron potential in the electrode. When the potential barrier is very high, this model can be reduced to the Julliére model. In addition, in M.B. Stearns' study [10], wave vector k was also introduced to describe the spin polarization.

3.1.4 Typical Structure of the Magnetic Sensing Unit

The room temperature TMR effect was first observed in this classic sandwich structure in 1995 [11, 12]. In order to realize parallel and antiparallel alignment of the magnetization direction of two FM layers through easily changing the external magnetic field, the coercivity of two FM layers should be different. This three-layer structure is similar to a spin valve, the magnetic moment of the low-coercivity FM layer rotates first with the increasing external magnetic field. In this case, antiparallel arrangement of the two FM layers is formed and the MTJs presents high resistance state. The transition between parallel and antiparallel states of two FM layers results in the transition between the low resistance state and high resistance state of the MTJs. For magnetic sensors, the quality of linear region during resistance transition is crucial. Although the core sandwich MTJ structure is simple and easy to prepare, there are some obstacles to be solved before practical application. For example, magnetization directions of two FM layers in this structure are easily affected by external factors, resulting in poor anti-external magnetic field interference ability and thermal stability. Moreover, magnetic-field response curve of this three-layer structure exhibits remarkable hysteresis. In other words, the measured resistance value and the external magnetic field cannot form a one-to-one correspondence relation, which is not suitable for the sensor application.

3.2 Material and Process

In order to improve the performance of MTJs, the choice and fabrication of the insulating layer and a high-spin-polarization ferromagnetic (FM) layer of high quality becomes the key point. In this section, the optimization of barrier layer and FM layer will be discussed in details.

3.2.1 TMR Barrier Materials

For the MTJs, the quality of the insulating layer is one of the key factors affecting its TMR ratio. A high-quality insulating layer requires to form a continuous film without pinholes in a few nanometers thickness, which is usually made of oxide materials. The insulating layer has evolved several decades, various materials are used as tunnel barrier in MTJs. MgO is currently regarded as the optimal insulating material. In 1975, a kind of germanium oxide was first used as the tunneling layer in a Fe/Ge/Co junction proposed by Julliére [3]. Then, NiO and GdO_x were used as tunneling layers in MTJs by Maekawa [13] and Nowak [14] et al. in 1982 and 1992, respectively. Maekawa et al. prepared three kinds of NiO-based MTJs, namely Ni/NiO/Ni, Ni/NiO/Fe and Ni/NiO/Co. Among them, the TMR value of Ni/NiO/Co reaches 2% at 4.2K, and an obvious change of resistance with magnetic field can still be observed when the temperature rises to 77K. For the other two MTJs, there are no significant MR effects observed when the temperature is higher than 4.2K. Nowak et al. prepared two kinds of MTJs Gd/GdO/Fe and Fe/GdO/Fe, and their TMR ratios reached 2.5%–7.7%. In 1994, T.S. Penlaskett [15] et al. studied the transport characteristics of spin polarized electrons tunneling through an insulating barrier in the NiFe/Oxide/Co structure. By comparing the output characteristics of the MTJs with different resistances, it is proved that the MTJs with significantly higher junction resistance than the electrode resistance indicates a better output curve and larger TMR ratio.

It was not until 1995 that Moodera et al. [11] prepared the CoFe/Al_2O_3/Co MTJs, which exhibited remarkable TMR effect with a value of 11.8% at room temperature. In the same year, Miyazaki [12] et al. fabricated the Fe/Al_2O_3/Fe junctions, in which, TMR value of 18% at room temperature and 30% at 4.2K were observed. Inspired by these reports, some research groups began to pay attention to the study of Al_2O_3-barrier MTJs, while others tried to look for other barrier materials for obtaining greater TMR values. In 1996 Yu Lu [16] et al. achieved a TMR value of up to 83% at 4.2K with a film stack of $La_{0.67}Sr_{0.33}MnO_3/SrTiO_3/La_{0.67}Sr_{0.33}MnO_3$. The MTJ of Co/$Ga_2O_3$/$Ni_{80}Fe_{20}$ structure shows a TMR ratio as high as18.2% at room temperature and 27.6% at 77K [17].

An important application scenario of MTJs is used as hard disk read heads [18], which requires the resistivity of the MTJs should be less than $10\Omega \cdot \mu m^2$. Reducing junction resistance means decreasing the barrier thickness or effective barrier height. However, the reduction of the insulator layer thickness tends to produce pinhole channels due to the unsophisticated film deposition technology. Therefore, the only solution is to find an insulating material with a small bandgap. Jianguo Wang [19] et al. adopted ZrO_x as the insulating layer and a maximum TMR value of 19.2% was observed. In addition, some groups used HfO_x material as an insulating layer to fabricate MTJs due to its small bandgap [20, 21].

In addition to the exploration of MTJs core materials, other impact factors of TMR values have also been studied. In 1997, Masashige Sato and Kazuo Kobayashi [22] first proved that annealing at an appropriate temperature is conducive to increase the TMR ratio. The MTJs with the core structure of Co/Al-AlO/Co was annealed at 300°C. As a result, the TMR value reached 24% at room temperature, which showed a significant change compared with the value before annealing. In 2000, Cardoso [23] and Xiufeng Han [24] et al. fabricated Al_2O_3-barrier MTJs, whose TMR ratio reaches up to 41% and 49.7% respectively at room temperature by annealing at 300°C. In the same year, the experiments of S. Yuasa [25] et al. confirmed the strong dependence of the TMR value on the crystal orientation of FM electrodes. In Fe/Al_2O_3/$Fe_{50}Co_{50}$ MTJs, the TMR ratio increases from 13% to 42% (8% to 26% at room temperature) when the crystal orientation of Fe electrode changes from (100) to (211). In addition, TiO_2-barrier MTJs have also received extensive attention. In 2003, M. Bibes et al. [26] obtained a high TMR ratio of up to 167% at low temperature in the $La_{0.7}Sr_{0.3}MnO_3/TiO_2/La_{0.7}Sr_{0.3}MnO_3$ MTJs.

Until 2001, W. H. Butler [27] and J. Mathon [28] et al. predicted that the TMR value of monocrystalline Fe/MgO/Fe MTJs might exceed 1000% through the method of first principle calculation. Simply, this huge TMR value results from spin coherent tunneling through monocrystalline MgO barrier, e.g., spin filter effect of MgO barrier. This prediction greatly encouraged research groups around the world to use MgO as a tunneling barrier. In 2004, Yuasa [29] et al. fabricated Fe/MgO/Fe MTJs by the molecular beam epitaxy, and high up to

180% TMR value at room temperature is observed. In the same year, Parkin [30] et al. prepared CoFeB/MgO/CoFeB MTJs by magnetron sputtering, and the TMR value at room temperature reached up to 220%. In Table 3.1, the research status of MTJs with different barrier materials is listed [31–34].

In addition to oxides, nitrides, and semiconductor compounds can be used as insulator layers. D. Chiba et al. [35] took the semiconductor multilayer film as a tunneling barrier, and their results indicated that the TMR effect in MTJ with semiconductor barrier resulted from spin-dependent scattering and tunneling. Then, in 2001, two semiconductor compounds, ZnS [36] and AlAs [37], were used as barrier layers. Xin Jiang et al. [38] used ZeSe as the intermediate layer and observed TMR value of 10.2% at room temperature and 15% at 15K. In 2004, S. Yoshimura et al. [39] used the nitride AlN instead of AlO_x as the insulator layer and obtained 49% TMR. Their study showed that the nitriding process of Al is slower than the oxidation process. Thus, the growth process of AlN film could be better controlled and beneficial for reducing the sensor gap and improving the quality of the tunneling layer.

In conclusion, high TMR ratio has always been the research target and there are some milestone-like achievements in the development of barrier materials. Amorphous Al_2O_3 is the first barrier material used in MTJs that can produce a large MR ratio at room temperature. By improving the surface smoothness and continuity of Al_2O_3 film, MR ratio can be further improved. At the same time, MTJs with crystallized Al_2O_3 barrier show a higher MR ratio than that in amorphous Al_2O_3-barrier MTJs by effectively reducing the disordered-atom-induced scattering. At last, the spin coherent tunneling effect in monocrystalline MgO-barrier MTJs opens the door to improve the TMR ratio, which makes MgO become the most concerned and popular barrier material till today.

Table 3.1 Development of insulator layer of MTJs.

Time	Material	TMR value (%)		Authors
		Low T	Room T	
1975	Ge	14	/	M. Julliére [3]
1982	NiO	2	/	S. Maekawa et al. [13]
1991	Al-Al_2O_3	3.5	2.7	T. Miyazaki et al. [31]
1992	GdO_x	7.7	/	J. Nowak et al. [14]
1994	MgO	/	0.2	T. S. Plaskett et al. [15]
1995	$Al2O_3$	30	18	T. Miyazaki et al. [12]
1996	$SrTiO_3$	83	/	Y. Lu et al. [16]
1996	HfO	30	/	C. L. Platt et al. [21]
2000	Ga_2O_3	27.6	18.2	Z. Li et al. [17]
2001	ZrO_x	/	19.2	J. Wang et al. [19]
2003	TiO_2	167	/	M.Bibes et al. [26]
2004	CoFeB/MgO/CoFeB	300	220	S. P. Parkin et al. [30]
2007	CoFeB/MgO/CoFeB	1010	500	Y. M. Lee et al. [32]
2008	CoFeB/MgO/CoFeB	1144	604	S. Ikeda et al. [33]
2009	Co_2FeAl/MgO/CoFe	700	330	W. Wang et al. [34]

3.2.2 Ferromagnetic Layers in TMR

In addition to the above discussed insulator materials used as tunneling barriers, the magnetic materials for fabricating ferromagnetic electrodes also play an important role in the TMR sensors. The Julliére model describes the theoretical maximum TMR value in MTJs with an amorphous barrier, which only depends on the spin polarization of the magnetic layer. Therefore, the magnetic layer material with high spin polarization is the key to improve the TMR effect. It is well known that Fe, Co, and Ni are the basic elements of ferromagnetic materials. Generally, the compounds and alloys stoichiometrically containing these elements are effective magnetic materials as electrodes in the TMR sensor configurations. The spin polarization of several common ferromagnetic materials used in MTJs is shown in Table 3.2 [40].

Although CoFe and CoFeB compounds are not the ferromagnetic materials with highest spin polarization, they are of superior compatibility in the process of preparing MgO-barrier MTJs on high quality. Therefore, CoFe and CoFeB become the most useful ferromagnetic materials in the MTJs nowadays. S. Ikeda et al. [41] fabricated a series of MgO-based MTJs by using CoFe and CoFeB materials in different compositions as ferromagnetic electrodes. The results are shown in Table 3.3.

According to Table 3.3, MTJs with two CoFeB FM electrodes present larger TMR value compared with other MTJs because MgO films deposited on CoFeB films are more easily to crystalize and form a highly (001) oriented crystallinity during annealing at an appropriated temperature. This kind of lattice-oriented MgO is beneficial for spin coherent tunneling, which can improve the TMR effect. Besides, S. Srivastava [42], et al. studied the influence of the composition of $(Co_xFe_{1-x})_{80}B_{20}$ alloys on the magnetic characteristics of the free layer. When the composition of Co is in the range of 35%–37.5%, the magnetization and anisotropy of free layer are minimally impacted by the annealing temperature, and the performance of the free layer is the most stable.

In addition to these common materials, the ferromagnetic materials with higher spin polarization are desired to develop the TMR sensors with high performance. Semimetal is a kind of materials with special band structure, in which electrons in one spin subband behave like a metal while electrons in another spin subband behave like a semiconductor or insulator. Near Fermi level, there are only electrons with one spin direction for semimetal, which indicates the nearly 100% spin polarization. The spin polarization of several semimetal materials have been extensively studied. As listed in Table 3.2, semimetal materials of $La_{0.7}Sr_{0.3}MnO_3$ has a spin polarization of nearly 100%. In 2002, MTJs with the $La_{0.7}Sr_{0.3}MnO_3$ FM electrode were fabricated, in which the TMR ratio as high as 1800% was determined at 4K [43]. In addition, the epitaxial Fe_3O_4 also possesses high spin polarization[44, 45]. Some studies have confirmed that full-Heusler alloys, e.g., Co_2MnSi, Co_2MnGe etc. possess semi-metallic properties, which have

Table 3.2 The spin polarization of some common FM materials.

Material	Spin polarization (%)	Materials	Spin polarization (%)
Ni	33	Fe_3O_4	98
Co	42	$La_{0.7}Sr_{0.3}MnO_3$	100
Fe	44	$Co_{40}Fe_{40}B_{20}$	58
$Ni_{80}Fe_{20}$	48	$Co_{20}Fe_{60}B_{20}$	58
$Co_{50}Fe_{50}$	55		

Table 3.3 Effect of CoFe and CoFeB FM electrodes in different compositions on the performance of MgO-based MTJs [41].

Pinning layer	Free layer	T_a^{op} (°C)	TMR value (%)
$Co_{40}Fe_{40}B_{20}$	$Co_{40}Fe_{40}B_{20}$	400	355
$Co_{20}Fe_{60}B_{20}$		400	351
$Co_{40}Fe_{40}B_{20}$	$Co_{50}Fe_{50}$	400	277
$Co_{40}Fe_{40}B_{20}$	$Co_{90}Fe_{10}$	350	131
$Co_{50}Fe_{50}$	$Co_{40}Fe_{40}B_{20}$	325	50
$Co_{50}Fe_{50}$	$Co_{50}Fe_{50}$	270	12
$Co_{90}Fe_{10}$	$Co_{40}Fe_{40}B_{20}$	300	75
$Co_{90}Fe_{10}$	$Co_{90}Fe_{10}$	270	53
$Co_{40}Fe_{40}B_{20}$	$Co_{40}Fe_{40}B_{20}$ / NiFe	300	65
CoFe / $Co_{40}Fe_{40}B_{20}$	$Co_{20}Fe_{60}B_{20}$ (pseudo-spin valve)	325	181
$Co_{40}Fe_{40}B_{20}$	$Co_{40}Fe_{40}B_{20}$	450	450
$Co_{20}Fe_{60}B_{20}$	**$Co_{20}Fe_{60}B_{20}$**	475	500

been widely used as ferromagnetic electrodes in MTJs [46–48]. Figure 3.3 shows the energy band structure of the constructed $Co_2MnSi/MgO/Co_{50}Fe_{50}$ MTJs by T. Ishikawa [49]. The observed TMR ratio is of 90% and 192% at room temperature and 4.2K, respectively. However, the deduced spin polarization of Co_2MnSi according to Julliére model was only 75% and 54% at 4.2K and room temperature, respectively, lower than the theoretical prediction. This may be due to the structure defect caused by the deviation of stoichiometric composition.

N. Tezuka et al. [50] prepared MTJs with $Co_2FeAl_{0.5}Si_{0.5}$ FM electrode, in which, TMR value of 76% at room temperature was observed. Their results also proved that the spin polarization of $Co_2FeAl_{0.5}Si_{0.5}$ with B2 phase is higher than that with L21 phase, benefiting

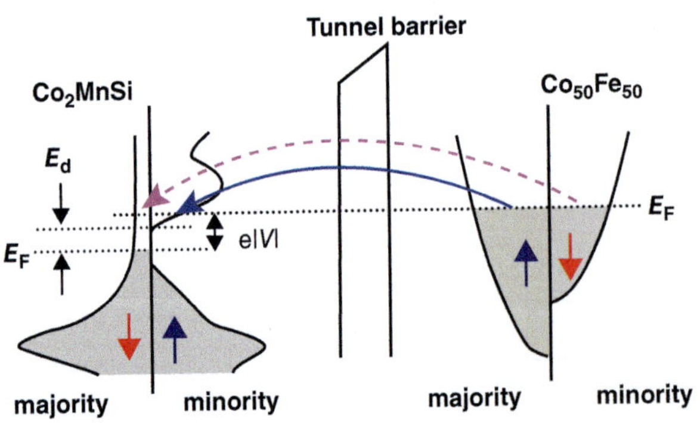

Figure 3.3 Schematic diagram of energy bands of $Co_2MnSi/MgO/Co_{50}Fe_{50}$ MTJs [49].

from the smooth surface of the B2 phase. In addition, the Ni_2MnX group Heusler alloys, such as Ni_2MnGa, Ni_2MnGe and Ni_2MnAl [51] are widely studied. Theoretically, these Heusler alloys possess spin polarization of 100%. However, the MTJs made up of these alloys presents much lower TMR values than expectation. D. Orgassa [52] et al. revealed that only a few percent atomic disorder can destroy the semi-metallic nature of the Heusler alloy and produce minority-spin states near the Fermi level. During the fabrication of Heusler alloy, it is inevitable to introduce defects or impurities, so the actual spin polarization of fabricated Heusler alloy could not reach 100%.

3.2.3 TMR Film Stack

The basic TMR film stack is a so-called tri-layered structure that consists of two ferromagnetic layers intermediated with an insulator spacer. One of the FM layer is pinned layer (PL) or reference layer (RL), another FM layer is free layer (FL) or sensing layer (SL), and the interlayer is a barrier. The magnetization direction of the reference layer is pinned in the direction of applied magnetic field during annealing. A large external magnetic field is needed to change its magnetization direction. The magnetization direction of sensing layer is free under an external magnetic field. In order to improve the performance of TMR sensor in application, the TMR film stack is much more complex than the trilayered structure. Figure 3.4 shows the film stack of MTJs utilized in most research and sensor applications, which consists of 13 thin films. The top CoFeB is ferromagnetic coupling with soft magnetic NiFe to construct free layer, which presents small coercivity. The antiferromagnetic (AFM) material of IrMn and synthetic antiferromagnetic (SAF) structure of CoFe/Ru/CoFe is used to pin the bottom CoFeB layer and enhance its coercivity. Compared with the basic structure of MTJs, this complex structure possesses more superior performance, which is suitable for realistic application.

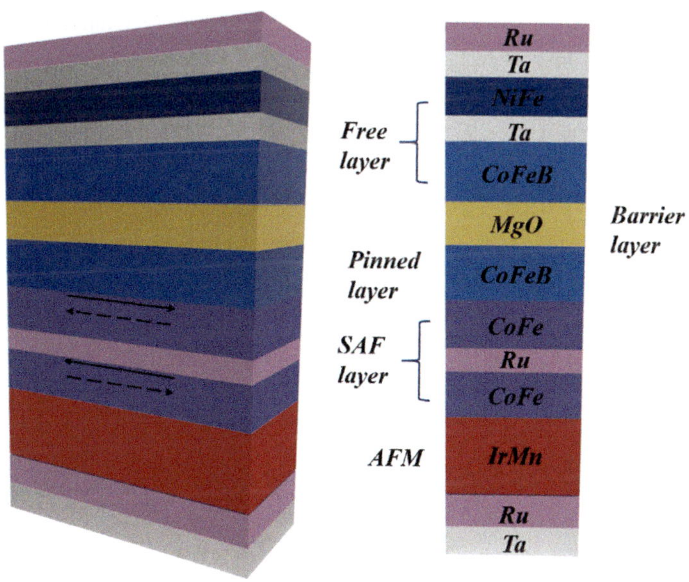

Figure 3.4 The structure of a MTJ film stack.

From the top side, the ruthenium (Ru) and tantalum (Ta) layers are used as the capping layer to prevent FM layer oxidation and improve the TMR value of MTJs. Some research indicates that the material type and thickness of capping layer is very important for the performance of MTJs. For example, P.W.T. Pong et al. [53] have demonstrated that as the thickness of the capping layer increases further, the stress or strain on the MgO layer also becomes stronger, which may distort the crystallinity of MgO (100) and adversely affects the TMR value. Therefore, as shown in Figure 3.5, the optimal TMR value of MTJs can be realized by precisely adjusting the thickness of Ru layer and Ta layer. In 2016, J. Zhou et al. studied the influence of different capping materials on the TMR value of the perpendicular magnetic anisotropic (PMA) MTJs. Their results proved that TMR value are sensitive to capping materials and tungsten (W) is an ideal capping material compared with Ta and other materials. Another research group [54] prepared 0.4 nm MgO film as the capping layer of the PMA MTJs. They find that the MgO capping layer can induce partially PMA at the CoFeB/MgO interface and further reduces the saturation magnetic field of the in-plane magnetic anisotropy (IMA) CoFeB under perpendicular field. This phenomenon can be used to reduce the linear region of the TMR sensor, and thus improve the sensitivity.

In order to improve the sensitivity of TMR sensor during measuring small magnetic field, we expect that the coercivity of free layer is as small as possible. Although CoFeB free layer presents high spin polarization and superior compatibility with MgO barrier, it also possesses a high magnetostriction coefficient, resulting in a large coercivity. To overcome this

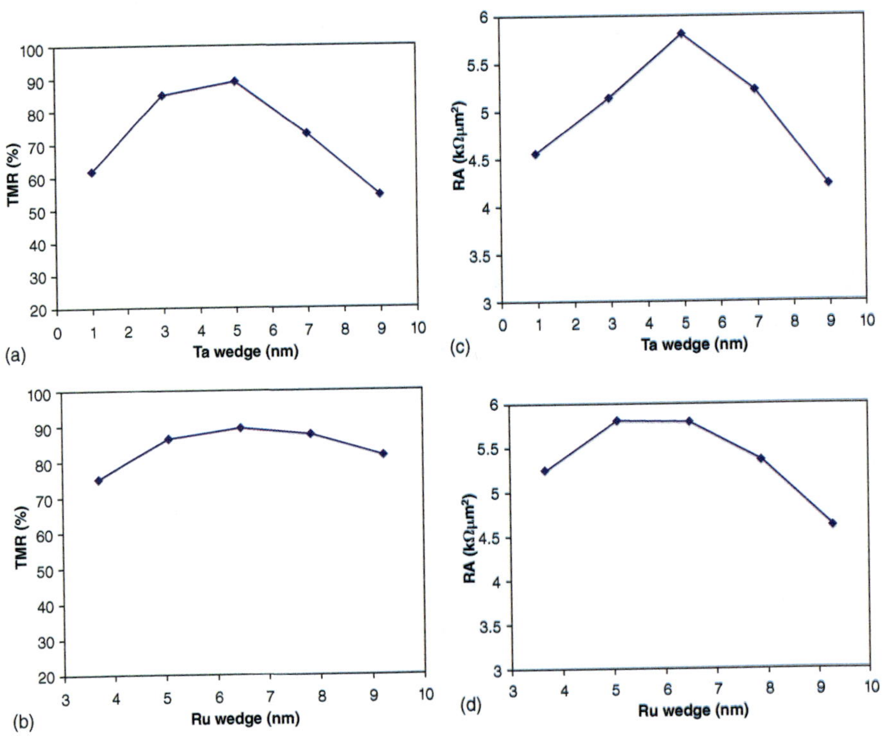

Figure 3.5 (a)(b) The TMR value variation with the thickness of Ta and Ru capping layer. (c)(d) The RA product variation with the thickness of Ta and Ru capping layer [53].

issue, an effective method is to integrate CoFeB layer and a soft magnetic NiFe layer with small magnetostriction coefficient [55] to form a composite free layer. In 2006, S. Ikeda et al. [56] found that the lattice structure of the free layer can influence the TMR value of MTJs. As shown in Figure 3.6, the TMR value of (001) oriented MgO-barrier MTJs with a free layer of a body-centered cubic (bcc) structure is higher than that of a face-centered cubic (fcc) structure.

As mentioned above, lattice structure is important for improving TMR value. There are some other points about lattice structure should be noted during the design and fabricating of MTJ film stack as shown in Figure 3.4. Before annealing, the sputtered CoFeB is

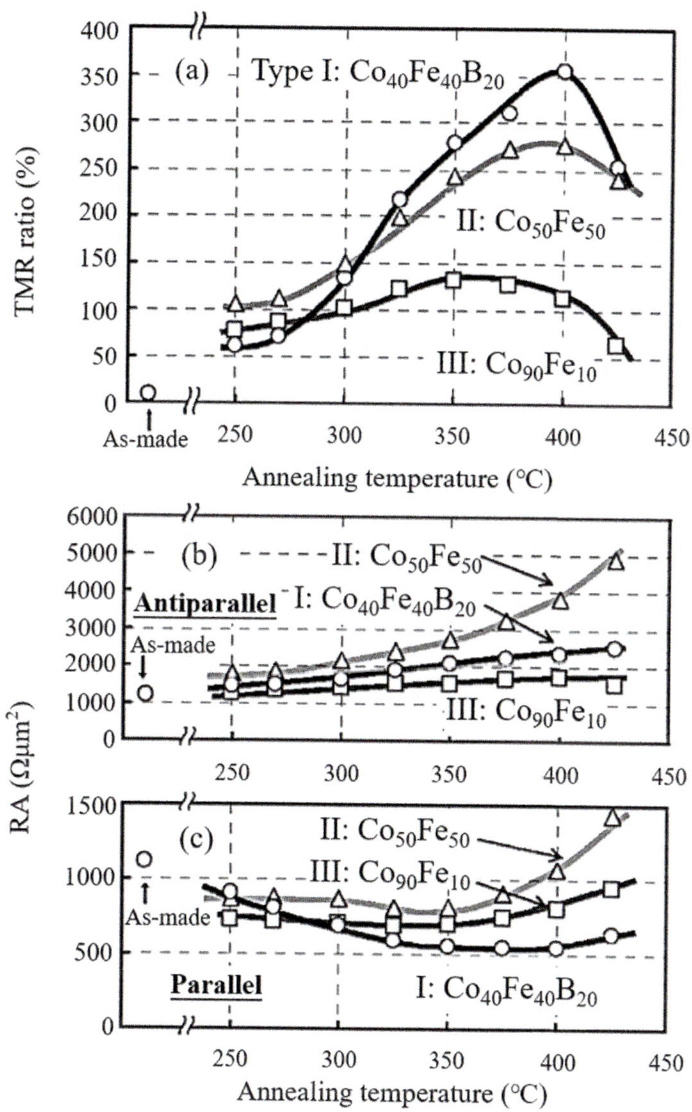

Figure 3.6 The influence of lattice structure of free layer on TMR ratio and RA values [56].

amorphous. (001)-oriented MgO with good crystallinity can grow on amorphous CoFeB because (001) the crystal phase is its preferred orientation. After high temperature annealing, the lattice structure of CoFeB tends to form bcc structure. The compatibility with MgO during the fabrication of high-quality core structure is an important factor to make CoFeB become a popular free-layer material. Furthermore, considering that the lattice structure of NiFe is fcc structure after annealing, a thin Ta layer is inserted between CoFeB and NiFe layers. Through this way, we can effectively avoid the lattice structure of CoFeB being influenced by the crystallization process of NiFe during annealing. Especially for complex MTJs film stacks, the lattice structure can be adjusted by inserting a thin insertion layer with bcc structure (e.g., Ta) to prevent the lattice dislocation at the interface. In summary, quite many results demonstrate that the performance of MTJs with core structure of CoFeB/MgO/CoFeB exhibits superior performance compared with other structures. From the perspective of materials, the MgO barrier possesses spin filter effect and CoFeB presents high spin polarization. From the perspective of process, high-quality CoFeB/MgO/CoFeB structure could be fabricated through the compatible growth mode of CoFeB and MgO.

Under the core structure of CoFeB/MgO/CoFeB, it is the synthetic antiferromagnetic (SAF) structure composed of a non-magnetic layer sandwiched between two FM layers. Through adjusting the thickness of non-magnetic layer, the two FM layers would present antiferromagnetic coupling, that is, the magnetization direction of the two FM layers is opposite. The commonly used non-magnetic layers is Ru layer, which can inhibit atomic interdiffusion.

The concept of SAF was first proposed in 1996 [57, 58] and then this structure was widely used in GMR and TMR film stacks. The SAF configuration presents excellent thermal stability, which can endure high temperature without losing its pinning ability. Moreover, through adjusting the thickness of FM layers in the SAF structure, the static magnetic coupling, and the demagnetizing field caused by the pinned CoFeB layer can be reduced effectively [59, 60]. In addition, some research has found that the thickness of Ru layer can also affect the TMR value, so optimizing the thickness of Ru layer is an effective way to improve the TMR value [61]. Except for the commonly used CoFe, CoFeB is also considered to use in SAF structure. When CoFeB replaces the top CoFe layer, the coupling strength of constructed CoFe/Ru/CoFeB multilayer is similar to that of the CoFe/Ru/CoFe multilayer [62]. However, when CoFeB replaces the bottom CoFe layer, the amorphous CoFeB will induce incompleted grain crystallization and disordered orientation in the following deposited Ru layer, leading to the reduction of coupling strength [63]. This also indicates that the exchange coupling strength in SAF structure is greatly dependent on the crystallinity of Ru.

Under the SAF structure, AFM material of IrMn layer is used to nail the magnetization direction of bottom FM layer in SAF structure through exchange coupling, which could further improve the thermal stability of SAF layer. Most AFM materials can be applied to pin magnetic layers, in which IrMn and PtMn are the commonly used. Because the characteristic thickness of IrMn is lower than that of PtMn, IrMn is commonly used as the nailing layer in the magnetic head [64].

The bottom Ta and Ru are used as buffer layers to induce the formation of (001) orientation grains in the InMn layer, which plays an induction role for the growth of the film stack with (001) lattice orientation. Furthermore, some research [65] has demonstrated that increasing the thickness of the Ta layer can effectively improve the TMR ratio. As mentioned above, Ta insertion in different parts of the membrane stack can optimize the growth of the lattice. Hence, Ta is a very popular buffer layer material. In addition to Ta and Ru, chromium [66] and tungsten [67] can also be used as buffer layers. Proper buffer layer should be selected based on the required lattice orientations of the upper layer materials, which can effectively improve the surface morphology and structure quality of MTJ stacks. It should be noted that copper nitride (CuN) (not shown in Figure 3.6) is widely used as an electrode material under the Ta/Ru buffer layer. On the one hand, the CuN exhibits excellent electrical conductivity so that the fabricated MTJ stack can perform current in-plane tunneling (CIPT) measurement. On the other hand, the deposited CuN possesses high flatness, which can further improve the performance of MTJ stack.

3.2.4 Perpendicular Magnetic Anisotropy (PMA) in TMR

To meet the demand of electronic products for minimization and high density, strong magnetic anisotropy material, which means high thermal stability, become an important consideration during fabricating MTJs. Therefore, perpendicular magnetic anisotropy film (PMA), in which the easy magnetization axis is perpendicular to the film surface, is used to overcome the superparamagnetic limit with the gradually decreasing size of free layer. Considering the compatibility with optimal core structure of MTJs, the X/CoFeB/MgO system is the widely used PMA FM structure, in which X represents heavy metal (HM), such as Pt, Ta, and W. The interfacial hybridization between CoFeB and its adjacent HM and MgO would induce an interfacial PMA. When CoFeB is thin enough, the interfacial PMA will exceed the bulk in-plane magnetic anisotropy (IMA) and make the magnetization perpendicular to the film plane. S. Ikeda et al. [68] reported a PMA MTJs with the core structure of Ta/CoFeB/MgO/CoFeB/Ta in 2010.

To obtain a better PMA, S. Peng et al. [69] calculated the magnetic anisotropy energy (MAE) at the MgO/CoFe interface and the CoFe/HM (Ru, Ta, and Hf) interface. According to Table 3.4, it is found that the sum of the MAE of MgO/CoFe and CoFe/X interfaces is approximately equal to that in the MgO/CoFe/HM structure, indicating that the PMA comes from the both interfaces of MgO/CoFe and CoFe/HM respectively. Therefore, the CoFeB/HM interface is critical to improve the PMA.

The thickness of free layer is also an important affecting factor of PMA. In 2019, Z. Lei et al. [70] studied the influence of free layer thickness on PMA and TMR value of MTJs with perpendicular SAF (p-SAF). When the thickness of the free layer is changed, the magnetization direction of MTJs exhibits different tendency towards horizontal component or vertical component. According to the variation of the magnetic moment of the free layer, it provides a theoretical basis for manufacturing three-dimensional (3D) TMR sensors.

3.2.5 Material Fabrication and Pattern Process

The fabrication of TMR sensors, similar to that of the GMR sensors, includes film deposition, pattern process, and encapsulation. Different from traditional semiconductor technology, the technical process of TMR sensors and GMR sensors are generally less than

Table 3.4 Calculated MAE values (erg/cm^2) for different structures [69].

X	CoFe surface	MgO/CoFe interface	CoFe/X interface	Sum of MAE in two interfaces	MgO/CoFe/X structure
Ru	0.41	0.57	0.52	1.09	0.98
Ta	0.41	0.57	1.13	1.70	1.77
Hf	0.41	0.57	1.65	2.22	2.28

five steps of photolithography in order to ensure the performance of sensors. Thus, the preparation process is simpler and the design process is more flexible. The technical process of TMR sensors is usually divided into two parts, material fabrication and pattern process. And this section will introduce the main processing steps of these two parts respectively.

The film deposition process is the basis of TMR sensor manufacturing. It refers to the preparation of MTJs film stacks by sputtering, epitaxy, evaporation, and other methods on silicon wafers. There are more than 10 layers of films in a MTJ film stack, which is required to possess uniform thickness and smooth surface for fabricating high-quality MTJs. Therefore, the film deposition process is very important for TMR sensors. These deposition techniques fall into two broad categories: physical vapor deposition (PVD) and chemical vapor deposition (CVD).

3.2.5.1 Magnetron Sputtering

PVD is a method of vaporizing the coating material by physical methods (such as evaporation, sputtering, etc.) and depositing the film on the substrate surface. The commonly used PVD method for MTJ film stack fabrication is magnetron sputtering. The working process of magnetron sputtering is described as follows:

1. High electric field would breakdown the inert gas and produce positive ions and electrons. Electrons fly to the substrate under the action of electric field, which would collide with atoms of inert gas, ionize them, and produce more positive ions and new electrons in the vacuum environment.
2. When the new electrons fly to the substrate, the ionized ions accelerate to the cathode target under the action of electric field, and then bombard the target surface with high energy to make the target material sputtering. Neutral target atoms or molecules are deposited on the substrate to form a thin film.
3. At the same time, the secondary electrons generated by the collision with the target will be bounded around the target by magnetic fields, which can improve the electron intensity and keep the collision ongoing.

Figure 3.7a is the schematic diagram of magnetron sputtering. Argon is a commonly used inert gas in magnetron sputtering. In addition to the collisions of high-speed electrons, the gas glow discharge to form plasma and electrons requires the entire process to take place in a low-pressure $\left(10^{-3} \sim 10^{-4}\, Torr\right)$ environment. At the same time, the low-pressure vacuum environment can effectively reduce the adsorption of water and oxygen on the chamber walls, providing a clean deposition environment and reducing pollution. After years of development, magnetron sputtering technology has been very mature and is suitable for fabricating large amounts of materials. Furthermore, the film prepared by this method shows better performance than other PVD technologies. As shown in Figure 3.7b, multiple target materials can be installed in the cavity of magnetron sputtering equipment, so multiple materials can be sputtered at the same time, which is known as co-sputtering technology. In addition, the deposition of all MTJs film stacks can be completed without opening the chamber through closed-loop control. Magnetron sputtering has become the most popular thin film deposition technology in the industry due to its advantages of high yield, high performance and closed-loop control [71, 72].

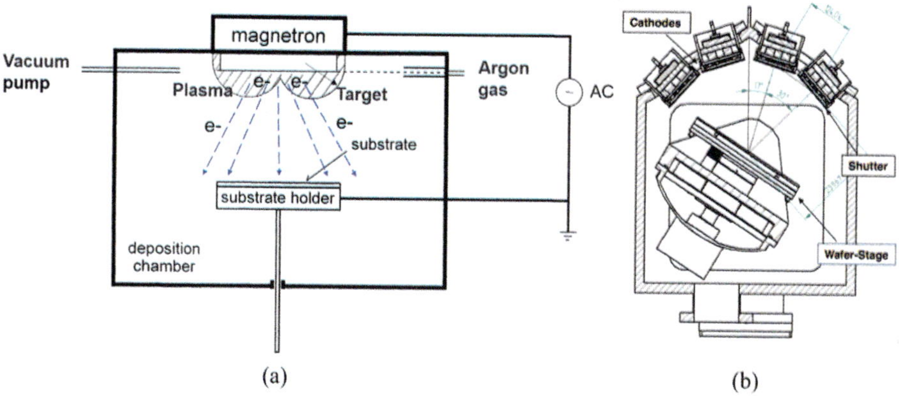

Figure 3.7 (a) Schematic of magnetron sputtering [73]. (b) Internal structure diagram of magnetron sputtering equipment.

3.2.5.2 Ion Beam Deposition (IBD)

Ion beam deposition is also a kind of thin film deposition technology which belongs to PVD. The principle is similar to sputtering, but this technique uses an ion source and then applies a voltage to the gate (graphite or tungsten) to accelerate the ion beam towards the target, depositing or sputtering the target materials (metal or dielectric) onto the substrates to form a metal or dielectric film. Because the ion beam is isoenergetic and highly collimated, it is able to precisely control thickness and deposit very dense and high-quality films compared with other PVD techniques. Since the ion beam is generated by the ion gun, the deposition parameters such as ion beam energy and sputtering angle are adjustable, which enables the operators to control and optimize the deposition process more independently.

3.2.5.3 Evaporation

Evaporation, generally refers to vacuum evaporation, is also a PVD technology. It is a deposition method that evaporates the material by heating under vacuum condition. The evaporated particles fly to the substrate surface and condense into film. And evaporation technology possesses the advantages of high deposition rate, large coating area, strong adhesion, simple equipment, and easy to operate, but the uniformity of the film is not as good as that of the film obtained by sputtering. Therefore, for the TMR sensor, evaporation technology is usually used to prepare the electrode of the device.

3.2.5.4 Chemical Vapor Deposition (CVD)

Chemical vapor deposition (CVD) is a kind of film deposition technology by chemical reaction on the surface of a substrate with several vapor phase compounds or elements. CVD technology possesses a high covering force, and it is widely used to cover deep defects, pores, and other complex three-dimensional structures. Furthermore, the deposition rate of CVD is high, which make it suitable to fabricate thick film. The CVD technology does not need ultra-high vacuum, and can adapt to a variety of technological changes [74]. Therefore, the equipment cost of CVD is much lower than that of PVD. However, CVD usually requires a high temperature to promote the chemical reaction of compounds, which is not suitable for magnetic film deposition. For MTJs film stacks CVD techniques are used to fabricate low cost, good quality insulating layers, such as silicon oxide and silicon nitride.

According to the working pressure and reaction type, CVD techniques could be subdivided to several types, such as atmospheric pressure CVD (APCVD), low pressure CVD (LPCVD), plasma-enhanced CVD (PECVD), and so on. PECVD is commonly used for the deposition of silicon nitride, which uses radio (RF) electric field to inspire gas glow discharge and generate plasmas in the deposition gas. These enhanced plasmas will increase the deposition energy, effectively reduce the reaction temperature, and improve the uniformity of the thin film. Figure 3.8 shows the schematic of a PECVD equipment. The silicon wafers are heated up to the working temperature through the heater. After the reaction gas enters, the RF voltage excites the plasma and deposition begins.

Pattern process transforms the fabricated multilayer stacks to functional devices (i.e., MTJs with support circuit), which is an essential part during the production of TMR sensors. Figure 3.9 shows the flow chart of the pattern process of TMR sensors, which includes photolithography, etching, lift off, backfilling, and other processing steps. The main purpose of pattern process is to transfer the designed graphs to the surface of a silicon wafer. Multiple photolithography and etching processes are performed because of the complex graphic structure.

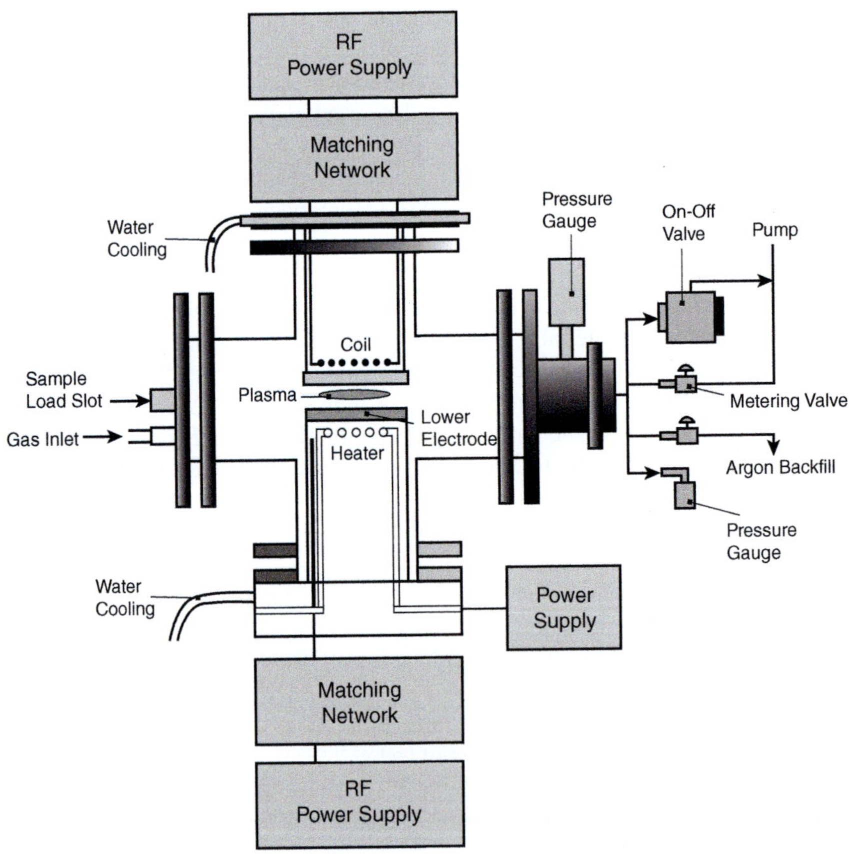

Figure 3.8 Schematic of a PECVD equipment [75].

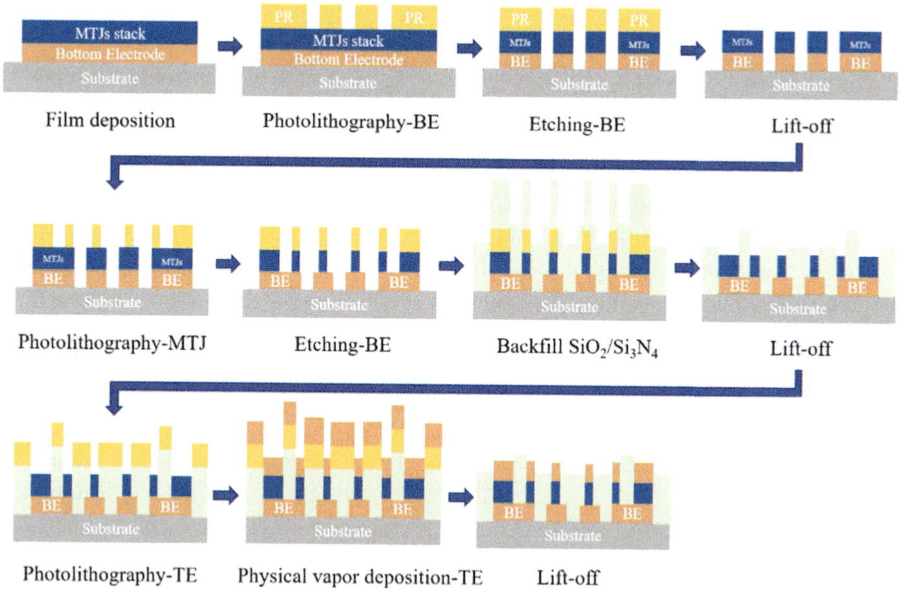

Figure 3.9 Flow chart of pattern process of MTJs. BE and TE are the bottom electrode and top electrode in shot respectively.

However, as the number of photolithography increases, the processing difficulties and risk also increase. Therefore, simplifying the process step and the number of photolithography can effectively reduce the cost and improve the survival rate of MTJs, which is also a research direction for the performance improvement of TMR sensor.

3.2.5.5 Photolithography

Photolithography is a graphic transfer process similar to photography. Two items of mask and the photoresist are very important for photolithography. The manufacture of mask plates is to split the design of a circuit into several parts to facilitate multi-step photolithography by dividing the drawn composite graph of size, shape, and surface alignment into individual mask layers. Photoresist is a photosensitive polymer solution and exposure could change its properties and structures. After exposure, the photoresist changes from a soluble substance to an insoluble substance, which is called negative photoresist. Conversely, after exposure, the photoresist changes from insoluble to soluble, which is called positive photoresist. Different photoresists are appropriate for different photolithography parameters and mask types, depending on the graphic design requirements.

3.2.5.6 Etching

Etching process is to remove the part of film stack without covering photoresist by chemical or physical methods, so as to obtain the pattern exactly corresponding to the mask pattern. There are two types of etching: wet etching and dry etching. Wet etching refers to etching the film and silicon wafer by chemical reaction between solution and film. Dry etching refers to the technique of thin film etching by plasma. The dry etching techniques, such as ion beam etching (IBE), inductively couple plasma (ICP) etching and reactive ion etching (RIE), are commonly used in the preparation of MTJs for TMR sensors.

IBE technology uses the accelerated plasma of inert gas generated by gas glow discharge to bombard the etched materials physically. Although the etching rate of IBE technology is stable and highly controllable, so IBE is suitable for the preparation of MTJs. RIE technology utilizes the plasma of active gas generated by an RF electric field to react with the etched material. Meanwhile, the applied electric field is perpendicular to the film plane, the accelerating ion could also play the role of physical bombardment. RIE technology is widely used to etch silicon oxide and silicon nitride by using the working gas of SF_6. Through selecting the appropriate reaction gas, air pressure, flow rate, and RF power, RIE technology can realize fast anisotropic etching rate. On the basis of RIE, ICP etching adds another RF electric field to generate alternating electromagnetic field, which could not only change the motion pass of ion to reduce surface damage, but also increase density of plasma to improve etching efficiency.

3.3 The Noise of TMR Sensors

TMR sensors is used to detect weak magnetic fields signal, its noise directly determines the resolution and application range of the sensors. One of effective solutions for improving signal-to-noise ratio of TMR sensors is to design and fabricate high-performance MTJ.

3.3.1 The Source of Noise from TMR Sensors

The noises in TMR sensor can be divided into white noise and low frequency noise according to the correlation with frequency. White noise refers to frequency independent noise, including thermal noise and shot noise. Low frequency noise is frequency dependent noise, including 1/f noise and random telegraph noise (RTN). According to different action mechanism, the noise in magnetic sensor can also be divided into magnetic noise and electrical noise. Electrical noise exists in all electrical materials and devices while magnetic noise is the unique noise in magnetic materials, which has the greatest impact on TMR sensors. In addition to the intrinsic noise of MTJs, an external circuit is required to amplify the signal in the process of sensor working and noise testing. Hence, the internal noise of the amplifier circuit will also affect the detection sensitivity [76].

These noise sources of TMR sensors have been widely studied over the past several decades. It is shown that these noise sources are unrelated to each other, and the total noise can be regarded as the superposition or sum of the noise components.

For magnetic materials, thermal noise can be divided into thermoelectric noise and thermomagnetic noise. Generally, thermal noise is referred to as thermoelectric noise, also known as Johnson-Nyquist noise. Thermoelectric noise originates from random fluctuation of carrier and belongs to balanced intrinsic noise of the material. It is a kind of white noise and only depends on resistance and temperature. Thus, the thermoelectric noise is constant when the resistance and temperature of the system keep fixed. And it usually exists in high frequency and low frequency band of signal. Thermomagnetic noise results from the magnetic disturbance of the magnetic material. Its main source in a TMR sensor is the noise generated by the random rotation of the magnetic moment of the free layer. Different from ordinary white noise, thermomagnetic noise varies with external magnetic field [77]. The generation mechanism of thermomagnetic noise follows the fluctuation dissipation theorem

(FDT), based on which, Egelhoff et al. [78] deduced the power spectral density formula of thermal magnetic noise

$$S_V^{therm-mag} = 4k_B T \mu_0 \alpha_G / \Omega \gamma M_S \tag{3.15}$$

where μ_0 is space permeability, α_G is the Gilbert damping coefficient, k_B is the Boltzmann constant, Ω is free layer volume, γ is electron spin magnetic ratio, and M_s is the saturation magnetization.

The shot noise comes from the particle characteristics of the carrier and is related to the dispersion of charge and the randomness of electron emission. In thin films, the particle noise is the result of charge quantization. For TMR films, the oxide layer in the film layer will cause the shot noise, so this noise is very weak in GMR films. The shot noise is also a kind of white noise, but the difference is that the shot noise is related to the membrane current. In the non-equilibrium system, it belongs to non-equilibrium intrinsic noise.

Unlike white noise, 1/*f* noise refers to the low frequency part of the noise and its intensity is proportional to $(1/f)^2$ (usually, $\beta = 1$). In magnetic films, 1/*f* noise can be divided into electrical 1/*f* noise and magnetic 1/*f* noise according to different action mechanisms. According to the reports, electrical 1/*f* noise is the disturbance noise caused by the electron capture charge in the defect and interface gap, and it is a kind of low-frequency noise prevalent in metal films. Its noise power spectral density decreases with the increasing frequency, which can be calculated by the following formula

$$S_V^{elec-1/f} = \alpha_{elec} V^2 / A f^{\beta} \tag{3.16}$$

where "V" is the voltage at both ends of the MTJs, "A" is the junction area, "*f*"is the frequency, usually $\beta = 1$, and α_{elec} is the Hooge constant of electrical 1/*f* noise. The Hooge constant is the empirical value, which is deduced from an approximate calculation of a large number of data obtained from Hooge's 1/*f* noise experiment [79]. At present, the Hooge constant is widely used to quantify the magnitude of 1/*f* noise [80]. For a TMR sensors, the influence factors of Hooge constant include the voltage applied to MTJs, TMR value and TMR performance parameters, such as the product of resistance and area (RA) [81].

The magnetic 1/*f* noise comes from the transition between metastable domains, i.e., the magnetic sequence rotation between the free layer and the pinning layer interface [82, 83]. It is a kind of noise produced by the magnetization disturbance inside the structure of the films, which is different from the magnetic moment rotation caused by the external magnetic field. The power spectral density of magnetic 1/*f* noise is [84]:

$$S_V^{mag-1/f} = 2B_{sat} \alpha_{mag} / \Omega f. \tag{3.17}$$

Here, B_{sat} is the saturation magnetic field of free layer, α_{mag} is a constant similar with Hooge constant in electrical 1/*f* noise, Ω is free layer volume.

Random telegraph noise, also known as popcorn noise, was first discovered in 1978 by Kandiah and Whiting [84] and they attributed the cause of RTN to the charging and discharging process at the center of the trap. For TMR sensors, RTN can be interpreted as a repetitive process in which a random single electron is captured and released by a single trap. When the sensing unit size of the TMR sensor is comparable with the size of a single defect, RTN becomes the main source of low-frequency noise. RTN can be effectively suppressed by improving annealing process to reduce film defects.

For a TMR sensor, the total noise is equal to the sum of the above noises. Thus, in the process of noise testing and analysis, we could figure out the magnitude of a kind of noise by eliminating it and calculating the noise spectrum difference, and further realize noises separation.

3.3.2 Methods to Suppress the Noise

It has also been shown that magnetic noise mainly comes from magnetic and thermal disturbance. This kind of noise, especially $1/f$ noise, will greatly deteriorate the linear response characteristics and low frequency detection capability of the sensors. Therefore, reducing sensor noise will greatly improve the detection accuracy and application range of magnetic sensor.

3.3.2.1 Increase the Number of MTJs in TMR Device

As mentioned in equation (3.16), electrical $1/f$ noise is inversely proportional to the area of free layer. In other words, we can effectively reduce the noise through increasing the area. In 1998, Tondra et al. [85] proposed that increasing the number of MTJs could effectively improve the signal-to-noise ratio of TMR sensor, which is equal to enlarge the area. In 2009, Guerrero et al. [86] experimentally achieved sensor detection sensitivity of $16.2\text{nT}/\sqrt{\text{Hz}}$ through series and parallel MTJ arrays. Considering the superposition calculation of thermoelectric noise and electrical $1/f$ noise, the detection accuracy formula is obtained by connecting number N of MTJs in series:

$$D=\sqrt{\frac{S_v^2}{(\Delta V/\Delta H)^2}}=\sqrt{\frac{1}{MR^2}\left(\frac{4Nrk_BT}{V^2}+\frac{\alpha}{NA}\frac{1}{f}\right)} \tag{3.18}$$

where MR is the TMR radio of a single MTJ, and r is the resistance value of a single MTJ. According to the formula, N junctions in series can reduce $1/f$ noise by $1/\sqrt{N}$ times. On this basis, number M of such series arms in parallel can obtain the detection accuracy formula:

$$D=\sqrt{\frac{1}{MR^2}\left(\frac{4Nrk_BT}{MV^2}+\frac{\alpha}{MNA}\frac{1}{f}\right)} \tag{3.19}$$

According to formula (3.19), the thermoelectric noise can be adjusted by the ratio of N/M, and the electrical $1/f$ noise decreases with the increase of the number of MTJs. Figure 3.10 shows the change of detection accuracy with the increase of the number of series junctions. The black point is the influence of $1/f$ noise on detectivity, while the white point is the influence of white noise on detectivity. Obviously, the $1/f$ noise decreases proportionally with the increase of N.

To sum up, increasing the number of MTJs in series and parallel can effectively reduce the electrical $1/f$ noise, but increasing the number of MTJ junctions also increases the output power consumption.

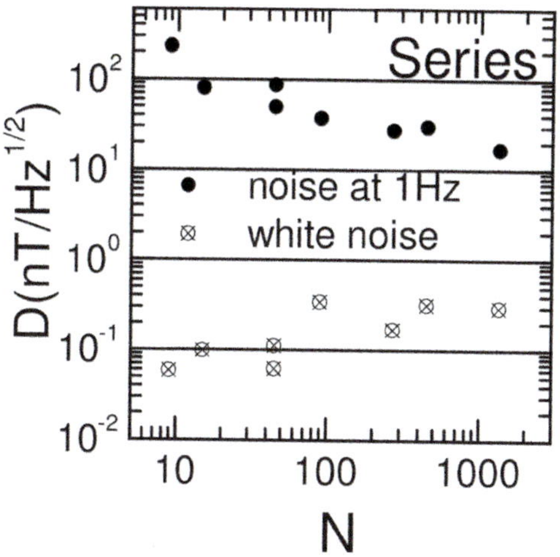

Figure 3.10 The detection rate varies with the number of junctions in a series structure [86].

3.3.2.2 Optimize Free Layer Volume

According to formula (3.15) and (3.17), increasing the volume of the free layer can reduce the magnetic noise, such as increasing the area or thickness of the free layer. Meanwhile, magnetic 1/*f* noise can also be suppressed by reducing the saturated magnetic field (B_{sat}) of the free layer, which is usually achieved by reducing the thickness of the free layer. However, the method of reducing thickness will leads to decrease in volume of free layer and TMR ratio of MTJ, both against the aim of decreasing the magnetic noise [87]. Therefore, it is necessary to comprehensively consider all impact factors and find a optimization parameters to reduce the noise [88].

3.3.2.3 Flux Concentrator

As mentioned above, the 1/*f* noise decreases with the increasing frequency. So, if the signal to be measured can be adjusted from the low-frequency range to the high-frequency range, the 1/*f* noise will be significantly reduced. Edelstein [89] and M. Pan et al. [90] proposed a horizontal motion flux concentrator and vertical motion flux modulation based on micro-electromechanical system (MEMS) respectively. In the following, we take horizontal motion flux concentrator as an example to introduce their working principle.

As shown in Figure 3.11, the horizontal motion flux modulation is constructed by depositing permalloy on the MEMS spring strip. Here, soft magnetic permalloy work as flux modulation film. The applied ac voltage to the electrostatic comb driver would drive the spring strip and deposited permalloy vibrates periodically in horizontal direction. The vibrating permalloy would induce a periodic magnetic field. So that the signal sensing by magnetic sensors is modulated to high frequency and 1/*f* noise is suppressed.

For flux modulation, in order to reduce the cost, the modulation structure needs to be simplified, but this will reduce the modulation efficiency. The horizontal motion flux modulation mentioned above has a low debugging efficiency. In order to improve the efficiency, the vertical motion flux modulation (VMFM) was proposed. Figure 3.12 is the principle diagram of the vertical motion flux modulation (VMM), soft magnetic film as the flux modulation film (FMF) was hung between two concentrators, the film can be driven up and down

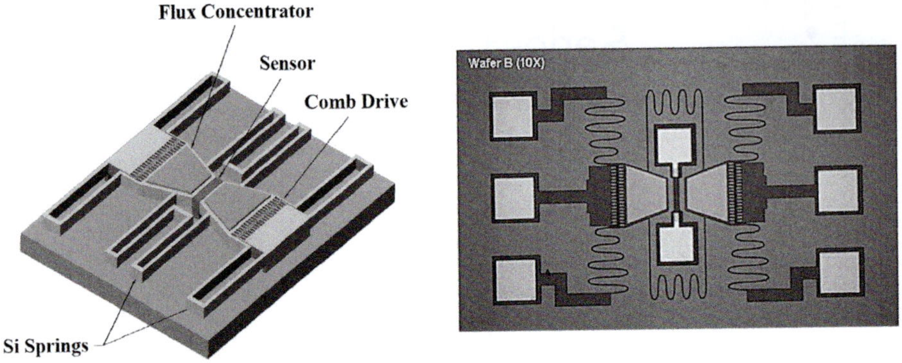

Figure 3.11 Illustrations of the concept of the MEMS flux concentrator and scanning electron micrograph picture of a MEMS flux concentrator [89].

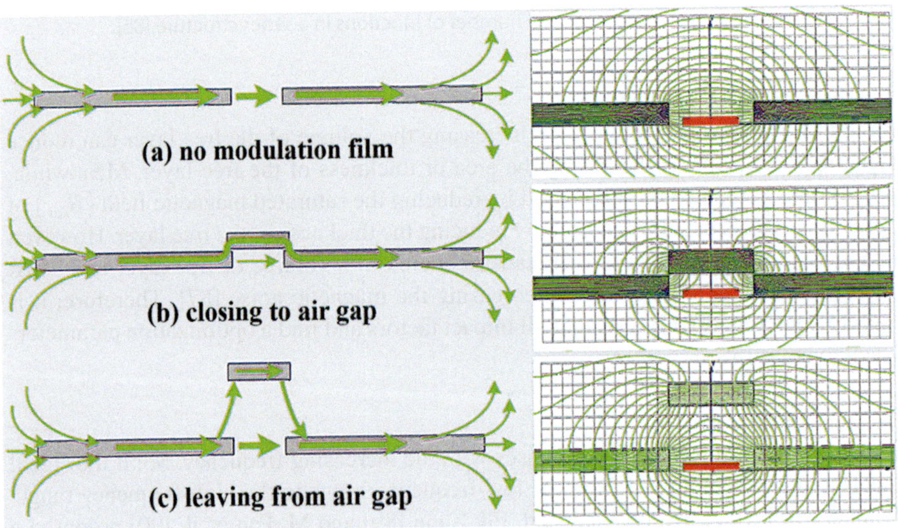

Figure 3.12 Illustration of magnetic flux distributions in VMM [90].

by MEMS drives along the vertical direction. The principle is the same as that of horizontal motion flux modulation. The magnetic field is changed by periodic vibration, and the magnetic field is modulated to high frequency to reduce 1/f noise. However, VMFM has higher modulation efficiency.

However, the flux concentrator needs high voltage to drive the concentrator vibration, leading to large power consumption. Moreover, the low modulation efficiency of MEMS cantilever beam is low and the complex process make it difficult to implement mass production.

3.3.2.4 Applying a Bias Magnetic Field

Considering the magnetic noise mainly comes from the disturbance of the magnetic moment of the free layer, adding a small bias field to keep the free layer in fixed direction when the external magnetic field is zero. By this way, the disturbance of magnetic moment could be remarkably suppressed and the noise of the TMR sensor can be effectively reduced. Meanwhile, the sensitivity of the sensor will also be reduced. Li Huang et al. [91] applied a bias field along the easy axis of the free layer to realize stable single domain and eliminate Barkhausen noise. Meanwhile, an excellent linear response to the external field is also realized by applying the bias field.

In addition to the noise suppression methods mentioned above, the single domain of the free layer can be realized by increasing the pinning strength of the pinning layer, and the shape anisotropy can be adjusted to reduce the noise of the sensor. However, these methods will reduce the sensitivity of the sensor [92]. There is a trade-off between reducing noise and improving sensitivity. How to adjust these parameters to maximize the detection accuracy of the sensor is an important direction to improve the sensor's performance in the future.

3.4 TMR Sensors and Applications

Due to the unique advantages of large MR value and high magnetic field sensitivity, TMR sensors exhibit an attractive application prospect. In recent years, TMR sensors have been widely used in many areas, including the internet of things (IoT), artificial intelligence (AI), military, medicine, etc. According to application scenarios, the existing TMR sensors can be classified into the following broad categories: read heads, switches, vector sensors, rotational speed measurements, micro-magnetic detection, nondestructive testing, and spin-MEMS sensors. One of the most important application is the read heads in the hard disk drive for magnetic recording. The TMR read heads are an important milestone in the process of improving the storage density and miniaturized the hard disk for high areal density recording. In addition to reading heads, TMR vector sensors are also widely used, including angle sensors and geomagnetic sensors. In 2017, Y. Fuji et al. reported a spintronic strain gauge sensor based on MTJs and applied it to the microphone, which is called the spin-MEMS microphone. Since then spin-MEMS sensors have received increasing attention. In order to give readers a comprehensive understanding of TMR sensor devices, several important TMR-sensors applications are described in details in this section.

3.4.1 TMR Read Heads

In recent years, TMR sensors have been widely used in many areas. Among them, the first commercial application of TMR sensor is the read head in hard disk drive, and it is also the most widely used TMR device at present. The main function of read head is to convert the magnetic information stored on the hard disk into electrical signal. With the increasing sensitivity of read head, the size and spacing of memory cell can be designed smaller, which means the disk record density can be improved. Therefore, the magnetic sensors with high sensitivity and large MR value can promote the hard disk head towards the direction of high precision, high speed, and miniaturization. AMR read heads replaced inductive read heads in the early 1990s, followed by GMR read heads in 1995 [93]. The TMR sensors are considered to have great potential because of their high MR value compared with GMR sensors. Since 2005, TMR read heads have been available on hard disks. Companies such as Western Digital and Seagate have been developing and manufacturing TMR read heads in high density perpendicular magnetic recording for years.

Figure 3.13 shows the typical structure of a TMR read head. The main part of this TMR sensor is a MTJ composed of top electrode, potential barrier layer and bottom electrode. The top and bottom shield are connected together through nonmagnetic lead to shield from the interference of external magnetic field. The "longitudinal bias" is a hard magnetic material, which is used to improve the linearity of TMR sensor's output signal through applying a bias magnetic field. Although the output performance of TMR read head is better, but its noise is higher than that of GMR sensor. Because the tunneling barrier layer of the MTJ increases the resistance, the thermoelectric noise of the TMR read head increases. Therefore, reducing its resistance and improving signal-to-noise ratio is one of the solutions to improve TMR read head performance, which is also the focus of the industry.

3.4.2 The TMR Angle Sensors

The angle sensor is used to measure the angle of rotation between the stationary component, i.e. the stator, and the rotating component, which converts the angle signal to electrical signal. For TMR sensors with spin valve structure, when the strength of external magnetic field is less than the strength of exchange bias field, the dependence of resistance on the angle $(\Delta\theta)$ between the magnetization direction of the free layer and the pinned layer is shown as follows

$$R(\Delta\theta) = R_P + \frac{(R_{AP} - R_P)}{2}(1 - \cos(\Delta\theta)) \tag{3.20}$$

where R_P *and* R_{AP} respectively are the resistance of MTJ with magnetizations of free layer and pinned layer under P and AP states. When the TMR sensor is saturated by external magnetic field, the resistance value is only related to the direction of the external magnetic field, and the value range of $\Delta\theta$ is from 0 degrees to 180 degrees. For TMR sensors, Wheatstone bridge structure is generally used. One of the main challenges for MR sensors in

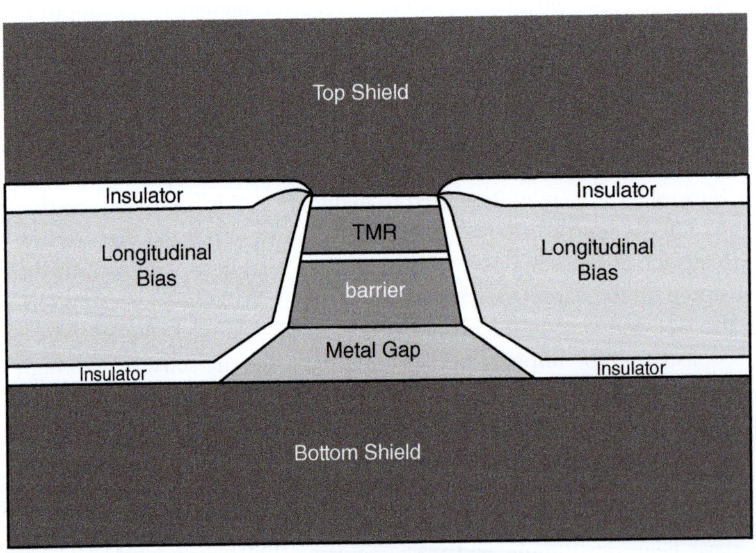

Figure 3.13 Typical structure of a TMR read head [94].

industrial applications is to overcome thermal drift. Considering that the ambient temperature of MR sensor is changing and the resistance and TMR value of MTJs decrease with the temperature increment. As a result, a single MTJ cannot be directly used as a sensor. In order to restrain the temperature drift and improve the stability of the sensor under different ambient temperatures, MTJs are usually connected into Wheatstone bridge structure. Wheatstone bridge is a special resistance network composed of four resistance elements. When the resistance value of the four bridge arms changes uniformly with the temperature, the bridge has no output, which is the principle of inhibiting the temperature drift. Therefore, the Wheatstone full bridge structure can effectively reduce the temperature drift, and increase the output voltage variation through differential circuit, and improve sensors' sensitivity. Each bridge arm in the structure can be composed of a single MTJ or multiple MTJs in series and parallel [95].

As shown in Figure 3.14, Wheatstone bridge structure consists of four resistors with same value, which can be divided into three types according to the number of MTJs, i.e., unique element, half bridge, and full bridge.

When the bridge sensor is at work, the bias Vin is applied at the two ends of diagonal, and the output voltage is measured with at the other ends. Among the three structures, the full bridge structure is composed of four MTJs bridge arms with the same range of resistance values. Figure 3.15 shows the circuit structure of full Wheatstone bridge, in which the output curves of each two adjacent bridge arms are opposite. Compared with other two structures, the whole bridge structure possesses the largest voltage change under the same external magnetic field change, indicating the highest sensitivity. Meanwhile, the output voltage of full bridge exhibits a complete linear relation with the external magnetic field. The sensitivity calculation formula of this structure is

$$S = \frac{\Delta V}{\Delta H} \tag{3.21}$$

where ΔV is the difference between the maximum and minimum resistance of the output linear curve, and ΔH is the linear range of the voltage output curve.

In order to realize the opposite change of resistance on the two adjacent bridge arms with an external magnetic field, the adjacent MTJs should have opposite magnetic pinning direction. But the pinning direction of films deposited on the same wafer is generally the same. Therefore, how to simply and efficiently prepare full-bridge sensors is one of the research hotspots in academic and industrial fields. In 2018, S. Yan et al. [96] proposed to use shape anisotropy and post-annealing process to set different pinning directions of adjacent bridge arms in a full bridge composed of GMR strip. In this method, the unannealed

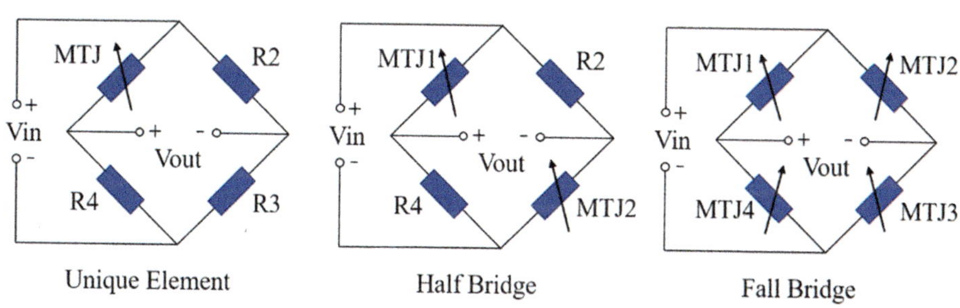

Figure 3.14 Schematic diagram of the Wheatstone bridge.

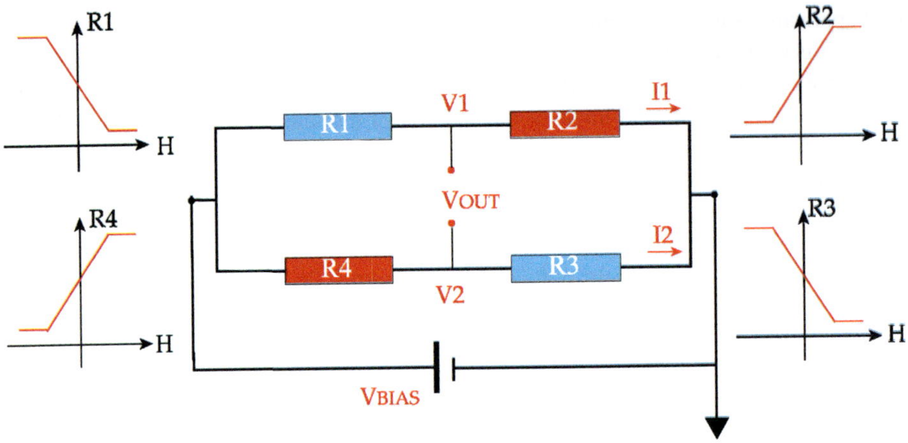

Figure 3.15 Circuit structure of full Wheatstone bridge and schematic diagram of magnetic field dependent resistance of four bridge arms [96].

GMR film is directly made into a full-bridge structure. The long axis of the GMR strip in the bridge arm is along the alternating 45°and 135° direction from X-axis, i.e., the long axis of adjacent GMR strip is perpendicular to each other. Then the external magnetic field in X-axis direction is applied during thermal annealing treatment. By using the interaction of demagnetizing field generated by shape anisotropy, induced anisotropy field and exchange bias field from the antiferromagnetic material, the easy magnetization axis of the pinned layer tends to the short axis direction (m_p) of each GMR strip, as shown in the Figure 3.16a. In this way, the linear output signal is enhanced under the Y-axis magnetic field while the signal is almost zero under X-axis magnetic field because of the full-bridge structure, as shown in Figure 3.16b. This method has been proved to suitable for mass production on 6-inch wafers. Furthermore, we can apply this technique to construct full bridge composed of TMR sensors.

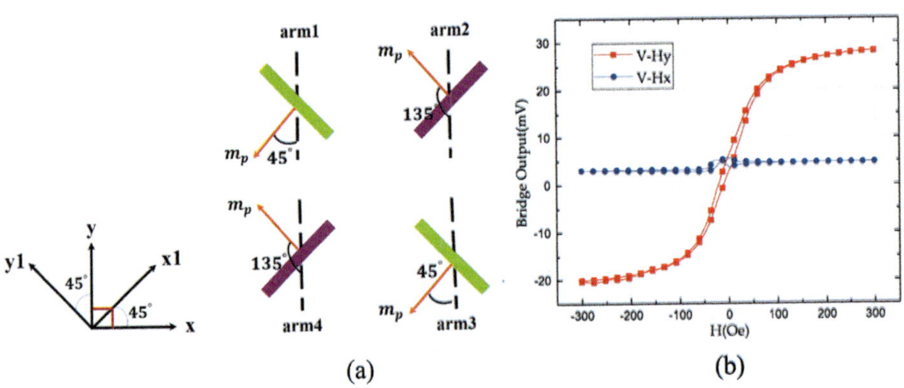

Figure 3.16 (a) The ideal direction of free layer and pinned layer of each GMR strip in a full-bridge structure after annealing. (b) The output of the Full Wheatstone bridge angular GMR sensor in the x-axes and y-axes.

When Wheatstone bridge is used in TMR sensors, it is assumed that the output curves of the four arms are completely consistent. In the case of saturation, the resistance relationship of the four arms can be expressed as follows:

$$R_1 = R_3 \; R_2 = R_4 \tag{3.22}$$

$$R_2\left(\Delta\theta\right) = R_1\left(\Delta\theta + 180\right). \tag{3.23}$$

Therefore, the output formula of the Wheatstone bridge can be obtained:

$$V_{out} = \left(\frac{R_1\left(\Delta\theta\right) - R_1\left(\Delta\theta + 180\right)}{R_1\left(\Delta\theta\right) + R_1\left(\Delta\theta + 180\right)}\right)V_b. \tag{3.24}$$

According to formula 3.23, the output signal range of a single TMR sensor based on Wheatstone bridge structure is only 0–180 degrees. The vertical placement of the two Wheatstone Bridges extends the angular measurement range of the TMR sensor from 0°–180° to 0°–360°. In 2019, J. Deak et al. [97] designed a new TMR angle sensor by placing a circular ferromagnetic attenuator on the TMR sensor element inside the magnetic angle sensor chip, extending the detection field range from 200 Oe to 1000 Oe.

3.4.3 Geomagnetic Measurement

Geomagnetic field is a typical weak magnetic field. It is also an important natural resource on earth and has important applications in military, medical, industrial, and other fields. The geomagnetic field can be equivalent to the magnetic field of a magnetic dipole located at the center of the Earth. It is a vector field with magnitude and direction. Therefore, geomagnetic measurement is a special angle measurement, which takes the center of the earth as the stator and measures the angle between geomagnetic field at different positions and the center of the Earth. The main application of the TMR sensor in geomagnetic measurement is the electronic compass. Something should be noted that electronic compass is the second largest application scenarios of the MR sensors after reading head. Represented by Honeywell and Multidimensional, the TMR electronic compasses have been widely produced and applied.

The working principle of electronic compass is to utilize a TMR sensor to measure the geomagnetic field and deduce the user's position through vector calculations. For a two-dimensional electronic compass, it only needs to measure the in-plane component of the geomagnetic field. According to the measured X-axes and Y-axes components, the azimuth angle can be figured out. Two in-plane TMR sensors with orthogonal sensing directions are usually used to realize two-dimensional electronic compass [98, 99].

Figure 3.17a shows an X-axis sensing TMR sensor with full bridge structure, which could measure X-axis component of the geomagnetic field. Meanwhile, a Y-axis sensing TMR sensor is needed to measure the Y-axis component of the geomagnetic field. As shown in Figure 3.17b, 360-degree angle measurement could be realized by combining these two TMR sensors.

In order to actually measure the three-dimensional geomagnetic field, a TMR sensor that can sense three-directional (X, Y, Z) magnetic field component is required. Three-dimensional TMR sensors can be constructed through adding a Z-axis inductive directive on the basis of

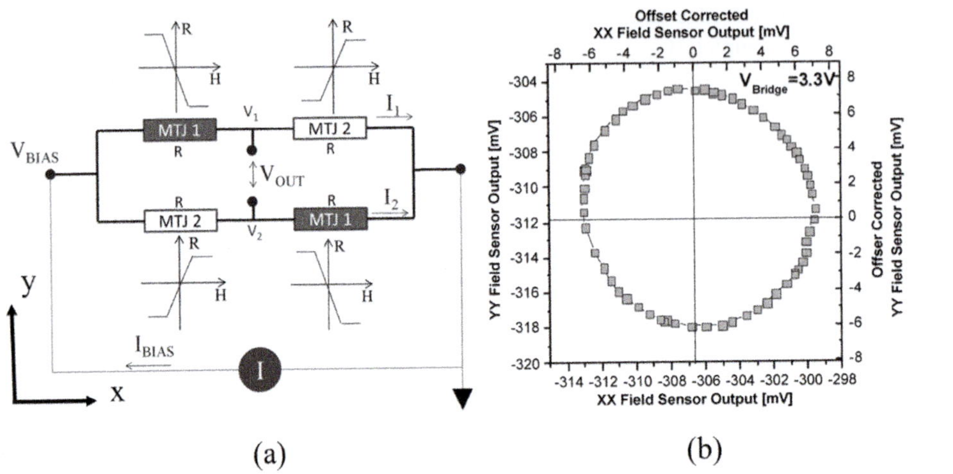

Figure 3.17 (a) Full Wheatstone bridge composed of 4 MTJs. In order to achieve a full bridge output, two types of MTJs with opposite dR/dH are required. (b) two-axis MTJ Full Bridge Magnetometer output measured by an electronic reading unit of digital compass during a 360 degree rotation [99].

two-dimensional sensors. To realize this objective, the research on the perpendicular TMR sensor has received a lot of attention.

In addition to the electronic compass, vehicle detection is also an application of geomagnetic detection. The magnitude and direction of geomagnetic field are very uniform and stable within a fixed area of the earth's surface. When a car made mainly of metal material passes by, which can be regarded as a series of bipolar magnets passing by, it will disturb the local geomagnetic field. A highly sensitive TMR sensor can detect this disturbance and obtain the vehicle passing information. With the popularity of intelligent traffic system, large amounts of TMR sensors connected with wireless sensor networks are installed on roads and cars to sense slight changes in geomagnetic field when vehicles pass, and obtain dynamic traffic information such as speed, vehicle position, and occupancy rate. It can be used to build large-scale traffic monitoring and management system, develop unmanned and anti-collision applications.

3.4.4 Spin-MEMS Combined Application

Magnetoresistance sensor has high sensitivity, small size, low power consumption, and other excellent performance, has been widely used in angle sensing, geomagnetic sensing, magnetic reading head, and other aspects. In addition, the application of spin-MEMS sensor combined with spin magnetoresistance sensor has received extensive attention. This combination can greatly improve the sensitivity and accuracy of traditional MEMS sensors and is a new application direction of TMR sensors. In 2017, Fuji, Y et al. [100] reported the spintronic strain gauge sensor based on MTJ and applied this sensor to the microphone, which is called Spin-MEMS microphone. The schematics of this microphone as shown in Figure 3.18. The MTJ series are located above the MEMS diaphragm, and deformation occurs with the diaphragm vibration, resulting in junctions resistance changes.

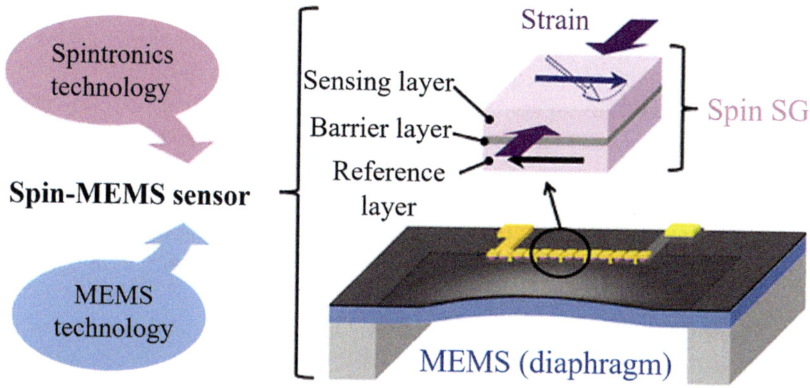

Figure 3.18 Schematics of Spin-MEMS sensors [101].

In recent years, new strain gauge sensors based on spintronic technology have been proposed. For example, a MTJ with high magnetostrictive free layer can act as strain gauge [102]. When the MTJ is subjected to compressive strain or tensile strain, the magnetization direction of free layer will rotate. As a result, the junction resistance of MTJs varies with the alternating compressive stress and tensile stress. Figure 3.19 shows the dependence of TMR value on applied stress.

For this kind of sensors, the higher the gauge factor (GF) is, the higher the sensitivity is. The TMR sensor is integrated into the MEMS diaphragm. When sound causes the vibration of the diaphragm, the direction and magnitude of the stress applying to MTJ is changed;

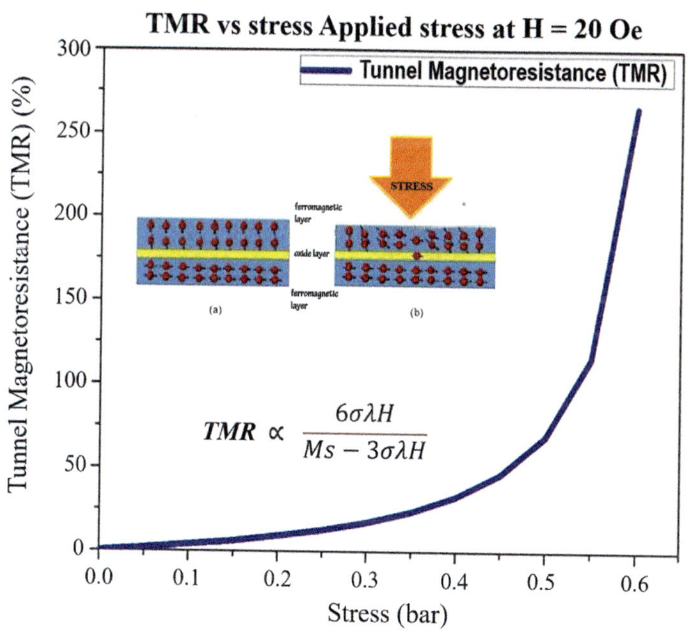

Figure 3.19 The dependence of TMR value on applied stress. Inset shows the shows the working mechanism [103].

thus, the resistance is changed, resulting in the change of the output voltage. Finally, the sound wave is converted into the change of the voltage to realize the sensing process. This is also the rationale behind this type of spin-MEMS microphone, as shown in Figure 3.20.

In order to improve the sensitivity of spin-MEMS microphones, Y. Fuji et al. [104] proposed to connect a large array of MTJs in series to reduce 1/*f* noise, which was explained in detail in the previous section. In the same year, this research group also proposed to improve the signal-to-noise ratio by optimizing the bias magnetic field [105]. According to a recent report by Y. Fuji et al. [101], the spin-MEMS microphone can realize a signal-to-noise ratio of 57 dB(A), benefiting from the high strain sensitivity of the MTJ-based strain gauge sensor.

In 2019, A.K. Ramesh et al. [103] reported that the MTJ with polycrystalline free layer (i.e. CoFeB) can be used to construct hypersensitive MEMS pressure sensor because of its higher stability than amorphous CoFeB.

Spin-MEMS sensor exhibits great potential and gradually becomes an important development direction of MR sensor in the future. In addition to the pressure sensor, the design of the microphone using the TMR sensor as the sensor unit will be future development direction that attracts much attention.

3.4.5 Nondestructive Testing (NDT)

Nondestructive testing (NDT) refers to the testing of materials or components in a way that does not impair or affect their future performance or application. Through NDT, we can obtain massive information about object under test, such as the defects, geometric features, and dimensions, internal composition, structure, physical properties, and states. There are many kinds of NDT techniques, among which eddy current testing (ECT) is based on electromagnetic induction, which is suitable for conductive materials. Figure 3.21 illustrates the working principle of ECT. When the alternating magnetic field generated by excitation coil is close to the measured object, the eddy current is induced in the conductor by the electromagnetic induction. This eddy current is influenced by many factors, such as the excited magnetic field, conductivity, and permeability of the conductor, and the defects reacting on the original excitation magnetic field. As a result, the change of some parameters (e.g., the impedance of the detection coil) indicates the existence of defects [106].

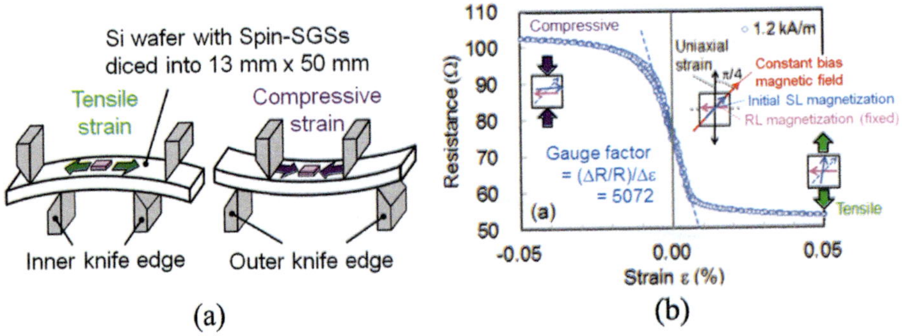

Figure 3.20 (a) The force of spintronic strain-gauge sensor (i.e. MTJ) under tensile and compressive strain. (b) The variation of MTJ magnetoresistance with tensile and compressive deformation [100].

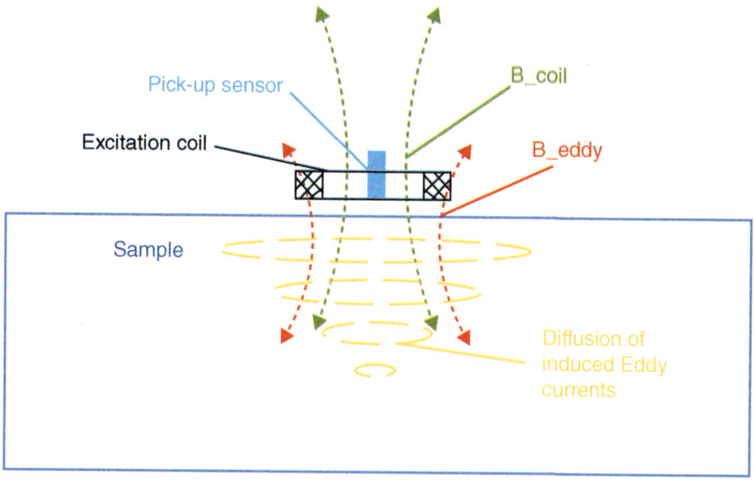

Figure 3.21 Illustration of the working principle of ECT [106].

The ECT technology based on MR sensor has also been proposed. When the eddy current flows through the defect, the obstruction of the defect would change the magnetic field generated by eddy current. The MR sensors can detect this magnetic field change and directly judge the existence of defects. In 2014, F.A. Cardoso et al. [107] fabricated an eddy current detection probe based on MTJ for the detection of surface defects. They prepared two kinds of TMR sensors by connecting six MTJs and 10 MTJs in series respectively, which can reduce noise and improve signal-to-noise ratio. D.W. Guo et al. [108] prepared a sensor for NDT by connecting 72 MTJs in series. Benefiting from the high TMR value of 159%, the obtained sensitivity is up to 2.9%/Oe. At 1V bias voltage, the field detectivity can reach $1.76\text{nT}/\sqrt{\text{Hz}}$ and $170\text{pT}/\sqrt{\text{Hz}}$, respectively. In the same year, F.A. Cardoso et al. [109] optimized the TMR-based eddy current detection probe based on finite element simulation, and verified that increasing the numbers of series MTJs can effectively improve the signal-to-noise of the sensors.

3.4.6 Ultra-low Magnetic Field Detection: Biosensor

Spintronic sensors have also received a lot of attention in the biomedical field, especially TMR sensors. As early as 2013, spintronic bioassay technology based on the GMR sensor was used to measure the concentration of target analyte molecules (e.g., DNA and proteins) in solution [110]. This biological detection uses magnetic tags to mark the target analyte molecules, and then uses high-sensitivity spintronic sensors to detect the magnetic signals generated by the tags. For example, TMR sensors are often used to detect ultra-low magnetic field signal from the brain and heart at room temperature. The TMR sensors can be made to the nanoscale to match the size of the target analyte molecules. However, the high inherent noise of TMR sensor is an inevitable challenge to achieve ultra-high sensitivity.

There are many methods for biological detection, sandwich assay method [111] is an effective and sensitive immunoassay. In 2018, Kim et al. [112] proposed a scalable MR biosensor array with an on-chip magnetic field generator and a high-speed data acquisition method. Using sandwich assay, magnetic nanoparticles were detected at signal levels as low as 6.92 PPM. Figure 3.22 shows the principle of biological detection using spintronic sensors. Antigens are quantified by capturing antibodies and detecting specific binding between the two layers of antibodies. The capture antibody is first attached to the surface of the spintronic sensors. The antibody captures an antigen in the solution, which attracts the detection antibody to form a sandwich structure, and then detects the magnetic nanoparticles (MNP) attached to the detection antibody. When the magnetic field is applied, the magnetization direction of MNP will follow the direction of magnetic field, and the number of MNP will be measured by detecting the change of magnetic field.

In addition to immunodetection, the electrical activity of various tissues of the human body will generate a weak magnetic field in the human heart and brain. The high sensitivity and low noise TMR sensors can be used to detect this weak magnetic field, and magnetocardiography (MCG) and magnetoencephalography (MEG) similar to electrocardiogram can be obtained. As shown in Figure 3.23a, QRS wave group reflects the changes of left and right ventricular potentials and time. The first downward wave is Q wave, the first upward wave is R wave, and the next downward wave is S wave. By averaging the signals for dozens of seconds, high signal-to-noise ratio QRS complex signals are realized. The real-time MCG is proven to be high sensitivity, which can help to improve the diagnostic accuracy of heart disease. However, the magnitude of MEG is much smaller than that of MCG, so measuring magnetoencephalography is still a big challenge. In the present study, MEG is achieved by using electroencephalography (EEG) signals as triggers to average the measured MEG signals, as shown in Figure 3.23b. The alpha waves produced by EEG are consistent with the decrease in MEG signals over time.

In 2019, Wang et al. measured the triaxial components of MCG signals in the male tester in the back position by rotating a highly sensitive TMR sensor. They obtained the R wave and QRS-Complex of MCG signals, which is related to cardiac cycles clearly without averaging calculation [114].

With the development of the artificial intelligence (AI) age and the popularization of the IoT, people's demand for smart life is increasingly strong. In 2020, Attayeb Mohsen et al. combined the IoT with TMR and proposed an IoT device based on the super-sensitive MTJs to measure the magnetic field generated by cardiovascular electromagnetic activity, namely MCG, with the goal to achieve remote heart health monitoring, which is shown in Figure 3.24.

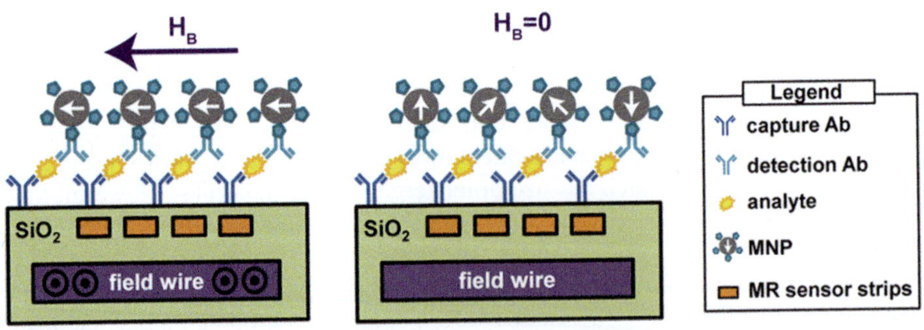

Figure 3.22 The principle of biological detection using spintronic sensors [111].

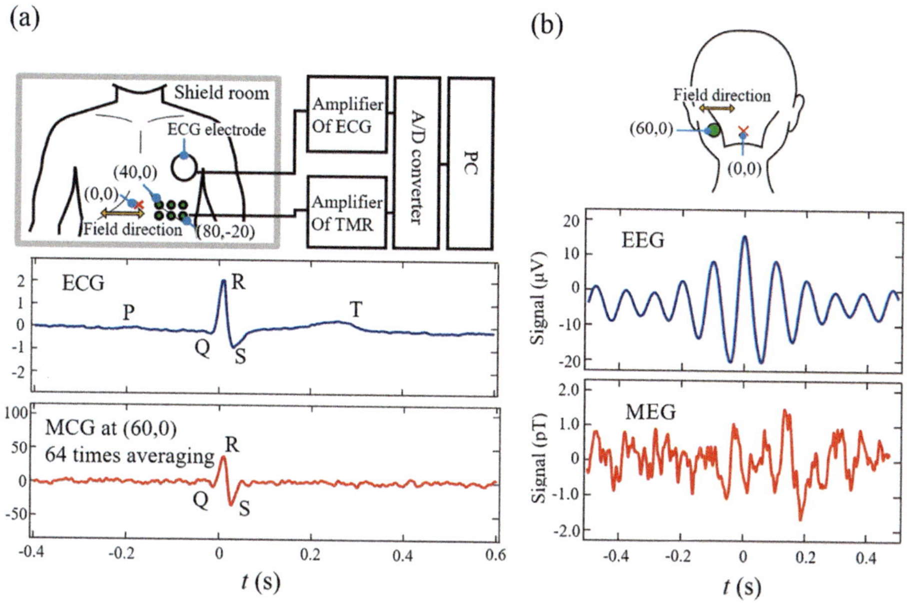

Figure 3.23 (a) Schematic of MCG test experimental apparatus. ECG reference and MCG signal. (b) Schematic of MEG test experimental apparatus. EEG reference and MEG signal [113].

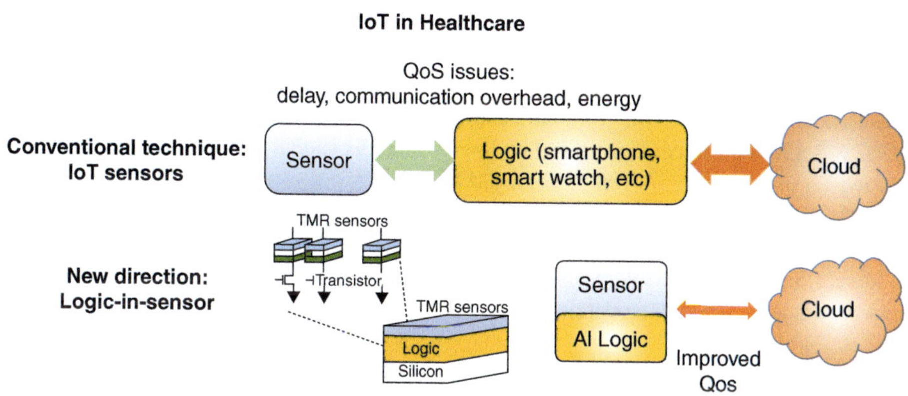

Figure 3.24 A health care system based on the Internet of Things, conceived by traditional and new logic sensors [115].

3.5 Conclusion

The sensitive detection of a magnetic field based on the magnetoresistance (MR) effect plays an important role in the solution of the next generation of high-precision sensing. After nearly 30 years of rapid development, the MR sensor has become an important member in academic fields and industries. Among the MR sensor family, as a new type of information sensors, TMR device will help to achieve magnetic field detection in the advantages of ultra-high sensitivity, potential miniaturization, high precision, good linearity, excellent thermal stability, and ultra-low power consumption. The application of TMR effect in read heads has increased the storage density of hard disk drive by more than a hundred thousand times in the past two decades, that directly opens the era of big data and cloud computing, and plays an irreplaceable role in promoting the development of intelligent industry. Moreover, due to its compatibility with CMOS technology, TMR device has widely application demands in nonvolatile memories, wearable electronic products, IoT, and biomedical applications.

References

1 Simmons, J.G. (1963). Generalized formula for the electric tunnel effect between similar electrodes separated by a thin insulating film. *J. Appl. Phys.* 34 (6): 1793–1803.

2 Tedrow, P.M. and Meservey, R. (1971). Spin-dependent tunneling into ferromagnetic nickel. *Phys. Rev. Lett.* 26 (4): 192–195.

3 Julliere, M. (1975). Tunneling between ferromagnetic films. *Phys. Lett. A* 54 (3): 225–226.

4 Valet, T. and Fert, A. (1993). Theory of the perpendicular magnetoresistance in magnetic multilayers. *Phys. Rev. B* 48 (10): 7099–7113.

5 Tedrow, P.M. and Meservey, R. (1973). Spin polarization of electrons tunneling from films of Fe, Co, Ni, and Gd. *Phys. Rev. B* 7 (1): 318–326.

6 Meservey, R. and Tedrow, P.M. (1972). Spin polarization of tunneling electrons from films of Fe, Co, Ni, and Gd. *Solid State Commun.* 11 (2): 333–336.

7 De Teresa, J.M., Barthélémy, A., Fert, A., Contour, J.P., Lyonnet, R., Montaigne, F. et al. (1999). Inverse tunnel magnetoresistance in Co/SrTiO3/La0.7Sr0.3MnO3: new ideas on spin-polarized tunneling. *Phys. Rev. Lett.* 82 (21): 4288–4291.

8 Slonczewski, J.C. (1989). Conductance and exchange coupling of two ferromagnets separated by a tunneling barrier. *Phys. Rev. B* 39 (10): 6995–7002.

9 Belashchenko, K.D., Tsymbal, E.Y., van Schilfgaarde, M., Stewart, D.A., Oleynik, I.I., and Jaswal, S.S. (2004). Effect of interface bonding on spin-dependent tunneling from the oxidized Co surface. *Phys. Rev. B* 69 (17): 174408.

10 Beth Stearns, M. (1977). Simple explanation of tunneling spin-polarization of Fe, Co, Ni and its alloys. *J. Magn. Magn. Mater.* 5 (2): 167–171.

11 Moodera, J.S., Kinder, L.R., Wong, T.M., and Meservey, R. (1995). Large magnetoresistance at room temperature in ferromagnetic thin film tunnel junctions. *Phys. Rev. Lett.* 74 (16): 3273–3276.

12 Miyazaki, T. and Tezuka, N. (1995). Giant magnetic tunneling effect in Fe/Al_2O_3/Fe junction. *J. Magn. Magn. Mater.* 139 (3): L231–L4.

13 Maekawa, S. and Gafvert, U. (1982). Electron tunneling between ferromagnetic films. *IEEE Trans. Magn.* 18 (2): 707–708.

14 Nowak, J. and Raułuszkiewicz, J. (1992). Spin dependent electron tunneling between ferromagnetic films. *J. Magn. Magn. Mater.* 109 (1): 79–90.

15 Plaskett, T.S., Freitas, P.P., Barradas, N.P., da Silva, M.F., and Soares, J.C. (1994). Magnetoresistance and magnetic properties of NiFe/oxide/Co junctions prepared by magnetron sputtering. *J. Appl. Phys.* 76 (10): 6104–6106.

16 Lu, Y., Li, X.W., Gong, G.Q., Xiao, G., Gupta, A., Lecoeur, P. et al. (1996). Large magnetotunneling effect at low magnetic fields in micrometer-scale epitaxial $La_{0.67}Sr_{0.33}MnO_3$ tunnel junctions. *Phys. Rev. B* 54 (12): R8357–R60.

17 Li, Z., de Groot, C., and Moodera, J.H. (2000). Gallium oxide as an insulating barrier for spin-dependent tunneling junctions. *Appl. Phys. Lett.* 77 (22): 3630–3632.

18 Song, D., Nowak, J., Larson, R., Kolbo, P., and Chellew, R. (2000). Demonstrating a tunneling magneto-resistive read head. *IEEE Trans. Magn.* 36 (5): 2545–2548.

19 Wang, J., Freitas, P.P., Snoeck, E., Wei, P., and Soares, J.C. (2001). Spin-dependent tunnel junctions with ZrO_x barriers. *Appl. Phys. Lett.* 79 (26): 4387–4389.

20 Wang, J., Liu, Y., Freitas, P.P., Snoeck, E., and Martins, J.L. (2003). Continuous thin barriers for low-resistance spin-dependent tunnel junctions. *J. Appl. Phys.* 93 (10): 8367–8369.

21 Platt, C.L., Dieny, B., and Berkowitz, A.E. (1996). Spin-dependent tunneling in HfO_2 tunnel junctions. *Appl. Phys. Lett.* 69 (15): 2291–2293.

22 Sato, M. and Kobayashi, K. (1997). Spin-valve-like properties and annealing effect in ferromagnetic tunnel junctions. *IEEE Trans. Magn.* 33 (5): 3553–3555.

23 Cardoso, S., Freitas, P.P., de Jesus, C., Wei, P., and Soares, J.C. (2000). Spin-tunnel-junction thermal stability and interface interdiffusion above 300°C. *Appl. Phys. Lett.* 76 (5): 610–612.

24 Han, X.-F., Daibou, T., Kamijo, M., Yaoita, K., Kubota, H., Ando, Y. et al. (2000). High-magnetoresistance tunnel junctions using $Co_{75}Fe_{25}$ ferromagnetic electrodes. *Jpn. J. Appl. Phys.* 39 (Part 2, No. 5B): L439–L41.

25 Yuasa, S., Sato, T., Tamura, E., Suzuki, Y., Yamamori, H., Ando, K. et al. (2000). Magnetic tunnel junctions with single-crystal electrodes: a crystal anisotropy of tunnel magneto-resistance. *Europhysics Letters (EPL)* 52 (3): 344–350.

26 Bibes, M., Bowen, M., Barthélémy, A., Anane, A., Bouzehouane, K., Carrétéro, C. et al. (2003). Growth and characterization of TiO_2 as a barrier for spin-polarized tunneling. *Appl. Phys. Lett.* 82 (19): 3269–3271.

27 Butler, W.H., Zhang, X.G., Schulthess, T.C., and MacLaren, J.M. (2001). Spin-dependent tunneling conductance of Fe/MgO/Fe sandwiches. *Phys. Rev. B* 63 (5): 054416.

28 Mathon, J. and Umerski, A. (2001). Theory of tunneling magnetoresistance of an epitaxial Fe/MgO/Fe(001) junction. *Phys. Rev. B* 63 (22): 220403.

29 Yuasa, S., Nagahama, T., Fukushima, A., Suzuki, Y., and Ando, K. (2004). Giant room-temperature magnetoresistance in single-crystal Fe/MgO/Fe magnetic tunnel junctions. *Nat. Mater.* 3 (12): 868–871.

30 Parkin, S.S.P., Kaiser, C., Panchula, A., Rice, P.M., Hughes, B., Samant, M. et al. (2004). Giant tunnelling magnetoresistance at room temperature with MgO (100) tunnel barriers. *Nat. Mater.* 3 (12): 862–867.

31 Miyazaki, T., Yaoi, T., and Ishio, S. (1991). Large magnetoresistance effect in 82Ni-Fe/Al-Al_2O_3/Co magnetic tunneling junction. *J. Magn. Magn. Mater.* 98 (1): L7–L9.

32 Lee, Y.M., Hayakawa, J., Ikeda, S., Matsukura, F., and Ohno, H. (2007). Effect of electrode composition on the tunnel magnetoresistance of pseudo-spin-valve magnetic tunnel junction with a MgO tunnel barrier. *Appl. Phys. Lett.* 90 (21): 212507.

33 Ikeda, S., Hayakawa, J., Ashizawa, Y., Lee, Y.M., Miura, K., Hasegawa, H. et al. (2008). Tunnel magnetoresistance of 604% at 300K by suppression of Ta diffusion in CoFeB/MgO/CoFeB pseudo-spin-valves annealed at high temperature. *Appl. Phys. Lett.* 93 (8): 082508.

34 Wang, W., Sukegawa, H., Shan, R., Mitani, S., and Inomata, K. (2009). Giant tunneling magnetoresistance up to 330% at room temperature in sputter deposited Co_2FeAl/MgO/CoFe magnetic tunnel junctions. *Appl. Phys. Lett.* 95 (18): 182502.

35 Chiba, D., Akiba, N., Matsukura, F., Ohno, Y., and Ohno, H. (2000). Magnetoresistance effect and interlayer coupling of (Ga,Mn)As trilayer structures. *Appl. Phys. Lett.* 77 (12): 1873–1875.

36 Guth, M., Dinia, A., Schmerber, G., and van den Berg, H.A.M. (2001). Tunnel magnetoresistance in magnetic tunnel junctions with a ZnS barrier. *Appl. Phys. Lett.* 78 (22): 3487–3489.

37 Lyu, P. and Moon, K. (2001). Tunneling magnetoresistance in diluted magnetic semiconductor tunnel junctions. *Phys. Rev. B* 64 (3): 035201.

38 Jiang, X., Panchula, A.F., and Parkin, S.S.P. (2003). Magnetic tunnel junctions with ZnSe barriers. *Appl. Phys. Lett.* 83 (25): 5244–5246.

39 Yoshimura, S., Shoyama, T., Nozawa, T., Tsunoda, M., and Takahashi, M. (2004). Nitridation Process of Al Layer by microwave-excited plasma and large magnetoresistance in Co-Fe/Al-N/Co-Fe tunnel Junctions-as a comparison with oxidization Process. *IEEE Trans. Magn.* 40 (4): 2290–2292.

40 Moodera, J.S., Nassar, J., and Mathon, G. (1999). Spin-tunneling in ferromagnetic junctions. *Annu. Rev. Mater. Sci.* 29 (1): 381–432.

41 Ikeda, S., Hayakawa, J., Lee, Y.M., Matsukura, F., Ohno, Y., Hanyu, T. et al. (2007). Magnetic tunnel junctions for spintronic memories and beyond. *IEEE Trans. Electron Devices* 54 (5): 991–1002.

42 Srivastava, S., Chen, A.P., Dutta, T., Ramaswamy, R., Son, J., Saifullah, M.S.M. et al. (2018). Effect of $(Co_xFe_{(1-x)})_{80}B_{20}$ composition on the magnetic properties of the free layer in double-barrier magnetic tunnel junctions. *Phys. Rev. Appl.* 10 (2): 024031.

43 Bowen, M., Bibes, M., Barthélémy, A., Contour, J.P., Anane, A., Lemaître, Y. et al. (2003). Nearly total spin polarization in $La_{2/3}Sr_{1/3}MnO_3$ from tunneling experiments. *Appl. Phys. Lett.* 82 (2): 233–235.

44 Aoshima, K.-I. and Wang, S.X. (2002). Epitaxial growth and characterization of Fe_3O_4 on Ru underlayer. *J. Appl. Phys.* 91 (10): 7146–7148.

45 Xiang, H., Shi, F., Rzchowski, M.S., Voyles, P.M., and Chang, Y.A. (2010). Epitaxial growth and magnetic properties of Fe_3O_4 films on TiN buffered Si(001), Si(110), and Si(111) substrates. *Appl. Phys. Lett.* 97 (9): 092508.

46 Kämmerer, S., Thomas, A., Hütten, A., and Reiss, G. (2004). Co2MnSi Heusler alloy as magnetic electrodes in magnetic tunnel junctions. *Appl. Phys. Lett.* 85 (1): 79–81.

47 Galanakis, I., Dederichs, P.H., and Papanikolaou, N. (2002). Slater-Pauling behavior and origin of the half-metallicity of the full-Heusler alloys. *Phys. Rev. B* 66 (17): 174429.

48 Lee, S.C., Lee, T.D., Blaha, P., and Schwarz, K. (2005). Magnetic and half-metallic properties of the full-Heusler alloys Co2TiX(X=Al,Ga;Si,Ge,Sn;Sb). *J. Appl. Phys.* 97 (10): 10C307.

49 Ishikawa, T., Marukame, T., Kijima, H., Matsuda, K.I., Uemura, T., Arita, M. et al. (2006). Spin-dependent tunneling characteristics of fully epitaxial magnetic tunneling junctions with a full-Heusler alloy Co_2MnSi thin film and a MgO tunnel barrier. *Appl. Phys. Lett.* 89 (19): 192505.

50 Tezuka, N., Ikeda, N., Miyazaki, A., Sugimoto, S., Kikuchi, M., and Inomata, K. (2006). Tunnel magnetoresistance for junctions with epitaxial full-Heusler $Co_2FeAl_{0.5}Si_{0.5}$ electrodes with B2 and $L2_1$ structures. *Appl. Phys. Lett.* 89 (11): 112514.

51 Lund, M.S., Dong, J.W., Lu, J., Dong, X.Y., Palmstrøm, C.J., and Leighton, C. (2002). Anomalous magnetotransport properties of epitaxial full Heusler alloys. *Appl. Phys. Lett.* 80 (25): 4798–4800.

52 Orgassa, D., Fujiwara, H., Schulthess, T.C., and Butler, W.H. (1999). First-principles calculation of the effect of atomic disorder on the electronic structure of the half-metallic ferromagnet NiMnSb. *Phys. Rev. B* 60 (19): 13237–13240.

53 Pong, P.W.T. and Egelhoff, W.F. (2009). Enhancement of tunneling magnetoresistance by optimization of capping layer thicknesses in CoFeB/MgO/CoFeB magnetic tunnel junctions. *J. Appl. Phys.* 105 (7): 07C915.

54 Yamamoto, H., Hayakawa, J., Miura, K., Ito, K., Matsuoka, H., Ikeda, S. et al. (2012). Dependence of magnetic anisotropy in $Co_{20}Fe_{60}B_{20}$ free layers on capping layers in MgO-based magnetic tunnel junctions with in-plane easy axis. *Appl. Phys. Express* 5 (5): 053002.

55 Reekstin, J.P. (1967). Zero magnetostriction composition of NiFe films. *J. Appl. Phys.* 38 (3): 1449–1450.

56 Ikeda, S., Hayakawa, J., Lee, Y.M., Tanikawa, T., Matsukura, F., and Ohno, H. (2006). Tunnel magnetoresistance in MgO-barrier magnetic tunnel junctions with bcc-CoFe(B) and fcc-CoFe free layers. *J. Appl. Phys.* 99 (8): 08A907.

57 Baumgart, P.M., Dieny, B., Gurney, B.A., Speriosu, V.S., and Wilhoit, D.R. (1996). Antiferromagnetic exchange coupling in magnetoresistive spin valve sensors. Google Patents.

58 van den Berg, H.A.M., Clemens, W., Gieres, G., Rupp, G., Schelter, W., and Vieth, M. (1996). GMR sensor scheme with artificial antiferromagnetic subsystem. *IEEE Trans. Magn.* 32 (5): 4624–4626.

59 Leal, J.L. and Kryder, M.H. (1998). Spin valves exchange biased by Co/Ru/Co synthetic antiferromagnets. *J. Appl. Phys.* 83 (7): 3720–3723.

60 Sun, J.J., Shimazawa, K., Kasahara, N., Sato, K., Saruki, S., Kagami, T. et al. (2000). Low resistance and high thermal stability of spin-dependent tunnel junctions with synthetic antiferromagnetic CoFe/Ru/CoFe pinned layers. *Appl. Phys. Lett.* 76 (17): 2424–2426.

61 Naganuma, H., Bae, I.-T., Miyazaki, T., Kubota, M., Inami, N., Kawada, Y. et al. (2012). Enhancement of magnetoresistance using CoFe/Ru/CoFe synthetic ferrimagnetic pinned layer in $BiFeO_3$ based spin-valves. *Appl. Phys. Lett.* 101 (7): 072901.

62 Lee, Y.M., Hayakawa, J., Ikeda, S., Matsukura, F., and Ohno, H. (2006). Giant tunnel magnetoresistance and high annealing stability in CoFeB/MgO/CoFeB magnetic tunnel junctions with synthetic pinned layer. *Appl. Phys. Lett.* 89 (4): 042506.

63 Guo, Z.B., Zong, B.Y., Qiu, J.J., Luo, P., An, L.H., Meng, H. et al. (2010). Tuning exchange coupling by replacing CoFe with amorphous CoFeB in the CoFe/Ru/CoFe synthetic antiferromagnetic structure. *Solid State Commun.* 150 (1): 45–48.

64 Childress, J.R., Schwickert, M.M., Fontana, R.E., Ho, M.K., Rice, P.M., and Gurney, B.A. (2001). Low-resistance IrMn and PtMn tunnel valves for recording head applications. *J. Appl. Phys.* 89 (11): 7353–7355.

65 Cao, J., Kanak, J., Stobiecki, T., Wisniowski, P., and Freitas, P.P. (2009). Effect of buffer layer texture on the crystallization of CoFeB and on the tunnel magnetoresistance in MgO based magnetic tunnel junctions. *IEEE Trans. Magn.* 45 (10): 3464–3466.

66 Sakuraba, Y., Nakata, J., Oogane, M., Kubota, H., Ando, Y., Sakuma, A. et al. (2005). Fabrication of Co_2MnAl Heusler alloy epitaxial film using Cr buffer layer. *Jpn. J. Appl. Phys.* 44 (9A): 6535–6537.

67 Kaidatzis, A., Bran, C., Psycharis, V., Vázquez, M., García-Martín, J.M., and Niarchos, D. (2015). Tailoring the magnetic anisotropy of CoFeB/MgO stacks onto W with a Ta buffer layer. *Appl. Phys. Lett.* 106 (26): 262401.

68 Ikeda, S., Miura, K., Yamamoto, H., Mizunuma, K., Gan, H.D., Endo, M. et al. (2010). A perpendicular-anisotropy CoFeB–MgO magnetic tunnel junction. *Nat. Mater.* 9 (9): 721–724.

69 Peng, S., Wang, M., Yang, H., Zeng, L., Nan, J., Zhou, J. et al. (2015). Origin of interfacial perpendicular magnetic anisotropy in MgO/CoFe/metallic capping layer structures. *Sci. Rep.* 5 (1): 18173.

70 Lei, Z., Yan, S., Cao, Z., Guo, Z., Song, P., Qiang, Y. et al. (2019). High TMR for both in-plane and perpendicular magnetic field justified by CoFeB free layer thickness for 3-D MTJ sensors. *AIP Adv.* 9 (8): 085127.

71 Bräuer, G., Szyszka, B., Vergöhl, M., and Bandorf, R. (2010). Magnetron sputtering – milestones of 30 years. *Vacuum* 84 (12): 1354–1359.

72 Kelly, P.J. and Arnell, R.D. (2000). Magnetron sputtering: a review of recent developments and applications. *Vacuum* 56 (3): 159–172.

73 Musil, J., Baroch, P., Vlček, J., Nam, K.H., and Han, J.G. (2005). Reactive magnetron sputtering of thin films: present status and trends. *Thin Solid Films* 475 (1): 208–218.

74 Pierson, H.O. (1999). Handbook of Chemical Vapor Deposition: Principles, Technology and Applications. William Andrew Publishing.

75 Meyyappan, M., Delzeit, L., Cassell, A., and Hash, D. (2003). Carbon nanotube growth by PECVD: a review. *Plasma Sources Sci. Technol.* 12 (2): 205–216.

76 Lei, Z.Q., Li, G.J., Egelhoff, W.F., Lai, P.T., and Pong, P.W.T. (2011). Review of noise sources in magnetic tunnel junction sensors. *IEEE Trans. Magn.* 47 (3): 602–612.

77 Hardner, H.T., Weissman, M.B., Salamon, M.B., and Parkin, S.S.P. (1993). Fluctuation-dissipation relation for giant magnetoresistive 1/f noise. *Phys. Rev. B* 48 (21): 16156–16159.

78 Egelhoff, W.F., Pong, P.W.T., Unguris, J., McMichael, R.D., Nowak, E.R., Edelstein, A.S. et al. (2009). Critical challenges for picoTesla magnetic-tunnel-junction sensors. *Sens. Actuator A Phys.* 155 (2): 217–225.

79 Hooge, F.N., Kleinpenning, T.G.M., and Vandamme, L.K.J. (1981). Experimental studies on 1/f noise. *Rep. Prog. Phys.* 44 (5): 479–532.

80 Jiang, L., Nowak, E.R., Scott, P.E., Johnson, J., Slaughter, J.M., Sun, J.J. et al. (2004). Low-frequency magnetic and resistance noise in magnetic tunnel junctions. *Phys. Rev. B* 69 (5): 054407.

81 Almeida, J.M., Wisniowski, P., and Freitas, P.P. (2008). Low-frequency noise in MgO magnetic tunnel junctions: Hooge's parameter dependence on bias voltage. *IEEE Trans. Magn.* 44 (11): 2569–2572.

82 Hardner, H.T., Hurben, M.J., and Tabat, N. (1999). Noise and magnetic domain fluctuations in spin-valve GMR heads. *IEEE Trans. Magn.* 35 (5): 2592–2594.

83 Ren, C., Liu, X., Schrag, B.D., and Xiao, G. (2004). Low-frequency magnetic noise in magnetic tunnel junctions. *Phys. Rev. B* 69 (10): 104405.

84 Kandiah, K. and Whiting, F.B. (1978). Low frequency noise in junction field effect transistors. *Solid State Electron.* 21 (8): 1079–1088.

85 Tondra, M., Daughton, J.M., Wang, D., Beech, R.S., Fink, A., and Taylor, J.A. (1998). Picotesla field sensor design using spin-dependent tunneling devices. *J. Appl. Phys.* 83 (11): 6688–6690.

86 Guerrero, R., Pannetier-Lecoeur, M., Fermon, C., Cardoso, S., Ferreira, R., and Freitas, P.P. (2009). Low frequency noise in arrays of magnetic tunnel junctions connected in series and parallel. *J. Appl. Phys.* 105 (11): 113922.

87 Shen, W., Schrag, B.D., Girdhar, A., Carter, M.J., Sang, H., and Xiao, G. (2009). Effects of superparamagnetism in MgO based magnetic tunnel junctions. *Phys. Rev. B* 79 (1): 014418.

88 Wisniowski, P., Almeida, J.M., and Freitas, P.P. (2008). 1/f magnetic noise dependence on free layer thickness in hysteresis free MgO magnetic tunnel junctions. *IEEE Trans. Magn.* 44 (11): 2551–2553.

89 Edelstein, A.S., Fischer, G.A., Pedersen, M., Nowak, E.R., Cheng, S.F., and Nordman, C.A. (2006). Progress toward a thousandfold reduction in 1/f noise in magnetic sensors using an ac microelectromechanical system flux concentrator (invited). *J. Appl. Phys.* 99 (8): 08B317.

90 Pan, M., Hu, J., Tian, W., Chen, D., and Zhao, J. (2012). Magnetic flux vertical motion modulation for 1/f noise reduction of magnetic tunnel junctions. *Sens. Actuator A Phys.* 179: 92–97.

91 Huang, L., Yuan, Z.H., Tao, B.S., Wan, C.H., Guo, P., Zhang, Q.T. et al. (2017). Noise suppression and sensitivity manipulation of magnetic tunnel junction sensors with soft magnetic $Co_{70.5}Fe_{4.5}Si_{15}B_{10}$ layer. *J. Appl. Phys.* 122 (11): 113903.

92 Moulin, J., Doll, A., Paul, E., Pannetier-Lecoeur, M., Fermon, C., Sergeeva-Chollet, N. et al. (2019). Optimizing magnetoresistive sensor signal-to-noise via pinning field tuning. *Appl. Phys. Lett.* 115 (12): 122406.

93 Freitas, P.P., Ferreira, R., Cardoso, S., and Cardoso, F. (2007). Magnetoresistive sensors. *J. Phys. Condens. Matter* 19 (16): 165221.

94 Araki, S., Sato, K., Kagami, T., Saruki, S., Uesugi, T., Kasahara, N. et al. (2002). Fabrication and electrical properties of lapped type of TMR heads for /spl sim/50 Gb/in/ sup 2/ and beyond. *IEEE Trans. Magn.* 38 (1): 72–77.

95 Freitas, P.P., Ferreira, R., and Cardoso, S. (2016). Spintronic Sensors. *Proc. IEEE* 104 (10): 1894–1918.

96 Yan, S., Cao, Z., Guo, Z., Zheng, Z., Cao, A., Qi, Y. et al. (2018). Design and fabrication of full Wheatstone-Bridge-based angular GMR sensors. *Sensors* 18: 6.

97 Deak, J. and Jin, I. (2019). High-field tunneling magnetoresistive angle sensor. *IEEE Trans. Magn.* 55 (10): 1–4.

98 Kammerer, J., Hebrard, L., Hehn, M., Braun, F., Alnot, P., and Schuhl, A. (2004). A two-axis magnetometer using a single magnetic tunnel junction. *IEEE Sens. J.* 4 (3): 313–321.

99 Ferreira, R., Paz, E., Freitas, P.P., Ribeiro, J., Germano, J., and Sousa, L. (2012). 2-Axis magnetometers based on full Wheatstone Bridges incorporating magnetic tunnel junctions connected in series. *IEEE Trans. Magn.* 48 (11): 4107–4110.

100 Fuji, Y., Hara, M., Higashi, Y., Kaji, S., Masunishi, K., Nagata, T. et al. (eds.) (2017). An ultra-sensitive spintronic strain-gauge sensor with gauge factor of 5000 and demonstration of a Spin-MEMS Microphone. 2017 19th International Conference on Solid-State Sensors, Actuators and Microsystems (TRANSDUCERS), 18–22 June 2017.

101 Fuji, Y., Hara, M., Higashi, Y., Kaji, S., Masunishi, K., Nagata, T. et al. (2019). An ultrasensitive spintronic strain-gauge sensor and a spin-MEMS microphone. *Electron. Commun. Jpn.* 102 (2): 48–54.

102 Tavassolizadeh, A., Hayes, P., Rott, K., Reiss, G., Quandt, E., and Meyners, D. (2015). Highly strain-sensitive magnetostrictive tunnel magnetoresistance junctions. *J. Magn. Magn. Mater.* 384: 308–313.

103 Ramesh, A.K., Rana, V., Das, P., and Singh, P. (eds.) (2019). Polycrystalline sense layer for Magnetic Tunnel Junction (MTJ) as ultrasensitive sensing element for MEMS pressure sensors. 2019 IEEE 32nd International Conference on Micro Electro Mechanical Systems (MEMS), 27–31 Jan. 2019.

104 Fuji, Y., Higashi, Y., Masunishi, K., Yuzawa, A., Nagata, T., Kaji, S. et al. (2018). Spin-MEMS microphone integrating a series of magnetic tunnel junctions on a rectangular diaphragm. *J. Appl. Phys.* 123 (16): 163901.

105 Higashi, Y., Fuji, Y., Kaji, S., Masunishi, K., Nagata, T., Yuzawa, A. et al. (eds.) (2018). SNR enhancement of a spin-MEMS microphone by optimum bias magnetic field and demonstration of operation sound monitoring of rotating equipment. 2018 IEEE Micro Electro Mechanical Systems (MEMS), 21–25 Jan. 2018.

106 Sophian, A., Tian, G., and Fan, M. (2017). Pulsed Eddy current non-destructive testing and evaluation: a review. *Chin. J. Mech. Eng.* 30 (3): 500–514.

107 Cardoso, F.A., Rosado, L., Ferreira, R., Paz, E., Cardoso, S., Ramos, P.M. et al. (2014). Magnetic tunnel junction based eddy current testing probe for detection of surface defects. *J. Appl. Phys.* 115 (17): 17E516.

108 Guo, D.W., Cardoso, F.A., Ferreira, R., Paz, E., Cardoso, S., and Freitas, P.P. (2014). MgO-based magnetic tunnel junction sensors array for non-destructive testing applications. *J. Appl. Phys.* 115 (17): 17E513.

109 Cardoso, F.A., Rosado, L.S., Franco, F., Ferreira, R., Paz, E., Cardoso, S.F. et al. (2014). Improved magnetic tunnel junctions design for the detection of superficial defects by Eddy currents testing. *IEEE Trans. Magn.* 50 (11): 1–4.

110 Li, L., Mak, K.Y., Leung, C.W., Ng, S.M., Lei, Z.Q., and Pong, P.W.T. (2013). Detection of 10-nm superparamagnetic iron oxide nanoparticles using exchange-biased GMR sensors in Wheatstone Bridge. *IEEE Trans. Magn.* 49 (7): 4056–4059.

111 Su, D., Wu, K., Saha, R., Peng, C., and Wang, J.-P. (2020). Advances in magnetoresistive biosensors. *Micromachines* 11 (1): 34.

112 Kim, K., Hall, D.A., Yao, C., Lee, J.-R., Ooi, C.C., Bechstein, D.J.B. et al. (2018). Magnetoresistive biosensors with on-chip pulsed excitation and magnetic correlated double sampling. *Sci. Rep.* 8 (1): 16493.

113 Fujiwara, K., Oogane, M., Kanno, A., Imada, M., Jono, J., Terauchi, T. et al. (2018). Magnetocardiography and magnetoencephalography measurements at room temperature using tunnel magneto-resistance sensors. *Appl. Phys. Express* 11 (2): 023001.

114 Wang, M., Wang, Y., Peng, L., and Ye, C. (2019). Measurement of triaxial magnetocardiography using high sensitivity tunnel magnetoresistance sensor. *IEEE Sens. J.* 19 (21): 9610–9615.

115 Mohsen, A., Al-Mahdawi, M., Fouda, M.M., Oogane, M., Ando, Y., and Fadlullah, Z.M. (eds.) (2020). AI aided noise processing of spintronic based IoT sensor for magnetocardiography application. ICC 2020–2020 IEEE International Conference on Communications (ICC), 7–11 June 2020.

4

Spin-Transfer Torque Materials and Devices for Magnetic Random-Access Memory (STT-MRAM)

Yan Cui and Jun Luo

Key Laboratory of Microelectronic Devices and Integrated Technology, Institute of Microelectronics, Chinese Academy of Sciences, Beijing, China

CHAPTER MENU

4.1 The Background and Mechanism of STT-MRAM

4.1.1 The Background of STT-MRAM

The first generation MRAM is called Toggle-MRAM, in which information writing is realized by the magnetization switching of an MTJ free layer induced by a magnetic field. However, there are two drawbacks for Toggle-MRAM: first, the coercive force of MTJ increases as the size decreases; second, as the magnetic field is non-localized, cells neighboring those selected are still affected by the magnetic field, which may induce operation error. Therefore, Toggle-MRAM is hard to scale down. In order to overcome these shortcomings, STT-MRAM technology has been developed [1–3].

4.1.2 The Mechanism of STT-MRAM

4.1.2.1 LLGS Equation

Before discussing the operating principle of STT-MRAM, one should first examine the Landau–Lifshitz–Gilbert–Slonczewski equation, which is used to describe the precession of

Spintronics: Materials, Devices and Applications, First Edition. Edited by Kaiyou Wang, Meiyin Yang, and Jun Luo.

magnetization [2]. The equation is as follows:

$$\frac{d\vec{m}_F}{dt} = -\gamma(\vec{m}_F \times (\vec{H}_{eff} + \vec{H}_{TH})) + \alpha(\vec{m}_F \times \frac{d\vec{m}_F}{dt}) - \vec{\tau}_{ST} \tag{4.1}$$

$$\vec{\tau}_{ST} = \underbrace{a_J(\vec{m}_F \times \vec{m}_F \times \vec{m}_P)}_{Slonczewski's} - \underbrace{b_J(\vec{m}_F \times \vec{m}_P)}_{FLT}, b_J = \eta_{FLT} a_J \tag{4.2}$$

m_F stands for the unit vector of ferromagnetic moment, τ_{ST} stands for the spin transfer torque of spin polarized current, which interacts on the ferromagnetic moment, γ and α are the gyromagnetic ratio and the damping coefficient, respectively. η_{FLT} is a field-like torque, which ranges from 0 to 1. $\vec{H}_{eff}$ is the effective magnetic field, including: the external field, the exchange field, the demagnetization field, and the anisotropy field. $\vec{H}_{TH}$ stands for the random fluctuation field under a finite temperature; i.e., above 0K. Generally, η_{FLT} ranges from 10%~100% in a magnetic tunneling junction.

By solving the LLGS equation we can see that, when the applied current exceeds the threshold current, the magnetic moment flips from one equilibrium state to the other. When the applied current is smaller than the threshold current, the magnetic moment flipping is random, as shown in Figure 4.1.

4.1.2.2 The Write Mechanism of STT-MRAM

The difference between STT-MRAM and Toggle-MRAM is the method of data writing [4]. As shown in Figure 4.2, STT-MTJ consists of ferromagnetic layer/barrier layer/ferromagnetic layer. One of the ferromagnetic layer's magnetizations is fixed, which is called the reference layer (FM1); the magnetization of another ferromagnetic layer, which is called the free layer (FM2), can be switched on or off by a current, depending on its direction. When the current flows from FM2 to FM1, electrons move from FM1 to FM2. Electrons with spin in the direction which is anti-parallel to the magnetization of FM1 will be reflected by the CoFeB/MgO interface and move into the free layer. This means that the magnetization of FM2 flips to the opposite direction of FM1, which is to say the magnetization of FM1 and FM2 is anti-parallel. Currently, the MTJ shows high resistance to data "1." On the other hand, when the current flows from FM1 to FM2, the electron moves from FM2 to FM1.

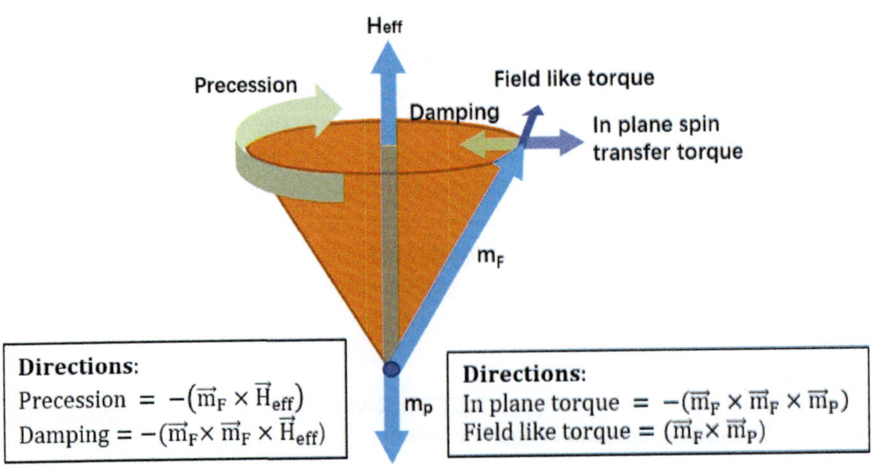

Figure 4.1 Precession process of single domain magnetic moment.

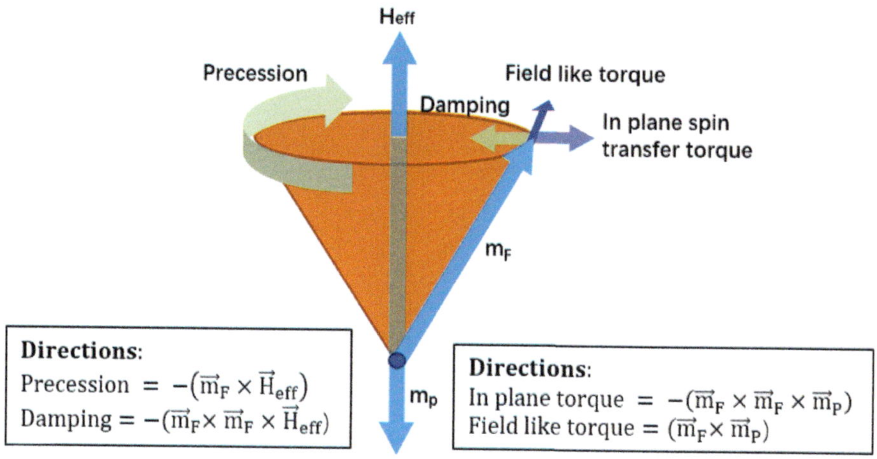

Figure 4.2 Spin transfer torque effect in a magnetic tunnel junction.

Electrons with spin in a direction parallel to the magnetization of FM1 will be passed into the free layer, which makes the magnetization of FM2 flip in the same direction as FM1: so, the magnetization of FM1 and FM2 is parallel. At this time, the MTJ shows low resistance which, indicated by "0" [5].

In the 1970s, Julliere devised a theoretical model to explain the magnetoresistance effect of MTJs. Figure 4.3 shows the energy band of an MTJ in parallel and anti-parallel states. The current transport mechanism in an MTJ obeys quantum mechanics; thus, the tunneling current has the following form:

$$I(V_{mtj}) \propto \int_{-\infty}^{\infty} N_{FM2}(E-eV_{mtj})N_{FM1}(E)[f(E-eV_{mtj})-f(E)]dE \tag{4.3}$$

$$I_{AP} \propto N_{FM1,up}\rho_{FM2,down} + N_{FM1,down}\rho_{FM2,up} \tag{4.4}$$

$$I_P \propto N_{FM1,up}\rho_{FM2,up} + N_{FM1,down}\rho_{FM2,down} \tag{4.5}$$

Here, N_{FM} is the effective number of injected electrons, ρ_{FM} is the effective density of states of the unfilled electrons in an electrode, V_{mtj} is the applied voltage on MTJ, I_P and I_{AP} is the tunneling current of MTJ in parallel- and anti-parallel states under applied voltage, respectively. The TMR (tunnel magnetoresistance ration) given by Julliere's model is as follows:

$$TMR = \frac{R_{AP}-R_P}{R_P} = \frac{2P_1P_2}{1-P_1P_2} \tag{4.6}$$

Here, R_P and R_{AP} stand for resistance of MTJ under parallel- and anti-parallel states, respectively. P_1 and P_2 are the spin polarizations of a reference layer and the free layer of MTJ.

After that, Slonczewski modifies Julliere's model, which considers the effective angle θ_{mtj} of magnetization between the reference layer and the free layer of MTJ. Slonczewski gives an opinion that the value of tunnel conductance depends on θ_{mtj} when the applied voltage is zero. It has the following form:

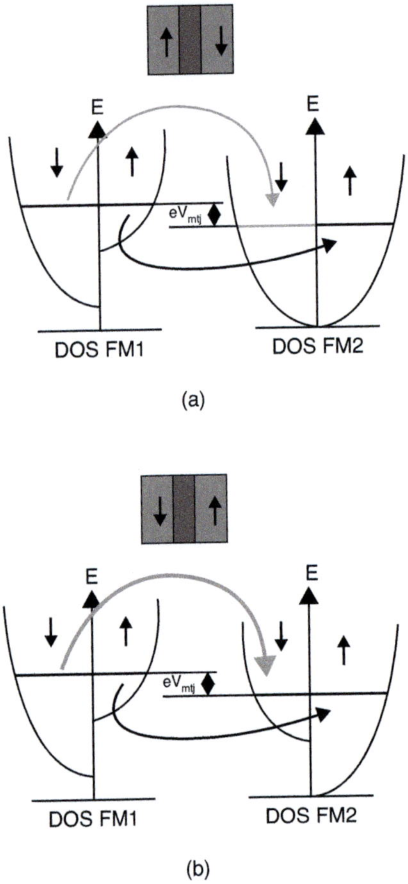

Figure 4.3 Spin-dependent electron tunneling in a magnetic tunnel junction in (a) parallel state and (b) anti-parallel state.

$$G_{mtj}(\theta) = G_0(1 + P_1P_2\cos\theta_{mtj}) \tag{4.7}$$

G_0 is the tunnel conductance at a temperature of absolute zero.

Based on this equation, Li et al. take the asymmetric tunneling barrier into account [6], which may affect tunnel conductance [7]:

$$G_{mtj}(V_{mtj}, \theta_{mtj}) = G_B[1 - TMR(V_{mtj})\cos\theta_{mtj} / (2 - TMR(V_{mtj}))] \tag{4.8}$$

Here, G_B is Brickman Conductance [8]:

$$G_B = G_0(1 - A0\Delta\varphi_b / 8\varphi_b{}^{3/2}eV_{bias} + 9A_0{}^2 / 128\varphi_b(eV_{mtj})^2),$$
$$A0 = 4E(2me)1/2xtox/3x\hbar \tag{4.9}$$

Here, φ_b is the barrier height, m_e is the effective mass of the electron, e is electron charge, t_{ox} is the thickness of the oxide, $\Delta\varphi_b$ stands for the coefficient of the asymmetric tunneling barrier, ħ is the Reduced Planck constant.

All the models mentioned above neglect the effect of the finite temperature. However, in the experiment the TMR of MTJ decreases as temperature increases. In order to explain this phenomenon, Stratton [9] gives a temperature-dependence model [10]. In this model, the tunnel conductance is divided into two parts: one is elastic scattering conductivity (G_E), which depends on the electron spin, another is non-elastic scattering conductivity (G_{SI}), which is independent of electron spin [11]. Generally, G_E doesn't change much with temperature while G_{SI} changes a lot. Taking this in consideration, the tunnel conductance has the following form:

$$G_{mtj} = \underbrace{\frac{G_0 CT}{\sin(CT)}[1 + P_1 P_2 \cos\theta_{mtj}]\sum_{i=0}^{2} C_i V^i}_{G_E} + \underbrace{ST^{1.33}}_{G_{SI}} \tag{4.10}$$

Here, $C = 1.387 \times 10^{-4} t_{ox} / \sqrt{\phi_b}$, t_{ox} is the thickness of oxide, φ_b is the height of the oxide tunnel barrier, T is the temperature, G_0 is the tunnel conductance at zero temperature.

$$G_{mtj} = \frac{3.16 \times 10^{10} \times \phi_b^{-1/2}}{t_{ox}} A_{mtj} \exp(-1.025 \times \phi_b^{-1/2} t_{ox}) \tag{4.11}$$

Based on the G_{mtj} mentioned above, the TMR under finite temperature has,

$$TMR = \frac{2G_E(V,T)P_1P_2}{G_E(V,T)\{1 - P_1(T)P_2(T)\} + G_{SI}(T)} \tag{4.12}$$

Here, $P_i(i=1,2)$ is the spin polarization of MTJ which depends on temperature. It has the following form:

$$P_i = P_0(1 - \beta_{FM} T^{3/2}) \tag{4.13}$$

P_0 is the spin polarization under zero temperature. β_{FM} is a coefficient which depends on material.

It can be seen that the bipolar current could make the magnetization of free layer switch into the opposite direction when the current is latger than I_c (critical switching current). This makes the MTJ exhibit high and low resistance, respectively. This is the operating mechanism of STT-MRAM.

Julliere first observed the TMR effect in the Fe/Ge/Co structure in 1975. After that, the MTJ with Al_2O_3 as the tunnel barrier was first reported in 1995, with the TMR ratio over 10% at room temperature [12]. Then, some researchers raised the TMR ratio to 70%, but it was still unable to meet application requirements. Until 2001, when Butler et al., Mathon and Umerski inferred from theory that there would be a significant improvement of the TMR ratio in MTJ consisting of Fe/MgO/Fe if the crystal phase was 001[13–14]. After that, experiments of MTJs with MgO as the tunnel barrier were reported, the TMR ratio was high enough for TMR=230% [15], TMR=260% [16], TMR=604% [17]. Since then, the magnetic tunnel junction has the possibility of practical application and has been widely considered by the industry. So far, many ferromagnetic and barrier materials have been studied, CoFeB/MgO/CoFeB has been introduced into mass production.

4.1.2.3 The Magnetism of STT-MTJ

Ferromagnetic film is the most important ingredient in MTJ. Generally, ferromagnetic material has at least one easy axis, in which direction the material is easy to be magnetized.

As shown in Figure 4.4, when the applied magnetic field is parallel to the easy axis (ϕ=0°or ϕ=180°), the magnetic film exhibits hysteresis behavior and is easy to saturate. On the other hand, when the applied magnetic field is vertical to the easy axis (ϕ=90°or ϕ=270°) the magnetic film exhibits paramagnetism and is hard to saturate. This axis is called the hard axis.

The magnetism of nano magnetic thin films is quite different from bulk materials. In the nano magnetic film, the magnetism depends on the properties of the shape and the interface. The magnetic anisotropy related to the shape of the film is called shape anisotropy, or demagnetization energy, which comes from magnetic dipole–dipole interaction. It tends to keep the magnetization direction in the plane [18, 19]. The shape anisotropy energy density U_{dem} has the following form:

$$U_{dem} = \frac{1}{2}\mu_0 M_S^2 \cos^2\theta_A \tag{4.14}$$

Here, μ_0 is vacuum permeability, M_S is saturation magnetization, θ_A is the angle between the magnetization and the normal plane of the film. When the magnetization direction is in the plane of the film, the shape anisotropy is the smallest. However, when other magnetic anisotropy is taken into account, the total magnetic anisotropy energy density under the applied magnetic field can be written as follows:

$$U_{total} = -\frac{1}{2}K_{eff}\cos^2\theta_A \pm \mu_0 M_S H\cos(\theta_A - \varphi) \tag{4.15}$$

Here, $K_{eff} = K_{vol} + \frac{2K_S}{t_f} - \frac{1}{2}\mu_0 M_S^2$ is the effective energy density which includes interface anisotropy energy density K_S/t_f, bulk anisotropy energy density K_{vol} and shape anisotropy energy density. The last term of the equation is Zeeman energy under an applied magnetic field. t_f is the thickness of magnetic film, ϕ is the angle between the applied magnetic field and the normal plane of the film.

It can be seen that if $K_{eff}>0$($K_{eff}<0$) the magnetization tends to go along with the out of plane direction (in plane). This can be further understood by the critical thickness (t_{cr}). When $t_f> t_{cr}$ the film exhibits in-plane magnetic anisotropy (IPA). When $t_f< t_{cr}$ the film exhibits perpendicular magnetic anisotropy (PMA). t_{cr} can be calculated by equation (4.15):

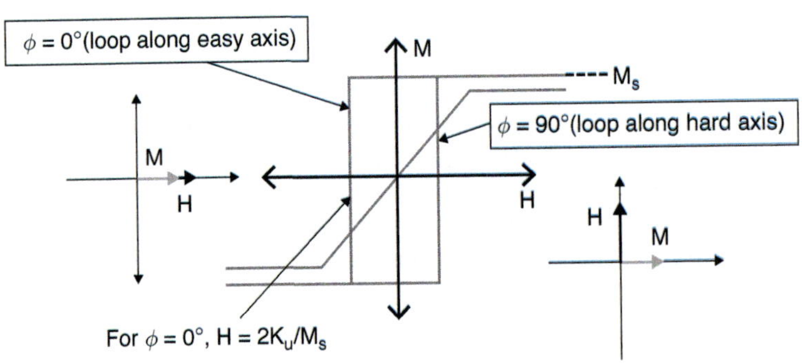

Figure 4.4 The magnetization curves along the hard-to-magnetize axis and the easy-to-magnetize axis of nano magnetic thin films.

$$t_{cr} = -\frac{2K_S}{K_{vol} - \frac{1}{2}\mu_0 M_S^2} \tag{4.16}$$

Figure 4.5 shows the relationship between the anisotropy energy of a magnetic film and the thickness of a magnetic film. Figure 4.6 shows the hysteresis loops of magnetic film with IMA and PMA.

4.1.2.4 The Switching Properties of STT-MTJ

Figure 4.7 shows the structure of the STT-MTJ device. The core structure of MTJ is CoFeB/MgO/CoFeB. But more function layers are needed to enhance the performance of MTJ, such as an anti-ferromagnetic pinned layer which is used to improve the thermal stability of the reference layer in MTJ. For MTJ with IPA, the commonly used materials are PtMn and

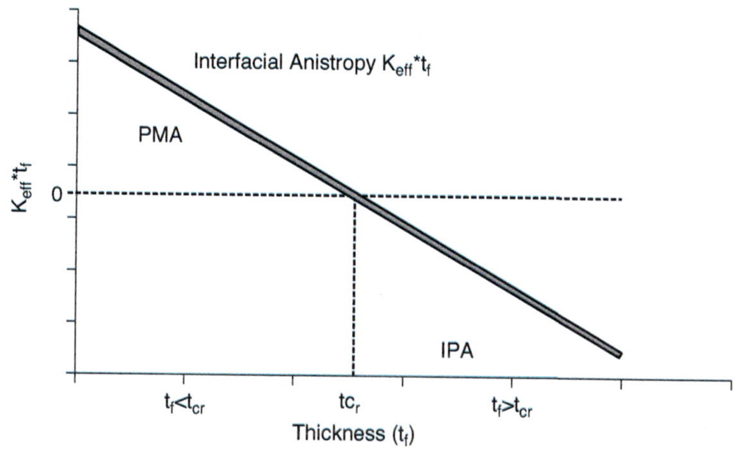

Figure 4.5 Relationship between magnetic anisotropy and film thickness of magnetic nano films.

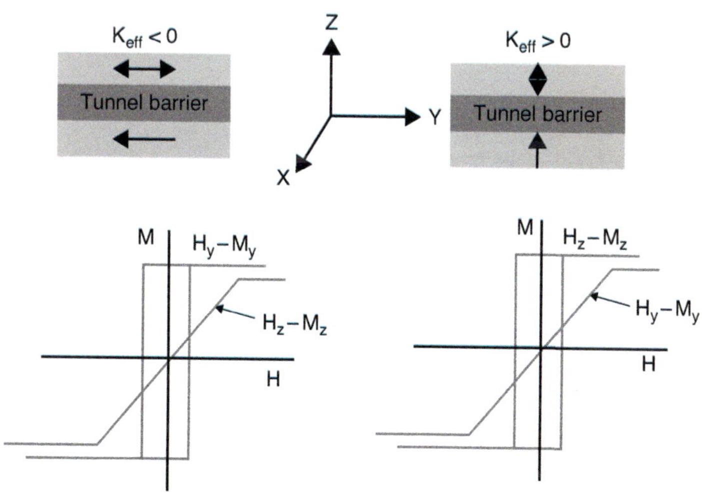

Figure 4.6 Hysteresis loops of in-plane and perpendicular magnetic anisotropy films.

IrMn. For MTJ with PMA, the commonly used materials are $[Co/Pt]_m/Ru/[Co/Pt]_n$. On the other hand, some heavy metals, such as W and Ta, are also used to improve thermal stability of the free layer. They deposit on the free layer as the capping layer to modify the texture of the free layer.

As mentioned above, the low- and high-resistance states of MTJ can be switched by a bipolar current as shown in Figure 4.7(b). Generally, the value of I_{HL0} (write high resistance to low resistance) and I_{LH0} (write low resistance to high resistance) is different.

By solving the LLGS equation of single domain approximation, one can get the critical switching current at T=0K. For MTJ with IPA, the critical switching current $I_{C0,IPA}$ has the following form [20],

$$I_{C0,IPA} = \frac{2e\alpha}{\eta(\theta_{mtj})\hbar} M_S V_f \left(H_{IPA} + \frac{1}{2}\mu_0 M_S\right) = \frac{4e\alpha}{\eta(\theta_{mtj})\hbar}\left(E_{B,IPA} + \frac{1}{4}\mu_0 M_S^2 V_f\right) \tag{4.17}$$

Here, η is the coefficient of spin transfer torque, θ_{mtj} is the angle between magnetization of the free layer and the reference layer. For the switching from anti-parallel state to parallel

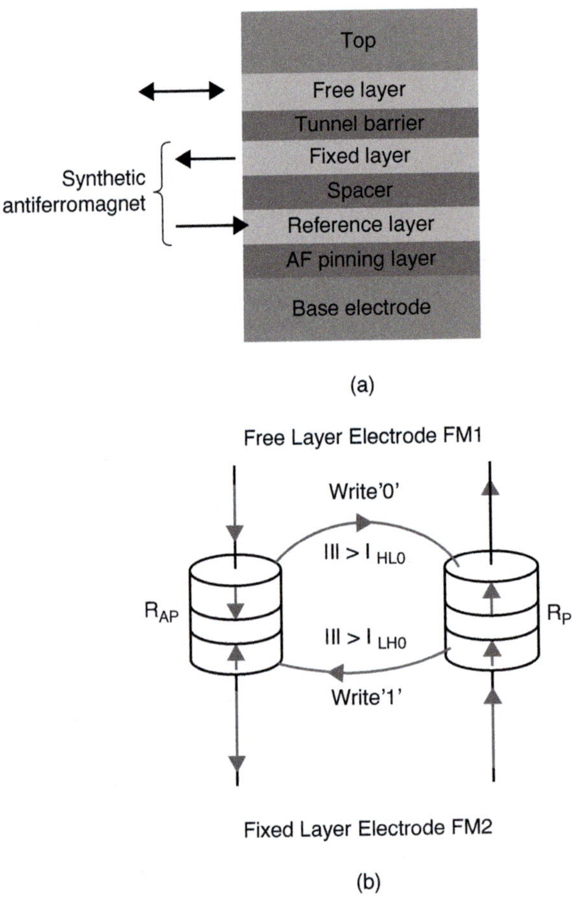

Figure 4.7 (a) The stack structure of STT-MTJ. (b) The switching operation of STT-MTJ.

state, θ_{mtj}=180° while for the switching from parallel state to anti-parallel state, 0°. η is the range from 0.2 to 0.6 which is experimental experience. $E_{B,IPA}$ is the switching energy barrier of parallel and anti-parallel states. H_{IPA}, M_S, V_f are the in-plane anisotropy field, saturated magnetization and the volume of the free layer. $\hbar, \alpha, \mu_0, e$ are the reduced Planck constant, damping coefficient, vacuum permeability and electronic charge. $\frac{1}{2}\mu_0 M_S$ is the demagnetization field.

On the other hand, for MTJ with PMA, the switching current $I_{C0,PMA}$ at T=0K has the following form,

$$I_{C0,PMA} = \frac{2e\alpha}{gh} M_S V_f (H_{PMA} - \mu_0 M_S) = \frac{4e\alpha}{gh} (E_{B,PMA}) \tag{4.18}$$

$$E_{B,PMA} = \frac{M_S V_f (H_{PMA} - \mu_0 M_S)}{2} \tag{4.19}$$

Here, H_{PMA}, $E_{B,PMA}$ stands for an out-of-plane anisotropy field and switching energy barrier, respectively. Other terms are the same as the in-plane MTJ. Looking at the equation of $I_{C0,PMA}$ and $I_{C0,IPA}$, it can be seen that $I_{C0,PMA}$ is smaller than $I_{C0,IPA}$. Therefore, MTJ with PMA is more compatible with small process nodes.

The switching time of MTJ with PMA can also be obtained by solving the LLGS equation; it has the following form [20],

$$\frac{1}{t_{Ps,PMA}} = \frac{2}{C + \ln(\frac{\pi^2 \Delta}{4})} \frac{\mu_B P}{e M_S V_f (1+P^2)} (I_{mtj} - I_{C0,PMA}) \tag{4.20}$$

Here, C=0.577 is the Euler constant, $\Delta = \frac{E_B}{k_B T}$ is the thermal stability coefficient, P is the spin polarization of the free layer, $\mu_B = 9.28 \times 10^{-24}$ is the Bohr magneton, $e = 1.6 \times 10^{-16}$ is the quantity of electric charge, M_S is the saturation magnetization of the free layer, V_f is the volume of the free layer. I_{mtj} and $I_{C0,PMA}$ are the supplied current and the critical switching current of MTJ.

It can be seen that, when the supplied current is larger than the critical switching current the switching time is inversely proportional to the supplied current.

On the other hand, when the supplied current is smaller than the critical switching current the switching process of the free layer is random. The relationship between switching probability and supplied current has the following form:

$$P_{SW} = 1 - \exp\left\{-\frac{t_{PW}}{\tau_0} \exp[-\frac{E_B^*}{k_B T}(1 - \frac{I}{I_{C0}})]\right\} \tag{4.21}$$

E_B^* is the effective magnetic anisotropy energy, which has the following form:

$$E_B^* = E_B (1 - \frac{H}{H_K}) \tag{4.22}$$

Here, H is the applied magnetic field, H_K is the anisotropy field at T=0K.

From equation (4.22) it can be seen that the switching probability of MTJ could reach 50% by strict control of the strength and pulse time of the applied current. At that time, the MTJ can be used as a true random number generator.

From equations (4.17) and (4.18) one can find that the perpendicular MTJ has lower critical switching current than in-plane MTJ. Moreover, in-plane MTJ suffers from the vortex effect when its size scales down, which seriously affects the thermal stability of the device. Thus, MTJ with perpendicular magnetic anisotropy is more compatible with the advanced technology node of the CMOS process. And it becomes the mainstream mass production technology route of MRAM. In 2010, the important experimental results of magnetic nanofilms were reported by Ikeda et al. [21]. The researchers proved that perpendicular anisotropy becomes stronger as the thickness of nanofilms decreases. For CoFeB, when the thickness is less than 1.5 nm it shows perpendicular anisotropy. Otherwise, it shows in-plane anisotropy. Based on this work, many works of CoFeB/MgO/CoFeB perpendicular MTJ have been reported recently [22–24]. And the TMR value of MTJ has made a huge leap [25–34]. For example, Jan et al. successfully fabricated a perpendicular MTJ device sized in 40 nm, RA=6.6Ωμm^2, Δ=74 and I_{C0}=25μA [23]. Then, Thomas et al reported devices with better performance, the size is 30 nm, RA=12Ωμm^2, Δ=77, TMR=150% [24].

Nevertheless, the proposal on STT-based switching for MRAM revived the research interest and the subsequent commercial interest [8, 35–37]. The first few proposals on STT-MRAM appeared in 2006 [38–40]. In August 2017, Everspin began sampling EMD4E001G, the 1Gb STT-MRAM, and has already mass-produced a 256Mb DDR3 STT-MRAM. The chip capacity is 256Mb with a 40 nm CMOS node and 1Gb with a 28 nm CMOS node. After that TSMC, Intel, Samsung, and GlobalFoundries launched their eMRAM production, based on 2Xnm Bulk, FFL and FDSOI technologies, respectively.

4.2 The Integrated Process of STT-MRAM

The memory cell of STT-MRAM is made up of one transistor and one MTJ (1T-1MTJ), which is integrated through BEOL. Generally, the BEOL of STT-MRAM concludes CMP, magnetic film deposition, photolithography, etching, dielectric isolation, metal contact and passivation.

4.2.1 CMP Technology

The film stack of perpendicular STT-MTJ is very complex and contains a 15~30 functional layer with the thickness of each layer about 1 nm. Therefore, the requirement of the surface roughness of the substrate is very high. Generally, the MTJ device is integrated between M4 and M5, that is to say the FEOL is stopped at Via4. At that time, the surface roughness of the wafer is quite poor, one should optimize it by CMP. It should be noted that a via is the W plug, which is fabricated by CVD or PVD at the CMOS process node over 0.13μm. However, the crystalline grain of W is pretty big, which results in a rough surface which is hard to polish. So a horizontal lead-out buffer layer is needed before MTJ can be integrated. On the other hand, for the advanced CMOS process, a via is fabricated by Cu Damascene technology. Combined with CMP technology, the surface roughness is greatly improved. Thus, the MTJ can be fabricated on the via directly, which improves the integration density of MRAM.

4.2.2 Magnetic Film Deposition Technology

Magnetic film deposition is one of the core technologies in the MRAM integration process. In the lab, one should use sputter, ALD, MBE and PLD to fabricate high property film. However, in the foundry, magnetron sputter is the preferred option in order to ensure production efficiency. As the performance of MTJ depends on the stack structure and interface texture of the film [41–53], the core stack needs to be deposited at one time in an ultra-high vacuum environment. On the other hand, the thickness of the core stack in MTJ with perpendicular magnetic anisotropy is about 1 nm; thus it is necessary to have strict controlability of the sputter rate and accuracy of the magnetron sputter. For now, CoFeB/MgO/CoFeB is the mainstream MTJ structure used in the industrial world. As the CMOS process scales down, the size of MTJ gets smaller and smaller. To keep the thermal stability of the MTJ device, one needs to add many kinds of functional layers to increase the PMA of the magnetic film. For example, TiN and Ta are used as a seed layer which could induce the bcc structure formed in CoFeB; [Co/Pt] n is used as a pinned layer, which increases the PMA of the CoFeB reference layer by SAF structure ($[Co/Pt]_n/Ru[Co/Pt]_m$); a dual MgO structure used in the CoFeB free layer increases the PMA through an extra MgO/CoFeB interface; Ta/Ru/Ta or TiN are often used to be the capping layer to protect the MTJ device and also to be a hard mask for MTJ etching.

4.2.3 Photolithography Technology

Photolithography technology is divided into UV lithography and electron beam lithography (EBL). EBL can directly write nano patterns, but for large scale patterns it needs a very long time and the overlay accuracy is not high. Therefore, the industry mainly uses UV lithography. UV lithography is also divided into DUV and EUV technology. DUV technology includes I-line lithography (wave length 365 nm), KrF DUV lithography (wave length 248 nm) and ArF DUV lithography (wave length 193 nm). For STT-MRAM mass production, ArF DUV lithography is used. And according to the different types of STT-MRAM, the critical dimension of an MTJ device ranges from 40 nm to 80 nm.

4.2.4 Etching Technology

Etching is another core technology of the MRAM integration process. Because of the complex constituents of MTJ, the photo resistance of DUV cannot satisfy the whole device etching. Thus, a hard mask layer is needed. First, the hard mask layer is etched with photoresist, then it is used as the barrier layer for MTJ etching. Generally, MTJ includes many magnetic elements and heavy elements such as Co, Fe, Ni, Pt. These elements are very sensitive to Cl and H_2O, which will affect the physical properties of such film and finally degrades the device's performance. To avoid the issue, C/F gas is used in the etching process. In the industry technology, RIE, IBE and RIE, combined with IBE, are used for MTJ etching. The advantages of RIE are: (a) It is widely used in the CMOS process; (b) Many kinds of gas can be used during the etching process and has the high etching rate which is helpful for nano-device etching. The disadvantages of RIE are: (a) For small size STT-MTJ, the issue of etching-induced redeposition on the device side wall is serious; (b) The energy distribution of the etching ion is quite large, which would damage the magnetic film of the MTJ device. For IBE technology, the advantage is that the profile is pretty good and can be used for small-scale array etching. The disadvantages are that: (a) There is no automatic cleaning function so the cavity needs to be opened frequently for maintenance; (b) The etching energy needs to be quite high in order to obtain a good profile, which could damage the device; (c) Insufficient capacity means it is not suitable for industry application. When RIE is combined with IBE etching technology it complements the characteristics of the two technologies: (a) Main etching of MTJ is done in the RIE chamber, which has an automatic cleaning function and is easy to maintain; (b) After main etching, the device can be transferred to an IBE chamber for shape optimization, which minimizes the etching damage to the device. What's more, OES or SIMS are used during etching MTJ in order to monitor the etched materials and the endpoint. On the other hand, as mentioned above, the MTJ device is sensitive to H_2O and O_2. To avoid this issue, one should transfer the MTJ device to a PECVD chamber for low temperature SiN protection after etching without breaking the vacuum. Thus, for industry application, the etching equipment for MRAM is usually integrated with a PECVD chamber.

4.2.5 Dielectric Isolation Technology

Although SiN is deposited after MTJ etching, the thickness of it is about several tens of a nanometer, which is not enough for electrical isolation. Thus, extra insulator is needed, such as SiO_x. However, high temperature deposition will degrade the performance of the MTJ device. One needs to use a low temperature (not over 400°C) deposition process.

4.2.6 Contact Technology

After dielectric isolation, one needs to lead out a metal electrode for device testing by contact technology. For the top electrode of an MTJ device, there are two ways of contacting. One way is using CMP to polish the dielectric layer until the top metal surface of the MTJ is exposed. This method needs the high accuracy and high uniformity of CMP. The other way is using RIE to etch the dielectric layer and form a via. This method needs the high selectivity and high accuracy of RIE. In the industry, the latter is widely used. For the bottom electrode of the MTJ device, RIE is used to fabricate the via. The high etching selectivity is also very important because the thickness of the bottom electrode layer is about 10 nm; one should stop on it during the etching process. After fabricating the top and electrode via, one should deposit metal material such as Ti/TiN or Cu for the fabricating test pad, which is the same with the CMOS process.

4.2.7 Passivation Deposition

Passivation deposition is the standard CMOS BEOL which is used to protect and enhance the lifetime of a chip. However, the annealing temperature of passivation is 400°C, which may degrade the performance of the MTJ device. Thus, the development of high temperature annealed magnetic tunnel junction materials is still one of the problems concerning scientific research and industry.

4.3 Testing of the STT-MTJ Device

The STT-MTJ device is a bipolar device and the resistance is about KΩ; thus one can use the two-point method to test its electronic properties. Routine performance tests in the lab include the R-H test, the R-I test while the industry is more attentive to reliability such as endurance tests, the BER test, a temperature shock test, a thermal stability test and so on. This chapter only introduces the R-H and R-I tests. For the R-H test, an external magnetic field is applied to the STT-MTJ device which makes the free layer magnetization switching and induces the change of resistivity. Then, a small given current is used to read out the resistivity. In this test, one can get the TMR, RA, coercivity field and stray field of the device. The R-I test divides into a dc test and an ac test. The dc test applies a dc bias voltage or current to the device. Under the STT effect mentioned above, the magnetization of the free layer could be switched. And the change of resistivity is read by the same current. From this test, one can get the critical switching current I_c of write "0" and write "1," the dependence of TMR on bias voltage, the breakdown current I_{BD} of device. The ac test applies an ac bias voltage or current to the device (the pulse time is ~ns). The test usually uses a GSG probe which has three terminals; thus the pad of the device needs a special design. The test mechanism is the same as the dc test. However, one can get more information from it. First, the highest writing speed and the critical current I_c in this fast write mode can be obtained. Then, the relationship between critical current and pulse time can be obtained. According to the relationship, one can get the thermal stability factor of the device by fitting calculation. Finally, one can also get the WER and write endurance of the device through the cycle test mode. Here, WER refers to "write error rate," defined as the number of writing errors/total number of write operations. Based on these test results, one can even set a STT-MTJ device model for a 1T-1MTJ design.

4.4 The Development Status of STT-MRAM

In the early days, an amorphous insulating layer such as AlO_x would be used as the tunneling barrier of MTJ. The TMR ratio depends mainly on the electronic band structure of the ferromagnetic materials. A maximum value of the TMR ratio could be achieved at 70% using AlOx [54–59]. However, when crystalline barriers such as MgO, with crystalline and band-matched magnetic materials such as Fe, were used in an MTJ, the TMR ratio was a qualitative improvement [10, 60–66]. In 2004, Yuasa et al. and Parkin et al. (2005) showed that the TMR of MgO-based MTJs can be greater than that of AlO_x barriers. In 2005, Djayaprawira et al. demonstrated a TMR ratio higher than 200% in CoFeB/MgO/CoFeB MTJ in the same year[67]. Over the same period, the Heuslar alloy was introduced into MTJ technology replacing CoFeB [68, 69]. MTJ with the Heuslar alloy shows a higher TMR ratio as the spin polarizability is almost equal to 1, while the dynamic power consumption in this system is higher than CoFeB/MgO/CoFeB [70–74]. After that, STT-MRAM entered the stage of mass production [35, 74–85]; TSMC, UMC, Intel, Samsung, and Global Foundries have launched eMRAM products, based on the 2Xnm CMOS process. The capacity is around the tens of megabits. In the 2020 IEEE International Electron Devices Meeting, Samsung exhibited their 28 nm eMRAM for frame buffer memory [84], which features the smallest macro size of 0.08 mm^2/Mb reported to date. Compared to SRAM, commonly used for frame buffer memory, eMRAM provides a 47% area saving. Global Foundries have demonstrated a 22 nm FD-SOI 40Mb embedded-MRAM for industrial grade (-40~125°C) applications, which passes all the JEDEC reliability tests [85]. TSMC presents an embedded STT-MRAM macro that can tolerate soldering reflow in a 16 nm FinFET CMOS logic process, which features a 9ns read access time with wide 128-IO, making it suitable for high-performance, general purpose MCU applications [86]. IBM have demonstrated the first Embedded STT-MRAM technology in a 14HP FinFET CMOS node. Using a 2 Mb eMRAM macro, they have achieved an integration at a tight MTJ pitch (160 nm) which fits vertically between M1 and M2, with a write performance down to 4ns [87].

In China, MRAM technology is in the state of research and development. Several research institutes are engaged in MRAM basic research, such as the Institute of Physics, the Chinese Academy of Sciences, the Institute of Semiconductors, the Chinese Academy of Sciences, the Institute of Microelectronics, the Chinese Academy of Sciences, Tsinghua University, Peking University, the Beijing University of Aeronautics and Astronautics, Huazhong University of Science and Technology, Xi'an Jiaotong University, and so on. The level of scientific research has reached the forefront of the world. That said, mass production technology is relatively backward.

In 2006, Xufeng Han's group at the Institute of Physics, Chinese Academy of Sciences introduced the first 4Kb STT-MRAM prototype chip in China, which is based on a nano-ring of MTJ. In 2017, The Institute of microelectronics, Chinese Academy of Sciences and the Beijing University of Aeronautics and Astronautics jointly launched the first sub-100-nanometer STT-MTJ device, based on 8-inch pilot technology in China; the electrical characteristics of the device attained international standards. In 2019, the Institute of Microelectronics, Chinese Academy of Sciences, developed 8-inch MTJ process which is compatible with CMOS BEOL. They are focusing on magnetic thin film deposition technology, deep ultraviolet lithography technology, tunnel junction etching technology, and metal interconnection technology, and have obtained wafer-level sub-100 nanometer STT-MTJ devices. The TMR is larger than 100%, writing speed reaches 2ns at the supplied voltage 0.7V, and the low resistance is about 2.6KΩ. Until now, The Institute of Microelectronics of the Chinese Academy of Sciences is the only scientific research institution with customized MRAM pilot research and development. Figure 4.8 shows the research results of the 8-inch STT-MRAM process at the Institute of Microelectronics, Chinese Academy of Sciences.

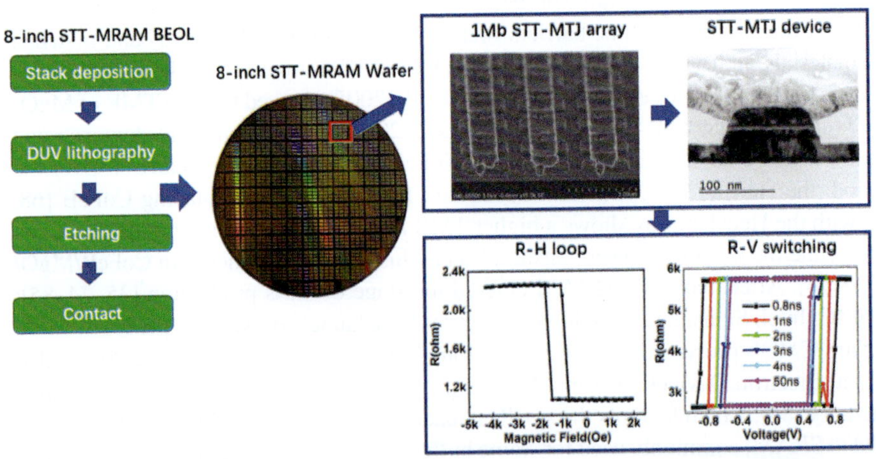

Figure 4.8 The research results of the 8-inch STT-MRAM process at the Institute of Microelectronics, Chinese Academy of Sciences.

References

1 Berger, L. (1978). Low-field magnetoresistance and domain drag in ferromagnets. *J. Appl. Phys.* 49 (3): 2156–2161.

2 Slonczewski, J.C. (1996). Current-driven excitation of magnetic multilayers. *J. Magn. Magn. Mater.* 159 (1–2): L1–L7.

3 Myers, E.B., Ralph, D.C., Katine, J.A., Louie, R.N., and Buhrman, R.A. (1999). Current-Induced Switching of Domains in Magnetic Multilayer Devices. *Science* 285 (5429): 867–870.

4 Zhang, S., Levy, P., and Fert, A. (2002). Mechanisms of Spin-Polarized Current-Driven Magnetization Switching. *Phys. Rev. Lett.* 88 (23): 236601.

5 Sankey, J.C., Cui, Y.-T., Sun, J.Z., Slonczewski, J.C., Buhrman, R.A., and Ralph, D.C. (2008). Measurement of the spin-transfer-torque vector in magnetic tunnel junctions. *Nat. Phys.* 4 (1): 67.

6 Li D, Feng J, Yu G, Guo P, Chen J, Wei HX, et al. (2013). Tunneling processes in asymmetric double barrier magnetic tunnel junctions with a thin top MgO layer. *Journal of Applied Physics* 2013; 114: 213909-213909.

7 Stiles, M.D. and Miltat, J. (2006). Spin transfer torque and dynamics. In: *Spin Dynamics in Confined Magnetic Structures III, ser. Topics in Applied Physics* (ed. B. Hillebrands and A. Thiaville), Vol. 101, 225–308. Berlin-Heidelberg, Germany: Springer-Verlag. Ch. 7.

8 Tsymbal, E.Y., Mryasov, O.N., and LeClair, P.R. (2003). Spin-dependent tunnelling in magnetic tunnel junctions. *J. Phys.: Condens. Matter* 15: 109.

9 Tiwari, R.K. et. al. (2014). Current-induced switching of magnetic tunnel junctions: effects of field-like spin-transfer torque, pinned-layer magnetization orientation, and temperature. *Appl. Phys. Lett.* 104 (2).

10 Brataas, A., Kent, A.D., and Ohno, H. (2012). Current-induced torques in magnetic materials. *Nat. Mater.* 11 (5): 372–381.

11 Skomski, R. and Zhou, J. (2006). Nanomagnetic models. In: *Advanced Magnetic Nanostructures* (ed. D.J. Sellmyer and R. Skomski), 41–90. New York: Springer-Verlag. Ch. 3.

12 Miyazaki, T. and Tezuka, N. (1995). Giant magnetic tunneling effect in Fe/Al2O3/Fe junction. *J. Magn. Magn. Mater.* 139 (3): L231–L234.

13 Butler WH, Zhang XG, Schulthess TC, MacLaren JM. (2001).Spin-dependent tunneling conductance of Fe|MgO|Fe sandwiches. *Physical Review B*2001; 63 (5): 054416.

14 Mathon J, Umerski A. (2001). Theory of tunneling magnetoresistance of an epitaxial Fe/MgO/Fe(001) junction. *Physical Review B*2001; 63 (22): 220403.

15 Djayaprawira, D.D. Tsunekawa K, Nagai M, Maehara H, Yamagata S, Watanabe N, et al. (2005). 230% room-temperature magnetoresistance in CoFeB/MgO/CoFeB magnetic tunnel junctions. *Appl. Phys. Lett.* 86 (9): 092502.

16 Hayakawa, J. (2005). Dependence of giant tunnel magnetoresistance of sputtered CoFeB/MgO/CoFeB magnetic tunnel junctions on MgO barrier thickness and annealing temperature. *Jpn. J. Appl. Phys.* 44 (19): 587.

17 Ikeda, S. Hayakawa J., Ashizawa Y., Lee Y.M., Miura K., Hasegawa H., Tsunoda M., Matsukura F. and Ohno H. (2008). Tunnel magnetoresistance of 604% at 300K by suppression of Ta diffusion in CoFeB/MgO/CoFeB pseudo-spin-valves annealed at high temperature. *Appl. Phys. Lett.* 93 (8): 082508.

18 Draaisma, H.J.G. and De Jonge, W.J.M. (1988). Surface and volume anisotropy from dipole-dipole interactions in ultrathin ferromagnetic films. *J. Appl. Phys.* 64 (7): 3610.

19 Johnson, M.T., Bloemen P.J.H., den Broeder F.J.A. and de Vries J.J. (1996). Magnetic anisotropy in metallic multilayers. *Rep. Prog. Phys.* 59: 1409.

20 Zhang, Y., Zhao W., Lakys Y., Klein J., Kim J., Ravelosona D., Chappert C. (2012). Compact Modeling of Perpendicular-Anisotropy CoFeB/MgO Magnetic Tunnel Junctions. *IEEE Trans. Electron Devices* 59 (3): 819.

21 Ikeda S, Miura K, Yamamoto H, Mizunuma K, Gan HD, Endo M, et al. (2010). A perpendicular-anisotropy CoFeB–MgO magnetic tunnel junction. *Nature Materials* 2010; 9 (9): 721–724.

22 Gajek M., Nowak J.J., Sun J.Z., Trouilloud P.L., O'Sullivan E.J., Abraham D.W., Gaidis M.C., Hu G., Brown S., Zhu Y., Robertazzi R.P., Gallagher W.J., and Worledge D.C. (2012). Spin torque switching of 20 nm magnetic tunnel junctions with perpendicular anisotropy. *Appl. Phys. Lett.* 100 (13): 132408.

23 Jan, G., Wang Y.J., Moriyama T., Lee Y.J., Lin M., Zhong T., Tong T.Y., Torng T. and Wang P.K. (2012). High Spin Torque Efficiency of Magnetic Tunnel Junctions with MgO/CoFeB/MgO Free Layer. *Appl. Phys. Express* 5: 093008.

24 Thomas, L., Jan G., Zhu J., Liu H., Lee Y.J., Le S., Tong R.Y., Pi K., Wang Y.J., Shen D., He R., Haq J., Teng J., Lam V., Huang K., Zhong T., Torng T., and Wang P.K. (2014). Perpendicular spin transfer torque magnetic random access memories with high spin torque efficiency and thermal stability for embedded applications (invited). *J. Appl. Phys.* 115 (17): 172615.

25 Kumagai, S., Tezuka, N., and Miyazak, T. (1997). Temperature Dependence of the Spin Tunneling Magnetoresistive Effect on NiFe/Co/Al2O3/Co/NiFe/FeMn Junctions. *Jpn. J. Appl. Phys.* 36: 1498.

26 R. C. Sousa, J. J. Sun, V. Soares, and P. P. Freitas. (1998). Large tunneling magnetoresistance enhancement by thermal anneal. *Appl. Phys. Lett.* 73: 3288.

27 Parkin, S.S.P. Roche, K. P., Samant, M. G., Rice, P. M., Beyers, R. B., Scheuerlein, R. E., Osullivan, E. J., Brown, S. L., Bucchigano, J., Abraham, D. W. (1999). Exchange-biased magnetic tunnel junctions and application to nonvolatile magnetic random. *J. Appl. Phys.* 85: 5828.

28 Han X.-F., Oogane M., Kubota H., Ando Y., and Miyazaki T. (2000). Fabrication of high-magnetoresistance tunnel junctions using $Co_{75}Fe_{25}$ ferromagnetic electrodes. *Appl. Phys. Lett.* 77: 283.

29 Kano H., Bessho K., Higo Y., Ohba K., Hashimoto M., Mizuguchi T. and Hosomi M. (2002). MRAM WITH IMPROVED MTJ MATERIAL. Magnetics Conference, 2002. INTERMAG Europe 2002, Digest of Technical Papers.

30 Wang, D., Nordman C., Daughton J.M., Qian Z.-H. and Fink J. (2004). 70% TMR at room temperature for SDT sandwich junctions with CoFeB as free and reference Layers. *IEEE Trans. Magn.* 40: 2269.

31 Bowen, M., Bowen M., Cros V., Petroff F., and Fert A. (2001). Large magnetoresistance in Fe/MgO/FeCo(001) epitaxial tunnel junctions on GaAs(001). *Appl. Phys. Lett.* 79: 1655.

32 Faure-Vincent, J. Faure-Vincent J., Tiusan C., Jouguelet E., Canet F., Sajieddine M., Bellouard C., Popova E., Hehn M., Montaigne F., and Schuhl A. (2003). High tunnel magnetoresistance in epitaxial Fe/MgO/Fe tunnel junctions. *Appl. Phys. Lett.* 82: 4507.

33 Yuasa, S., Matsumoto R., Fukushima A., Kubota H., Nagahama T., Djayaprawira D.D., Tsunekawa K., Maehara H., Nagamine Y., Nagai M., Yamagata S., Suzuki Y., Mizuguchi M., Deac M. and Ando K. (2006). Giant tunneling magnetoresistance in MgO-based

magnetic tunnel junctions and its industrial applications. Nanotechnology Materials and Devices Conference.

34 Yuasa, S., Fukushima A., Nagahama T., Ando K., Suzuki Y. (2004). High Tunnel Magnetoresistance at Room Temperature in Fully Epitaxial Fe/MgO/Fe Tunnel Junctions due to Coherent Spin-Polarized Tunneling. *Jpn. J. Appl. Phys.* 43: 588.

35 Hosomi, M., Yamagishi H., Yamamoto T., Bessho K., Higo Y., Yamane K., Yamada H., Hachino H., Fukumoto C., Nagao H. and Kano H. (2005). A novel nonvolatile memory with spin torque transfer magnetization switching: spin-ram. Electron Devices Meeting, IEDM Technical Digest 459.

36 Parkin, S., Xin J., Kaiser C., Panchula A., Roche K., Samant M. (2003). Magnetically engineered spintronic sensors and memory. *Proc. IEEE* 91: 661.

37 Yang T., Hirohata A., Kimura T. and Otani Y. (2007). *J. Nanosci. Nanotechnol.* 7: 259.

38 Zhao W., Belhaire E., Mistral Q., Chappert C., Javerliac V., Dieny B., et al. (2006). Magnetically engineered spintronic sensors and memory. IEEE International Behavioral Modeling and Simulation Workshop.

39 Huai, Y., Nguyen, P.P., and Albert, F. (2006). Three-terminal magnetostatically coupled spin transfer-based MRAM cell. Google Patents.

40 Nam, K.-T., Oh S.C., Lee J.E., Jeong J.H. and Moon J.T. (2006). Switching Properties in Spin Transfer Torque MRAM with sub-5Onm MTJ size. Non-Volatile Memory Technology Symposium, NVMTS 2006. 7th Annual.

41 Girgis, E., Schelten, J., Shi, Janesky, et al. (2000). Switching characteristics and magnetization vortices of thin-film cobalt in nanometer-scale patterned arrays. *Appl. Phys. Lett.* 76: 3780.

42 Shi, J., Tehrani, S., and Scheinfein, M. (2000). Geometry dependence of magnetization vortices in patterned submicron NiFe elements. *Appl. Phys. Lett.* 76: 2588.

43 Zheng, Y. and Zhu, J.-G. (1997). Switching field variation in patterned submicron magnetic film elements. *J. Appl. Phys.* 81: 5471.

44 Shi, J. and Tehrani S. (1999). Magnetization vortices and anomalous switching in patterned NiFeCo submicron arrays. *Appl. Phys. Lett.* 74: 2525.

45 Schrag, B., Anguelouch A., and Xiao G. (2000). Magnetization reversal and interlayer coupling in magnetic tunneling junctions. *J. Appl. Phys.* 87: 4682.

46 Carey, M.J., Fontana, R.E., Jr., and Gurney, B.A. (2001). Magnetic sensors having antiferromagnetically exchange-coupled layers for longitudinal biasing. Google Patents.

47 Nishimura N., Hirai T., Koganei A., Ikeda T., Okano K., Sekiguchi Y. and Osada Y. (2002). Magnetic tunnel junction device with perpendicular magnetization films for high-density magnetic random access memory. *J. Appl. Phys.* 91: 5246.

48 Nakayama N., Hirai T., Koganei A., Ikeda T., Okano K., Sekiguchi Y., and Osada Y. (2008). Magnetic tunnel junction device with perpendicular magnetization films for high-density magnetic random access memory. *J. Appl. Phys.* 103: 07A710.

49 Carcia, P., Meinhaldt, A., and Suna, A. (1985). Perpendicular magnetic anisotropy in Pd/Co thin film layered structures. *Appl. Phys. Lett.* 47: 178.

50 Broeder F.D., Donkersloot H.C., Draaisma H. and Jonge W.D. (1987). Magnetic properties and structure of pd/co and pd/fe multilayers. *J. Appl. Phys.* 61: 4317.

51 Brad N. Engel, Craig D. England, Robert A. Van Leeuwen, Michael H. Wiedmann, and Charles M. Falco. (1991). Magnetocrystalline and magnetoelastic anisotropy in epitaxial Co/Pd superlattices. *J. Appl. Phys.* 70: 5873.

52 Sugimoto, T., Katayama T, Suzuki Y, Hashimoto M, Nishihara Y, Itoh A, et al. (1992). Temperature dependence of perpendicular magnetic anisotropy in Co/Au and Co/Pt multilayers. *J. Magn. Magn. Mater.* 104: 1845.

53 Joo, H. et al. (2006). *J. Appl. Phys.* 99.

54 Martin and Jourdan. (2014). Revival of Heusler compounds for spintronics. *Mater. Today* 17: 362.

55 Diao, Z., Panchula A., Ding Y., Pakala M., Wang S., Li Z., Apalkov D., Nagai H., Driskill-Smith A., Wang L.-C., Chen E. and Huai Y. (2007). Spin transfer switching in dual MgO magnetic tunnel junctions. *Appl. Phys. Lett.* 90: 132508.

56 Hu, G.,, Lee J.H., Nowak J.J., Sun J.Z., Harms J., Annunziata A., Brown S., Chen W., Kim Y.H., Lauer G., Liu L., Marchack N., Murthy S., O'Sullivan E.J., Park J.H., Reuter M., Robertazzi R.P., Trouilloud P.L., Zhu Y., Worledge D.C. (2015). STT-MRAM with double magnetic tunnel junctions. Electron Devices Meeting (IEDM).

57 Liu, H., Honda Y., Taira T, Matsuda KI, Arita M, Uemura T, et al. (2012). Comment on "Water-driven programmable polyurethane shape memory polymer: Demonstration and mechanism." *Appl. Phys. Lett.* 101 (13): 1488-R.

58 Lu H, Liu Y, Leng J, Du S. (2010). Magnetization switching by spin-torque effect in off-aligned structure with perpendicular anisotropy. *Appl. Phys. Lett.* 97 (5): 056101.

59 Sbiaa, R., Piramanayagam, S.N., and Liew, T. (2013). High speed in spin-torque-based magnetic memory using magnetic nanocontacts. *Phys. Status Solidi* 7: 332.

60 Sbiaa, R. (2013). Magnetization switching by spin-torque effect in off-aligned structure with perpendicular anisotropy. *J. Phys. D: Appl. Phys.* 46: 395001.

61 Kamberský, V. (2007). Spin-orbital Gilbert damping in common magnetic metals. *Phys. Rev. B* 76: 134416.

62 Fähnle, M. and Illg, C. (2011). Electron theory of fast and ultrafast dissipative magnetization dynamics. *J. Phys.: Condens. Matter* 23: 493201.

63 Mizukami, S. et al. (2009). *J. Appl. Phys.* 105: 7D306.

64 Lenz, K., Wende H, Kuch W, Baberschke K, Nagy K, Jánossy A. (2006). Two-magnon scattering and viscous Gilbert damping in ultrathin ferromagnets. *Phys. Rev. B* 73: 144424.

65 Lagae L, Wirix-Speetjens R, Eyckmans W, Borghs S, De Boeck J. (2005). Increased Gilbert damping in spin valves and magnetic tunnel junctions. *J. Magn. Magn. Mater.* 286: 291.

66 Smith, N. and Arnett, P. (2001). White-noise magnetization fluctuations in magnetoresistive heads. *Appl. Phys. Lett.* 78: 1448.

67 Djayaprawira DD, Tsunekawa K, Nagai M, Maehara H, Yamagata S, Watanabe N. (2005). 230% room-temperature magnetoresistance in CoFeB/MgO/CoFeB magnetic tunnel junctions. *Applied Physics Letters*2005; 86 (9): L231.

68 Piramanayagam, S.N., Sbiaa, R., and Tahmasebi, T.. (2012). Magnetoresistance device and memory device including the magnetoresistance device. Google Patents.

69 Ma, Q. et al. (2015). *Sci. Rep.* 5.

70 Bai, Z. Shen L., Han G., Feng Y.P. (2012). *Data Storage: Review of Heusler Compounds Spin. SPIN*2012; 02 (04): 1230006.

71 Chumak, A., Serga A.A. and Hillebrands B. (2015). Magnon spintronics. *Nat. Phys.* 11: 453.

72 Husain, S., Akansel S., Kumar A., Svedlindh A.P., and Chaudhary S. (2016). *Sci. Rep.* 6.

73 Galanakis, I. and Dederichs, H. (2005). Half-metallicity and Slater–Pauling behavior in the ferromagnetic Heusler alloys. In: *Half-metallic Alloys*, 1. Springer.

74 Hirohata A, Kikuchi M, Tezuka N, Inomata K, Claydon JS, Xu YB, et al. (2006). Heusler alloy/semiconductor hybrid structures. *Curr. Opin. Solid State Mater. Sci.* 10: 93.

75 Sato, J., Oogane M, Naganuma H, Ando Y. (2011). Large Magnetoresistance Effect in Epitaxial $Co_2Fe_{0.4}Mn_{0.6}Si/Ag/Co_2Fe_{0.4}Mn_{0.6}Si$ Devices. *Appl. Phys. Express* 4: 113005.

76 Tezuka N., Ikeda N., Mitsuhashi F. & Sugimoto S. (2009). Improved tunnel magnetoresistance of magnetic tunnel junctions with heusler $co_2FeAl_{0.5}Si_{0.5}$ electrodes fabricated by molecular beam epitaxy. *Appl. Phys. Lett.* 94: 162504.

77 Hirohata, A., Sukegawa H, Yanagihara H, Žutić I, Seki T, Mizukami S, et al. (2015). Roadmap for Emerging Materials for Spintronic Device Applications. *IEEE Trans. Magn.* 51: 1.

78 Åkerman, J., Brown P, DeHerrera M, Durlam M, Fuchs E, Gajewski D, et al. (2004). Demonstrated reliability of 4-mb MRAM. *IEEE Trans. Device Mater. Reliab.* 4: 428.

79 Zhao, W., Zhang Y, Devolder T, Klein JO, Ravelosona D, Chappert C, et al. (2012). Failure and reliability analysis of STT-MRAM. *Microelectron. Reliab.* 52: 1848.

80 Xiangyu D, Xiaoxia W, Guangyu S, Yuan X, Li H, Yiran C. (2008). Circuit and microarchitecture evaluation of 3D stacking magnetic RAM (MRAM) as a universal memory replacement. Design Automation Conference, DAC 2008. 45th ACM/IEEE.

81 Sun G, Dong X, Xie Y, Li J, Chen Y. (2009). A novel architecture of the 3D stacked MRAM L2 cache for CMPs. IEEE 15th International Symposium on High Performance Computer Architecture.

82 Faber L, Zhao W, Klein J, Devolder T, Chappert C. (2009). Dynamic compact model of Spin-Transfer Torque based Magnetic Tunnel Junction (MTJ). 4th International Conference on Design & Technology of Integrated Systems in Nanoscale Era, DTIS'09.

83 Panagopoulos, G., Augustine, C., and Roy, K. (2011). 28-nm 0.08 mm^2/Mb Embedded MRAM for Frame Buffer Memory. Device Research Conference (DRC), 2011 69th Annual.

84 Han SH, Lee JM, Shin HM, Lee JH, Suh KS, Nam KT, et al. (2020). Electron Devices Meeting (IEDM).

85 Naik V.B., Yamane K., Lee T.Y., Kwon J., Chao R., Lim J.H., Chung N.L., Behin-Aein B., Hau L.Y., Zeng D., Otani Y., Chiang C., Huang Y., Pu L., Jang S.H., Neo W.P., Dixit H., Goh S.K.L.C., Toh E.H., Ling T., Hwang J., Ting J.W., Low R., Zhang L., Lee C.G., Balasankaran N., Tan F., Gan K.W., Yoon H., Congedo G., Mueller J., Pfefferling B., Kallensee O., Vogel A., Kriegerstein V., Merbeth T., Seet C.S., Ong S., Xu J., Wong J., You Y.S., Woo S.T., Chan T.H., Quek E. and Siah S.Y. (2020). JEDEC-Qualified Highly Reliable 22nm FD-SOI Embedded MRAM For Low-Power Industrial-Grade, and Extended Performance Towards Automotive-Grade-1 Applications. Electron Devices Meeting (IEDM).

86 Shih Y.C., Lee C.F., Chang Y.A., Lee P.H., Lin H.J., Chen Y.L., Lo C.P., Lin K.F., Chiang T.W., Lee Y.J., Shen K.H., Wang R., Wang W., Chuang H., Wang E., Chih Y.D. and Chang J. (2020). A Reflow-capable, Embedded 8Mb STT-MRAM Macro with 9nS Read Access Time in 16nm FinFET Logic CMOS Process. Electron Devices Meeting (IEDM).

87 Edelstein D, Rizzolo M, Sil D, Dutta A, DeBrosse J, Wordeman M, et al. (2020). A 14 nm Embedded STT-MRAM CMOS Technology. Electron Devices Meeting (IEDM).

5

Spin-Orbit Torque (SOT) Materials and Devices

Yucai Li[1,2], *Kevin William Edmonds*[3], *and Kaiyou Wang*[1,2,4]

[1]*State Key Laboratory of Superlattices and Microstructures, Institute of Semiconductors, Chinese Academy of Sciences, Beijing, China*
[2]*College of Materials Science & Opto-Electronic Technology and Center for Excellence in Topological Quantum Computation, University of Chinese Academy of Science, Beijing, China*
[3]*School of Physics and Astronomy, University of Nottingham, Nottingham, United Kingdom*
[4]*Beijing Academy of Quantum Information Sciences, Beijing, China*

CHAPTER MENU

5.1 Spin-Orbit Coupling in Materials

SOT originates from the interaction between local magnetization and spin currents, induced by SOC effects in materials. SOC is a relativistic effect of a particle's spin with its motion in an electric field.

In a hydrogen-like atom, considering the rest frame of an electron with momentum $\boldsymbol{P}$, the nucleus undergoes a circular motion around the electron, which induces a magnetic field

$$\boldsymbol{B} = -\frac{\boldsymbol{v} \times \boldsymbol{E}}{c^2} \tag{5.1}$$

where $\boldsymbol{v}$ is the velocity of the electron, c is the speed of light, $\boldsymbol{E} = Ze\boldsymbol{r} / 4\pi\varepsilon_0 r^3$ is the electrical field, Z is the atomic number, $\boldsymbol{r}$ is the position vector, ε_0 is the permittivity of vacuum and e is the elementary charge. Due to the angular momentum $\boldsymbol{L} = \boldsymbol{r} \times \boldsymbol{P}$:

Spintronics: Materials, Devices and Applications, First Edition. Edited by Kaiyou Wang, Meiyin Yang, and Jun Luo.

$$\boldsymbol{B} = \frac{Ze}{4\pi\varepsilon_0 mc^2 r^3}\boldsymbol{L} \tag{5.2}$$

The spin magnetic moment of the electron is:

$$\boldsymbol{\mu}_s = -\frac{g_s \mu_B}{\hbar}\boldsymbol{S} \tag{5.3}$$

where g_s is the electron-spin Lande g factor, μ_B is the Bohr magneton, $\hbar$ is the reduced Planck constant and $\boldsymbol{S}$ is the spin angular momentum of the electron.

SOC is the interaction between the electron spin magnetic moment and the magnetic field, given by $H = -\mu_s.B$. Considering the relativistic effect in coordinate transformation, the contribution of SOC to the Hamiltonian can be expressed as [1]

$$E = Z^4 \frac{\mu_0 g_0 \mu_B^2}{8\pi n^3 a_0^3} \frac{\left(j(j+1) - l(l+1) - s(s+1)\right)}{l(l+1/2)(l+1)} \tag{5.4}$$

where l, s and j are the orbital, spin, and total angular momentum quantum numbers, respectively, n is the principal quantum number, a_0 is the Bohr radius. It is noticed that the energy of SOC is directly proportional to the fourth power of Z; the atoms further down a particular column of the periodic table (including heavy metals, such as Pt, Ta, W) have larger SOC.

Although the contribution of SOC to the total energy is a small perturbation, it has an important influence on the band structure and Fermi surface in the solids. Inversion symmetry breaking, which could induce a built-in electric field in materials, generates SOC terms in the Hamiltonian of charge carriers. Various types of SOC effects have been observed in solids, which are dependent on the origin of inversion symmetry breaking. The main types related to SOT are the Dresselhaus effect, the Rashba effect, and the Dzyaloshinskii–Moriya interaction (DMI).

The Dresselhaus effect was first observed by Dresselhaus [2] in noncentrosymmetric crystalline zinc-blende semiconductors in 1955. It originated from the bulk inversion asymmetry. The Hamiltonian of cubic Dresselhaus SOC could be written down as [3]

$$H_d = (\gamma / \hbar)((p_y^2 - p_z^2)p_x\sigma_x + c.p) \tag{5.5}$$

where γ is the gyromagnetic ratio, p_x, p_y, p_z are the components of momentum along [100], [010], and [001] crystal orientation, respectively; σ_x, σ_y, σ_z are the Pauli matrices for the spin, and *c.p.* denotes circular permutations of indices.

The deformation introduced by strain could also result in the breaking of symmetry. For convenience, considering the strain to be along the [001] direction, the cubic Dresselhaus SOC can be written as:

$$H_d = (\gamma p_z^2 / \hbar)(p_x\sigma_x - p_y\sigma_y). \tag{5.6}$$

The interface and surface of materials also can induce inversion symmetry breaking, which can generate a built-in internal electric field at the interface or surface. When the itinerant electrons flow in the built-in internal electric field, they experience an effective magnetic field, the contribution to the Hamiltonian called Rashba SOC [4], which can be written as [3, 5]

$$H_R = (\alpha_R / \hbar)(p_x\sigma_y - p_y\sigma_x) \tag{5.7}$$

where α_R is the Rashba parameter.

Both Dresselhaus and Rashba effects can induce spin-momentum locking and have an influence on the energy band. The spin textures at the Fermi surface, induced by the energy band spin-orbit splitting, are shown in Figures 5.1a and 5.1b. The spin-to-charge conversion, induced by the Rashba effect at the interface or surface, is called the Rashba-Edelstein effect, in which the charge current is transferred to spin current by Rashba SOC, as illustrated in Figure 5.1c.

For a ferromagnet, the SOC induced by the inversion symmetry breaking can influence the chirality of the magnetic structure [5]. The DMI, also known as the antisymmetric exchange, favors spin canting and influences the magnetization dynamics. The DMI was first introduced to explain weak ferromagnetism in antiferromagnetic insulators such as Fe_2O_3 [6, 7]. More recently, the DMI in the non-centrosymmetric magnetic metals and magnetic multilayers has been investigated. The contribution of DMI to the Hamiltonian can be written as $D_{ij}\cdot(S_i \times S_j)$, where S_i and S_i are the neighbor spins, and the orientation of the D_{ij} vector is determined by the electron wave function and the symmetry of the crystal structure [8, 9]. The typical DMI in a ferromagnet (FM)/heavy metal (HM) bilayer is displayed in Figure 5.1e.

The spin Hall effect (SHE) resulting from the SOC is another important source of SOT. The charge-to-spin conversion induced by the SHE can generate a spin-polarized current. When the electrons flow in the materials with strong SOC, such as heavy metals, the electrons with opposite sign spins will accumulate at the opposing interface boundaries [10], as illustrated in Figure 5.1f. This is an effective way to generate and inject a spin-polarized

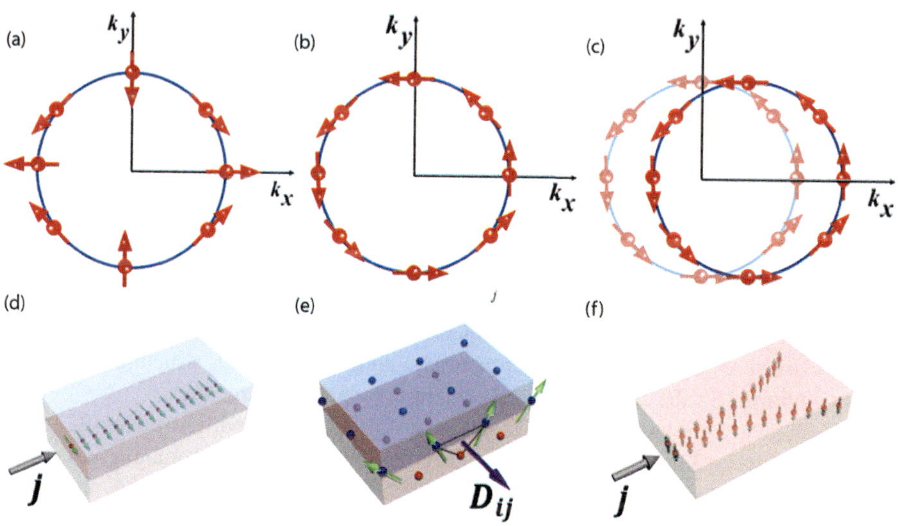

Figure 5.1 Illustration of spin-orbit coupling phenomena. (a), (b), Spin texture induced by Dresselhaus SOC when strain is along [001] direction and Rashba SOC in the momentum space. (c), The nonequilibrium spin distribution due to electrical current and the Rashba-Edelstein effect. (d), Schematic of the Rashba-Edelstein effect. (e), Schematic of DMI at the interface of the bilayer. (f), Schematic of the spin Hall effect.

current. Dyakonov and Perel [11] first predicted the SHE in 1971. More than 30 years later, the experimental observation of the SHE was reported by Awschalom et al. [12] and Wunderlich et al. [13] independently. The origin of the SHE can be attributed to either an extrinsic or intrinsic origin. The extrinsic mechanisms of skew scattering [14] and side jump [15] are related to spin-dependent scattering by impurities. The intrinsic mechanism is related to a Berry phase acquired by the electron trajectories in a complex spin-orbit split band structure [13]. The charge-to-spin conversion efficiency of materials can be quantified by the spin Hall angle θ_{SH}, which is defined as

$$\theta_{SH} = \frac{|j_s|}{|j_e|} \tag{5.8}$$

where j_s is the spin current density and j_e is charge current density.

To achieve high density SOT devices, samples with perpendicular magnetic anisotropy are needed. The anomalous Hall effect (AHE) could be used to measure the magnetic properties electrically, where the anomalous Hall resistance is directly proportional to the out-of-plane magnetization and caused by SOC.

5.2 Manipulation of Magnetic Materials by SOT

5.2.1 The Mechanism of SOT in Ferromagnets

Miron et al. [16] and Liu et al. have reported the magnetization switching induced by SOT in magnetic multilayers. However, the interpretations of the two groups about the origin of SOT were different. Miron et al. [16] attributed the switching to the Rashba effect yielding an effective magnetic field in Pt/Co/AlO_x multilayers due to the structural asymmetry. The orbital angular momentum from the crystal lattice is transferred to the local magnetization; thus, the magnetization was reversed by current. Liu et al. considered the SHE as the origin of SOT-induced magnetization switching in CoFeB/β-Ta bilayers. The spin-polarized current generated by SHE in the heavy metal was injected into the ferromagnet and the spin angular momentum of the spin-polarized current is transferred to the ferromagnetic CoFeB layer. What's more, Liu et al. [17] also reported SOT magnetization switching in Pt/Co/AlO_x, where the structure was the same as in Miron's work. Liu et al. concluded that the Rashba effect offers no measurable contribution to the magnetization switching. Many works have focused on making clear the contribution of the Rashba effect and SHE to SOT [18–21]. Yang et al. [19] demonstrated a strong damping-like torque from the spin Hall effect and an unmeasurable field-like torque from the Rashba effect in a symmetric Pt/Co/Ni/Co/Pt multilayer. The switching efficiency in the symmetric structure was not decreased compared with the Pt/Co/AlO_x structure. The spin Hall effect is dominant for the current-induced switching in HM/FM structures.

Magnetization switching induced by SOT in ferromagnetic layers with in-plane [22] and out-of-plane easy axes [23] have both been widely investigated. But ferromagnetic layers with perpendicular magnetic anisotropy are favorable for higher density MRAM, and thus have attracted more attention due to greater potential for practical applications. Because of the qualitatively similar magnetization switching mechanism in ferromagnetic layers with in-plane and out-of-plane easy axes, we mainly introduce SOT magnetization switching in materials with perpendicular magnetic anisotropy.

The precessional magnetization dynamics of SOT on magnetization can be described by the Landau–Lifshitz–Gilbert (LLG) equation with an additional SOT term

$$\frac{d\boldsymbol{M}}{dt} = -\gamma \boldsymbol{M} \times \boldsymbol{H}_{eff} + \frac{\alpha}{M_s} \cdot \left(\boldsymbol{M} \times \frac{d\boldsymbol{M}}{dt} \right) + \frac{\gamma}{\mu_0 M_s} \boldsymbol{\tau}_{SOT} \tag{5.9}$$

where γ is the absolute value of the gyromagnetic ratio, $\boldsymbol{H}_{eff}$ is the effective magnetic field, α is the Gilbert damping coefficient, and $\boldsymbol{\tau}_{SOT}$ is SOT. The SOT term consists of a field-like torque $\boldsymbol{\tau}_{FL} \sim \boldsymbol{m} \times \boldsymbol{y}$ and a damping-like torque $\boldsymbol{\tau}_{DL} \sim \boldsymbol{m} \times (\boldsymbol{y} \times \boldsymbol{m})$, where $\boldsymbol{m}$ is the magnetization unit vector and $\boldsymbol{y}$ is the unit vector along the direction of spin polarization. The torque $\boldsymbol{\tau}_{FL}$ acts on the magnetization as an effective magnetic field and the torque $\boldsymbol{\tau}_{DL}$ acts as an effective magnetic damping, which is the origin of their names [24]. The damping-like term of STT is parallel or antiparallel to the damping torque, so that the competition between STT and damping induce many precessions around the equilibrium position. Unlike STT, the damping-like term of SOT is perpendicular to the damping torque; thus, SOT does not compete with the damping terms. Therefore, the magnetization switching induced by SOT is faster than that induced by STT.

The in-plane current-induced SOT in typical SOT bilayers (HM/FM) with perpendicular magnetic anisotropy is illustrated in Figure 5.2. The SOT can only drive the magnetization in the in-plane direction, so that after removing the writing current, the magnetization will switch to up or down randomly, due to the energy degeneracy. Thus, the deterministic magnetization switching by SOT often requires an in-plane magnetic field to break the symmetry. However, the requirement for an in-plane magnetic field restricts the application of SOT devices. For the application of SOT device, many works have focused on developing field-free SOT-induced magnetization switching.

5.2.2 Measurement Techniques of SOT

To measure SOT quantitatively, harmonic Hall voltage and spin-torque ferromagnetic resonance (ST-FMR) techniques have been applied to probe the effect of the current on the magnetization. The SOT can induce resonant and non-resonant oscillations of magnetization. Thus, the effective field could be extracted as a function of the magnetization angle, the amplitude, and the phase of the applied current.

The harmonic Hall voltage technique is an electrical measurement to detect SOT. The method utilizes the harmonic response of magnetization in an applied low frequency ac

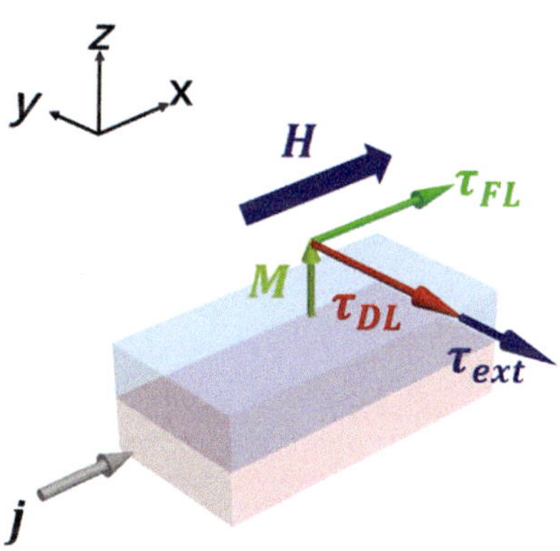

Figure 5.2 Schematics of the magnetic torques in a HM/FM bilayer with external field H and current density j. $\boldsymbol{\tau}_{FL}$ is the field-like SOT and $\boldsymbol{\tau}_{DL}$ is the damping-like SOT. τ_{ext} is the torque induced by the external magnetic field.

current. In 2010, Pi et al. [25] devised the method to measure the field-like torque 40 under the assumption that the damping-like torque and the planar Hall effect (PHE) are negligible. Soon after, this method was improved [26, 27] to investigate a system with AHE and PHE, and it has been an effective way to measure SOT. The schematic illustration of the experimental setup is shown in Figure 5.3. The magnetization oscillations driven by the low frequency current modulate the Hall resistance, and the homodyne mixing of the ac current with the Hall resistance can generate a second harmonic Hall voltage v_{2f}. The effective magnetic field of SOT can be extracted from the second harmonic Hall voltage. The transverse effective magnetic field can be calculated using the following equation when a magnetic field is swept along the y axis

$$\Delta H_T = \frac{-2\dfrac{\partial v_{2f}}{\partial Hy}}{\dfrac{\partial^2 v_f}{\partial H_y^2}} \tag{5.10}$$

where H_y is the magnetic field along the y axis and is the first harmonic Hall voltage. The longitudinal effective magnetic field can be calculated using the following equation when a magnetic field is swept along the x axis

$$\Delta H_L = \frac{-2\dfrac{\partial v_{2f}}{\partial H_x}}{\dfrac{\partial^2 v_f}{\partial H_x^2}} \tag{5.11}$$

where H_x is the magnetic field along the x axis. The planar Hall resistance induced by the PHE also contributes to the Hall voltage. Considering the effect of the PHE, the equation can be modified as follows

$$H_{DL(FL)} = -\frac{\Delta H_{L(T)} \pm 2\xi \Delta H_{T(L)}}{1-4\xi^2} \tag{5.12}$$

where $H_{DL(FL)}$ is the damping-like effective field (field-like effective field), and ξ is the ratio of the planar Hall resistance and the anomalous Hall resistance.

ST-FMR uses a radio-frequency current to excite the magnetization oscillation. The magnetization driven by the radio-frequency (rf) current exhibits ferromagnetic resonance. This was utilized to measure the STT in magnetic tunnel junctions (MTJ) [28], and was more

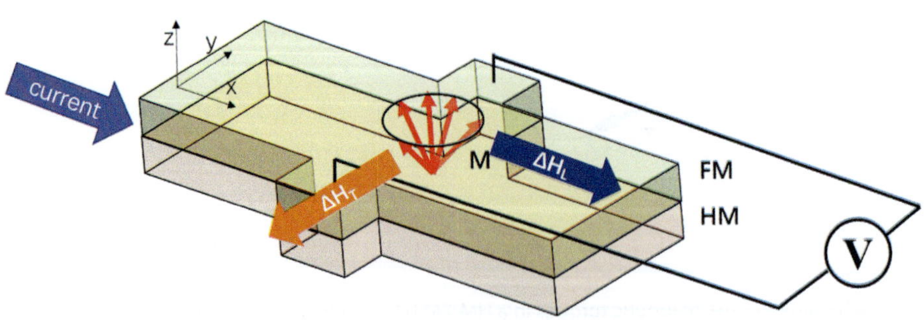

Figure 5.3 The schematic illustration of harmonic Hall voltage measurement.

recently extended to measure SOT in magnetic multilayers [29]. The circuit used for ST-FMR is illustrated in Figure 5.4. When a radio-frequency (RF) current is applied to the device, there is an RF anisotropic magnetoresistance (AMR) due to the oscillating magnetization. The applied RF current is assumed to have the form $I_{rf} = I\cos(\omega t)$ with frequency $\omega / 2\pi$. Then the time-dependent resistance can be written as $R(t) = R_0 + \Delta R\cos^2\theta(t)$, where R_0 is the resistance when the RF current is perpendicular to the magnetization, ΔR is the AMR. $\theta(t)$ is the angle between the RF current and the magnetization, given by $\theta(t) = \theta_H + \theta_c(\omega t + \delta)$, where θ_H is the angle between the magnetic field and the RF current, θ_c is the ferromagnetic resonance angle, and δ is the spin-resonance phase between magnetization and SOT. The voltage could be calculated as follows:

$$\begin{aligned} V(t) = I(t)R(t) &\approx (IR_0 + I\Delta R\cos^2\theta_H)\cos(\omega t) \\ &- I\Delta R\sin 2\theta_H . \frac{\theta_c\cos(\omega t + \delta)}{2} \\ &- I\Delta R\sin 2\theta_H . \theta_c \frac{\cos\delta}{2} \end{aligned} \tag{5.13}$$

It should be noted that the last term of the voltage is independent with time. This rectified voltage induced by the mixing of RF current and RF AMR includes the information of SOT. The symmetric Lorentzian part of the rectified voltage is proportional to the damping-like torque and the antisymmetric part includes the field-like torque and Oersted field.

5.2.3 Field-Free SOT Magnetization Switching in Ferromagnets

In 2014, Yu et al. [30] first observed the field-free SOT-induced magnetization switching by introducing a lateral structural asymmetry in $Ta/Co_{20}Fe_{60}B_{20}/TaO_x$. They used a wedge-shaped TaO_x capping layer to produce a new torque induced by the Rashba effect. An effective magnetic field along z axis determined the final states of the current-induced magnetization switching. This work provides a way to achieve field-free magnetization-switching by breaking the symmetry of the material. Analogously, You et al. [31] broke the perpendicular magnetic anisotropy and made the magnetization have a small in-plane component, and field-free magnetization switching due to the exchange interaction was also observed. Recently, taking advantage of the chiral coupling between out-of-plane and in-plane magnets, Luo et al. achieved field-free magnetization switching [32].

Although the structure symmetry breaking of the device is an effective way to achieve field-free magnetization switching, it presents a new problem for practical application because of the

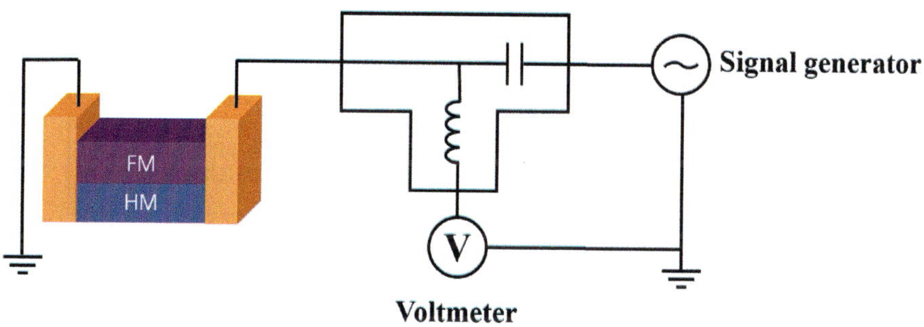

Figure 5.4 The schematic illustration of ST-FMR measurement.

difficulties in fabrication or growth of high quality homogeneous structures. In 2016, many works demonstrated field-free SOT magnetization switching by introducing an antiferromagnetic layer [21, 33]. The exchange coupling between ferromagnetic and antiferromagnetic layers induces an exchange bias field, which could be used to replace the external in-plane magnetic field.

Combining the STT and SOT in MTJ, Wang et al. [34] demonstrated the field-free magnetization switching. On the one hand, SOT could lower the switching current densities and improve the lifetime of the MTJ. On the other hand, the additional STT removes the requirement for an in-plane assisting magnetic field.

However, the chirality of SOT-induced magnetization switching is hard to change once the multilayers are deposited. Programmable devices, in which the switching chirality can be reset by an external electric field, may help to meet the rapidly growing demand for big data processing. In 2017, Cai el al. [35] reported voltage-controlled SOT magnetization switching in a PMN-PT/Pt/Co/Ni/Co/Pt heterostructure, as shown in Figure 5.5a–c. The chirality of SOT magnetization switching depends on the polarization of the ferroelectric substrate. Before the PMN-PT was polarized, the magnetization switching under an in-plane magnetic field is the same as in similar samples grown on a Si/SiO_2 substrate. After polarizing the PMN-PT with 500V voltage along x direction in zero magnetic field, as shown in Figure 5.5b, a clockwise magnetization switching loop was reproducibly observed. The loop could be tuned to counterclockwise by applying a –500V voltage to change the polarization of the PMN-PT, which is shown in Figure 5.5c. The origin of the voltage-controlled SOT magnetization switching is that the gradient of spin densities along the current direction, caused by the polarization of the ferroelectric substrate, could induce a new torque, which would realize the deterministic switching. By shrinking the device

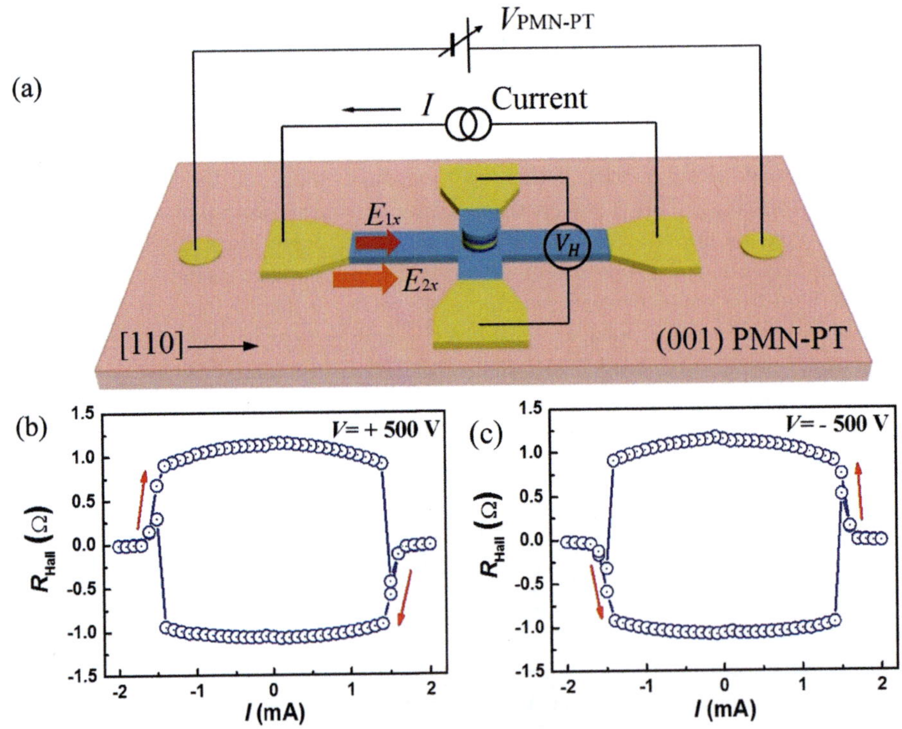

Figure 5.5 Magnetization switching induced by the polarization of a PMN-PT substrate. (a), The structure of the device; (b), (c) The deterministic magnetization switching after applied 500V and –500V voltages [35].

size, the spin logic functions were demonstrated using more than one order of magnitude lower of the electric field controlled current-induced magnetization switching [36]. The spin current gradient created using the wedge shape of the Pt/Co/Pt structure was used to realize current-induced magnetization switching, which further confirmed that the gradient spin current is an effective method for introducing controllable field-free switching [37].

By annealing part of the sample with a focused laser beam after deposition of a multilayer, a lateral Pt/Co composition gradient can be formed within the ferromagnetic layer. The deterministic magnetization switching using lateral SOT was observed in such structures by Cao et al. [38], where the switching direction is sensitive to the position of the laser spot. The lateral SOT originates from the lateral Pt–Co asymmetry, which may offer a pathway to all electrical manipulation of spins without external out-of-plane spin current injection, and provides further insights into the in-plane current-induced SOT in such asymmetric structures. Based on these devices, Boolean logic gates of AND, OR, NAND, and NOR as well as a spin-orbit half adder containing an exclusive-OR gate are obtained [39].

In the magnetic trilayers, SOTs induced by the spin-orbit precession could also be used to manipulate the magnetization [40]. When the spin currents generated in the bottom ferromagnetic layer, with in-plane magnetization, transit the spacer layer, the spin currents will precess around the spin-orbit field at the interface and the orientation will be changed. Due to the spin-orbit precession, the in-plane polarized spin current transferred to the top ferromagnet layer will have the out-of-plane spin polarization component, which could switch the top ferromagnet without an external magnetic field. Because the direction of spin current polarization depends on the magnetization of the bottom layer, the deterministic switching direction of the magnetization could be tuned by changing the magnetization of the bottom layer.

Another programmable magnetization switching method was also developed using interlayer exchange coupling [41]. The effective magnetic field induced by the Ruderman–Kittel–Kasuya–Yosida (RKKY) interaction between two ferromagnetic layers was used to achieve field-free SOT magnetization switching [42]. In the FM1(in plane)/HM/FM2(out of plane) structure, the interlayer exchange coupling between the two ferromagnetic layers could generate an effective field, and both the ferromagnetic layers could be switched independently without any applied magnetic field [41]. The direction of an effective magnetic field depends on the magnetization of the ferromagnetic layers, which determines the chirality of magnetization switching. The magnetization of the in-plane ferromagnetic layers could be reversed by SOT. As a result, the effective magnetic field induced by the interlayer coupling on the ferromagnetic layers with perpendicular magnetic anisotropy could be tuned electrically and a programmable device could be achieved in this structure.

5.2.4 Domain Wall and Skyrmion Motion Driven by SOT

In addition to the magnetization switching, SOT can be used to drive magnetic domain wall and skyrmion motion. Magnetic domain wall is the transitional interface between two magnetic domains with different magnetization [43–45]. When a magnetic field is applied to the ferromagnet, the domain expands and results in domain wall motion. Usually, the field-driven domain wall motion has no preferred direction. However, SOT can drive domain wall motion in a specific direction, which depends on the direction of the current, the spin polarization, and the chirality of the domain wall [46–48]. In 2010, Miron et al. [16] demonstrated domain wall nucleation and propagation in Pt/Co/AlO_x nanowires. The velocity of the domain wall motion could reach 400 m/s. In this asymmetric structure, the DMI induced by the interface could stabilize Néel domain walls and determine the chirality of the domain wall [49, 50].

As shown in Figure 5.6, the magnetization of the Néel domain walls is orthogonal to the direction of the spin polarization by the SHE, which is the origin of the efficient SOT-driven domain wall motion. Multilayer stacks of Pt/CoFe/MgO and Ta/CoFe/MgO show opposite directions of domain wall motion [51, 52], which could be explained by the different signs of the spin Hall angles in Pt and Ta. The magnetization dynamics of domain walls driven by SOT has been observed by time-resolved scanning transmission X-ray microscopy [53].

Magnetic skyrmions are quasiparticles with special topological structure. Skyrmion motion, driven by STT with an ultralow critical current density of 65Acm^{-2}, was observed in FeGe in 2012 [54]. Due to the potentially high density and ultralow power consumption [54], magnetic skyrmions have attracted lots of attention. The nucleation and propagation of skyrmions induced by the spin-polarized current from SHE was predicted [55]. The direction of SOT-driven skyrmion motion is not parallel to the current direction. There is a transverse component of the velocity, which is called the skyrmion Hall effect. The longitudinal component is determined by the chirality of the skyrmion and the spin Hall angle, and the transverse component depends on the central magnetization of the skyrmion. SOT-induced skyrmion motion was first observed in Ta/CoFeB/TaOx multilayers by magnetic optical Kerr microscopy [56]. Although the skyrmion trajectory is consistent with the theory, the relation between the velocity and the size of the skyrmion, and the relation between the skyrmion Hall angle [56–58] and the velocity are not explained by theory [57, 59]. More investigations are needed to understand the skyrmion motion.

The skyrmion Hall effect restricts the application of the skyrmion. The antiferromagnetic skyrmion was expected to solve the problem [60–62]. Because in a different magnetic lattice, the central magnetization is the opposite, thereby the transverse components are also opposite. The transverse components of the coupled skyrmion could be counteracted.

5.2.5 Manipulation of Antiferromagnets by SOT

SOT can also be used to manipulate the staggered magnetization of antiferromagnets effectively. The net magnetization of antiferromagnets is zero and is hard to influence with an

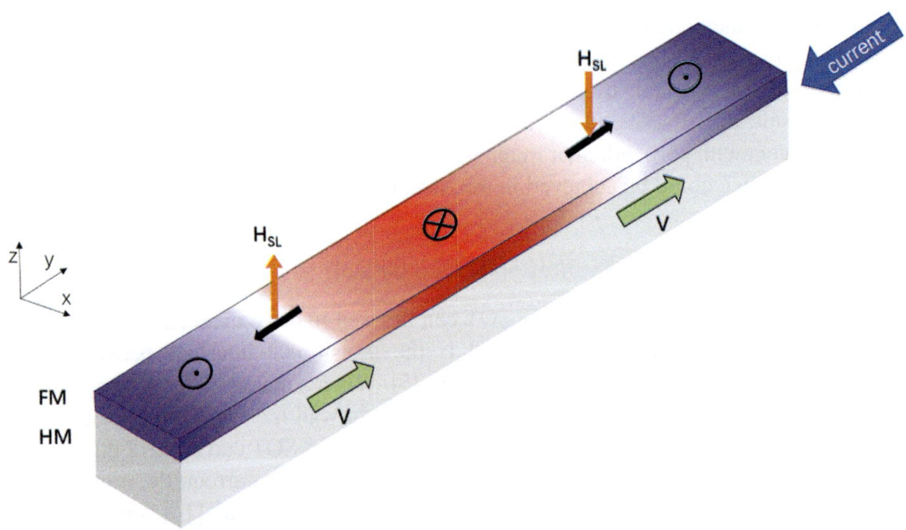

Figure 5.6 The dynamics of current-driven domain wall motion. The Néel domain walls in Pt/CoFe/MgO have left-hand chirality and the current drives the domain walls' motion in the same direction.

external magnetic field. Thus, the antiferromagnets are hard to detect and manipulate. But from another point of view, the antiferromagnets also have potential application in higher density and field-insensitive memory devices. The antiferromagnets have great advantages in high speed and high frequency devices, where the precession frequency is in the THz range, much higher than in ferromagnets.

Synthetic antiferromagnets, consisting of two ferromagnets with antiparallel coupling, are easier to manipulate and combine the advantages of natural antiferromagnets and ferromagnets. In fact, synthetic antiferromagnets have been widely used in high density storage, because the influence of stray field could be greatly reduced. The speed of the domain wall motion driven by SOT in synthetic antiferromagnets is much faster than in ferromagnets, which has been demonstrated in Co/Ni/Co/Ru/Co/Ni/Co multilayers [63]. The exchange coupling between the two ferromagnetic layers could stabilize a chiral Néel domain wall, and therefore much higher velocities could be achieved. The SOT-induced magnetization switching in synthetic antiferromagnets also has been demonstrated and the mechanism is similar to in ferromagnets [64, 65].

The SOT-induced switching of natural antiferromagnets with a special crystal structure was predicted by Zelezny et al. [66]. The two spin sublattices of the Mn_2Au have broken inversion symmetry and form inversion partners, so that a charge current can induce a local nonequilibrium spin polarization of opposite sign at each sublattice site. The total spin polarization and SOT is zero due to the antiferromagnetic structure. However, SOT with staggered components could be utilized to change the antiferromagnetic order. Later, Wadley et al. [67] experimentally demonstrated the antiferromagnetic order switching by SOT in CuMnAs, whose structure shares the combined time-reversal and space-inversion symmetry of Mn_2Au. The antiferromagnetic order could be switched reversibly between orthogonal easy axes by SOT, which was detected by anisotropic magnetoresistance as well as x-ray photoemission electron microscopy. Similar results were later observed in Mn_2Au [68, 69]. In addition, electrical switching using THz electrical pulses in CuMnAs was demonstrated by Olejnik et al. [70].

The manipulation of the antiferromagnetic order by SOT in a heavy-metal/antiferromagnet heterostructure has been the focus of many works. Reichlova et al. [71] measured SOT in Ta/IrMn/CoFeB, using the second harmonic resistance technique and indicating the complex interplay of antiferromagnetic order and SHE. In contrast to the case of antiferromagnets, such as Mn_2Au and CuMnAs, the damping-like torque orients the Néel order towards the direction of the writing current in the biaxial antiferromagnetic insulators/heavy bilayer system (NiO(001)/Pt) [72]. The topological antiferromagnetic state of antiferromagnetic Weyl metal Mn_3Sn [73] could be manipulated by SOT in the Mn_3Sn/Pt structure.

Besides the antiferromagnetic order, the exchange bias between antiferromagnet and ferromagnet also could be manipulated by SOT. Lin et al. [74] demonstrated concurrent switching of ferromagnetic magnetization and exchange bias in a Pt/Co/IrMn multilayer, which was induced by the interfacial spins tuned by SOT [75].

5.3 SOT Materials

5.3.1 Traditional Materials

Most of the reported studies of SOT concern a ferromagnetic layer on a nonmagnetic metal layer, with an oxide or metal capping layer. Enhancing SOT-induced magnetization switching efficiency to decrease power consumption is particularly important for applications. The key point is to search for materials with high charge-to-spin conversion efficiency, which could be quantified by the spin Hall angle. As we introduced before, SOC in heavy metals is usually strong, and was used to generate spin-polarized current for SOT in the majority of studies to date [76]. The spin Hall angle of heavy metals, such as Pt [77], Ta, Pd [78], W [79], Hf [80], etc., has been widely studied. The thickness and crystalline structure both have impact on the spin Hall angle [81]. Moreover, impurity scattering can enhance the spin Hall angle in materials due to the extrinsic SOC. For example, Cu with Ir [82] or Bi [83] doping shows a strong spin Hall effect. Sandwiching a FM between two heavy metals with opposite spin Hall angles is also an effective way to enhance SOT efficiency, as the two heavy metal layers give rise to parallel SOT at the interfaces [84].

Not limited to nonmagnetic materials, the ferromagnets and antiferromagnets were also investigated for spin generation. In 2013, Miao et al. [85] reported the inverse spin Hall effect in a permalloy/yttrium iron garnet, and Du et al. [86] measured the spin Hall angle of 3d transition metals, such as antiferromagnetic Cr and FeMn. Zhang et al. [87] reported that the spin Hall angle of 5d metal alloys, such as antiferromagnetic IrMn and PtMn, is larger than the 4d metal antiferromagnetic alloy PdMn and 3d alloy FeMn. Therefore, the spin Hall angle in antiferromagnetic alloys is determined by the nonmagnetic elements.

5.3.2 Interfacial Engineering

Interfacial engineering is an effective way to tune the spin Hall angle and SOT efficiency. The interfacial asymmetry could be modified by inserting interfacial layers on capping layers, to change the magnitude and direction of SOT [88–90]. In 2014, Fan et al. [88] reported the influence of a Cu spacer layer on the ratio of the damping-like and field-like SOT in Pt/Cu/CoFeB. Inserted layers of other metals, such as Hf, were also reported to enhance the damping-like torque effectively due to the decrease of the spin-mixing conductance [91]. In 2016, Qiu et al. [92] demonstrated that SOT in HM/FM/Ru is larger than that of HM/FM/MgO. In the HM/FM/MgO, if the thickness of the FM layer is thinner than the spin dephasing length, the spin current will be reflected by the FM/MgO interface, which will lower SOT efficiency. By replacing the MgO layer with Ru, SOT efficiency could be enhanced due to the absorption of spin current by the Ru layer.

Oxidation also is an effective way to tune SOT. In 2015, Qiu et al. [93] reported the effect of the top SiO_2 layer on SOT. By controlling the oxidation of the CoFeB in Pt/CoFeB/MgO/SiO_2 heterostructures, the sign and magnitude of SOT could be tuned. When the thickness of the SiO_2 layer was up to 1.5 nm, the sign of the effective field and magnetization switching sequence changed. This means interfacial SOT could be controlled by oxidation effectively. Emori et al. [94] demonstrated the enhancement of damping-like torque through controlling the oxidation in Pt/Co/GdO_x by voltage gating. The migration of the O^{2-} ion from GdO_x could be driven by an electrical field and thus the oxidation in the interface could be modified by voltage, which could have an influence on SOT. The magnetization anisotropy could be affected by the voltage (known as voltage-controlled magnetic anisotropy, VCMA), which is usually induced by the interface modification by voltage. Combining VCMA and magnetization switching by SOT, the critical current and power consumption could be effectively lowered. What's more, through VCMA, a complementary logic operation was demonstrated [95–97].

5.3.3 Oxide Heterostructures

The investigation of SOT is not only restricted to conventional magnetic materials such as Co, CoFeB. At the interface between the two kinds of insulator $LaAlO_3$ and $SrTiO_3$, there is a strong Rashba SOC because of the symmetry breaking and prominent spin-charge interconversion effects which can be observed [98]. Through spin-pumping, Lesne et al. [99] demonstrated the spin-to-charge conversion in the conductive 2D electron gas at the interface. Soon after, Wang et al. [100] demonstrated the charge-to-spin conversion at the $LaAlO_3$/$SrTiO_3$ interface. The spin Hall angle measured by spin-torque ferromagnetic resonance is up to 6.3 at room temperature.

In 2019, Liu et al. [101] reported SOT-induced magnetization switching in all-oxide heterostructures. In $SrIrO_3$/$SrRuO_3$ bilayers grown on a $SrTiO_3$ substrate, because the easy axes of the $SrIrO_3$ have an in-plane component, the spin current generated by the $SrRuO_3$ could switch the magnetization of the $SrIrO_3$ in the absence of any external magnetic field.

5.3.4 The van der Waals Materials and Topological Materials

The emergence of the van der Waals ferromagnets, with high quality interfaces and reduced thickness, provides a new platform for investigation of SOT. Wang et al. [102] and Alghamdi et al. [103] reported SOT-induced magnetization switching of Fe_3GeTe_2, while Ostwal et al. reported SOT-induced magnetization switching of $Cr_2Ge_2Te_6$ [104]. This efficient SOT-induced switching proves that van der Waals ferromagnets have the potential for low-power spintronic applications.

Also, 2D materials have the potential to produce a large spin Hall angle: transition metal dichalcogenides with strong SOC were investigated for spin generation. The Edelstein effect in WTe_2 implies a potential to convert charge current to spin current, which could induce SOT. MacNeill et al. [105] and Shi et al. [106] reported magnetization switching in WTe2/ferromagnet heterostructures. The critical current density to switch magnetization in a permalloy layer is $2.96 \times 10^5 A\,cm^{-2}$, which demonstrates the capability of WTe_2 to efficiently manipulate magnetization.

Topological materials with strong SOC can convert charge current to spin-polarized current efficiently [107]. Topological insulators, a class of topological material, are insulating in the interior but metallic on the surface, induced by strong SOC [108, 109]. Due to the Rashba spin texture in the surface state of 3D topological insulators, the momentum of the electrons and the spin are locked [110, 111]. When a voltage is applied to the topological insulators, most of the current will flow through the surface states, so that nonequilibrium spin will be accumulated due to spin-momentum locking. In 2014, Mellnik et al. [112] reported the spin current generated by the Rashba-Edelstein effect in topological insulators Bi_2Se_3, and they measured SOT in a Bi_2Se_3/Py structure. Soon after, Fan et al. [113] showed magnetization switching in a $(Bi_{0.5}Sb_{0.5})_2Te_3/(Cr_{0.08}Bi_{0.54}Sb_{0.38})_2Te_3$ heterostructure at 1.9K, where $(Cr_{0.08}Bi_{0.54}Sb_{0.38})_2Te_3$ is a ferromagnetic topological insulator. The effective field to current ratio is almost three orders of magnitude larger than in typical heavy metal/ferromagnet structures. The high charge to spin conversion efficiency of topological insulators was verified by N.H.D. Khang et al. [114], who demonstrated ultralow-power SOT-induced magnetization switching in a BiSb/MnGa bilayer with perfect perpendicular magnetic anisotropy at room temperature. The BiSb spin Hall angle was estimated to be up to 52, and critical currents for magnetization switching down to 1.5×10^6 Acm^{-2} were obtained.

However, the high quality topological insulators are usually grown by molecular beam epitaxy, which is expensive and inefficient. D. Mahendra et al. [115] demonstrated the use of the industry-friendly sputtering method to deposit BiSe/CoFeB, and the critical current was even reduced to 4.3×10^5 Acm^{-2} at room temperature, which indicates the topological insulators are good candidates to optimize SOT efficiency.

We summarize the charge-to-spin conversion efficiency of various materials in Table 5.1. Comparison of material properties may provide clues to find better materials to act as spin generation sources in SOT devices with high efficiency.

Table 5.1 *T* is the temperature for the measurement with RT for room temperature, λ_{Sd} is the spin diffusion length, σ_{NM} is the conductivity, θ_{SH} is the spin Hall angle, Method is the measuring method, NLD = nonlocal detection, SP = spin pumping, ST-FMR = spin torque ferromagnetic resonance, SMR = Spin Hall magnetoresistance, SHM = second harmonic method.

Materials	*T*(k)	λSd(nm)	σNM($10^6\Omega^{-1}m^{-1}$)	θSH(%)	Method	Reference
Al (12 nm)	4.2	455 ± 15	10.5	0.032 ± 0.006	NLD	[77, 116]
Al (25 nm)	4.2	705 ± 30	17	0.016 ± 0.004	NLD	[77, 116]
Cr	RT	13.3	/	–5.1 ± 0.5	SP	[86]
Mn	RT	10.7	/	–0.19 ± 0.01	SP	[86]
Ti	RT	13.3	/	-0.036 ± 0.004	SP	[86]
V	RT	14.9	/	1 ± 0.1	SP	[86]
Au	4.5	65	48.3	<2.3	NLD	[117]
	≤10	40 ± 16	25	1.4 ± 0.4	NLD	[83]
	RT	86 ± 10	37	11.3	NLD	[118]
	RT	83	37	3	NLD	[119]
	RT	35 ± 4	28	7 ± 2	NLD	[120]
	RT	35 ± 3	25.2 ± 0.13	0.35 ± 0.03	SP	[121]
	RT	35 ± 3	5.26	1.6 ± 0.1	SP	[122]
	RT	35 ± 3	7	0.335 ± 0.006	SP	[122]
	RT	35	/	1.1 ± 0.3	SP	[123]
$Au_{99.58}Fe_{0.42}$	RT	33 ± 3	23.3	7 ± 1	NLD	[120]
$Au_{99.05}Fe_{0.95}$	RT	27 ± 3	14.3	7 ± 3	NLD	[120]
$Au_{98.6}PT_{1.4}$	RT	25 ± 3	14.5	12 ± 4	NLD	[124]
$Au_{98.6}Pt_{1.4}$	RT	50 ± 8	16.7	0.8 ± 0.2	NLD	[124]
$Au_{93}W_7$	RT	1.9	1.75	>10	NL and SP	[125]
Bi (interface)	RT	/	2.4 ± 0.3	–(7.1 ± 0.8)	SP	[126]
Bi(volume)	RT	/	50 ± 12	1.9 ± 0.2	SP	[126]
Cu	RT	500	16	0.32 ± 0.03	SP	[76]
CuIr	10	5–30		2.1 ± 0.6	NLD	[82]
$Cu_{99.7}Bi_{0.3}$	≤10	86 ± 17	31.25	–26 ± 11	NLD	[83]
$Cu_{99.5}Bi_{0.5}$	≤10	45 ± 14	19.6	–24 ± 9	NLD	[83]
$Cu_{99.5}Pb_{0.5}$	≤10	53 ± 15	18.52	–13 ± 3	NLD	[83]
Ag		700	15	0.7 ± 0.1	SP	[76]
$Ag_{99}Bi$	≤10	29 ± 6	14.7	–2.3 ± 0.6	NLD	[83]

(Continued)

Table 5.1 (Continued)

Materials	T(k)	λSd(nm)	σNM($10^6\Omega^{-1}m^{-1}$)	θSH(%)	Method	Reference
Mo	10	8.6 ± 1.3	2.8	–(0.8 ± 0.018)	NLD	[127]
	RT	/	4.66	–(0.05 ± 0.01)	SP	[121]
Nb	10	5.9 ± 0.3	1.1	–(0.87 ± 0.2)	NLD	[127]
Pd	10	13 ± 2	2.2	1.2 ± 0.4	NLD	[127]
	RT	5.5	5	1.2 ± 0.3	SP	[128]
	RT	2.0 ± 0.1	3.7	0.8 ± 0.2	ST-FMR	[129]
Pt	10	11 ± 2	8.1	2.1 ± 0.5	NLD	[127]
	RT	1.2	/	8.6 ± 0.5	SP	[130]
	RT	1.4	/	12 ± 4	SP	[123]
	RT	3.4 ± 0.4	6	5.6 ± 1	SP	[131]
	RT	2.1 ± 0.2	3.6	2.2 ± 0.8	ST-FMR	[132]
	RT	2.1 ± 0.2	3.6	8.5 ± 0.9	ST-FMR	[132]
	RT		2.4	4	SP	[133]
	RT	1.5 ± 0.05	0.5–3	11 ± 8	SMR	[134]
	RT	0.9	3.2	14	SMR	[135]
Ta	10	2.7 ± 0.4	0.3	–0.37 ± 0.11	NL	[127]
	RT	1.8 ± 0.7	/	$-2 \pm^{0.008}_{0.015}$	SP	[136]
β-Ta	RT	/	/	–(12 ± 4)	ST-FMR	
	RT	1.5 ± 0.5	/	–(3 ± 1)	SP, SMR	[137]
W	RT	2.1	0.55	–14 ± 1	SP	[76]
	RT	1.4	0.78	–22	SMR	[135]
β-W	RT	/	0.38 ± 0.006	–(33 ± 6)	ST-FMR	[79]
Hf(amorphous)	RT	0.3	0.25	11(Abs)	SMR	[81]
Hf(hexagonal close packed)	RT	1.3	0.64	7(Abs)	SMR	[81]
Re	RT	1.5	3.57	4(Abs)	SMR	[81]
PtMn	RT	0.5 ± 0.1	60.1	6 ± 1	SP	[138]
IrMn	RT	0.7 ± 0.2	37.1	2.5 ± 0.5	SP	[138]
$Ir_{20}Mn_{80}$	RT	/	/	2.9 ± 1.5	ST-FMR	[71]
$Ir_{25}Mn_{75}$	RT	/	0.7	2	ST-FMR	[139]
PdMn	RT	1.3 ± 0.1	44.9	1.5 ± 0.5	SP	[138]
FeMn	RT	1.8 ± 0.5	59.8	0.8 ± 0.2	SP	[138]
Bi_2Se_3	RT	/	/	200–350	ST-FMR	[112]
	RT	/	/	43	SP	[107]
$Bi_{0.9}Sb_{0.1}$	RT	/	25	5200		[114]
$LaAlO_3$/ $SrTrO_3$	/	/	/	630	ST-FMR	[100]
WTe_2	RT	/	/	17	NLD	[140]
WSe_2	RT	/	/	0.3	NLD	[141]
$PtTe_2$	RT	/	0.3–3	5–15	ST-FMR	[142]

5.4 Devices and Application

With the growth of the information era, more and more data need to be stored and processed. Traditional silicon devices are facing big challenges. There are promising opportunities for spintronics devices based on SOT with nonvolatility, high speed, low power consumption, and unlimited endurance. SOT devices have shown the advantages of in-memory computing, neuromorphic computing, probabilistic computing avionics and the internet of things.

5.4.1 SOT-MTJ and SOT-MRAM

SOT could be used for writing information through switching the free layer of MTJ, which is the memory cell of the MRAM [143]. The relative magnetization direction between the free layer and the reference layer in the MTJ determines the tunneling probability, and thereby the tunneling magnetoresistance, which could be used to read the states of the MTJ. The parallel or antiparallel alignment of the magnetizations correspond to the low resistance and high resistance states of the TMR, respectively. The writing endurance of SOT-MTJ of up to 10^{12} cycles is observed experimentally and is predicted up to 10^{16} in theory [53], which satisfies the requirements of information processing applications. As shown in Figure 5.7, SOT-MTJ is a three-terminal device, where the writing channel and the reading channel are separated. The writing current only passes in the plane and does not need to pass through the tunneling barrier. Only a very small current is needed to read the TMR. Thus, the separated channels could improve the endurance and reliability of the device.

MRAM is one of the best candidates in non-volatile memory, but need to improve high density and low cost are required. Currently, the capacity of commercial STT-MRAM reaches 1Gb [144], but with less density than DRAM and slower speed than SRAM. The development of stackable three-dimensional (3D) memory devices is a promising approach to improve the capacity of STT-MRAM. The 3D cross-point STT-MRAM has been

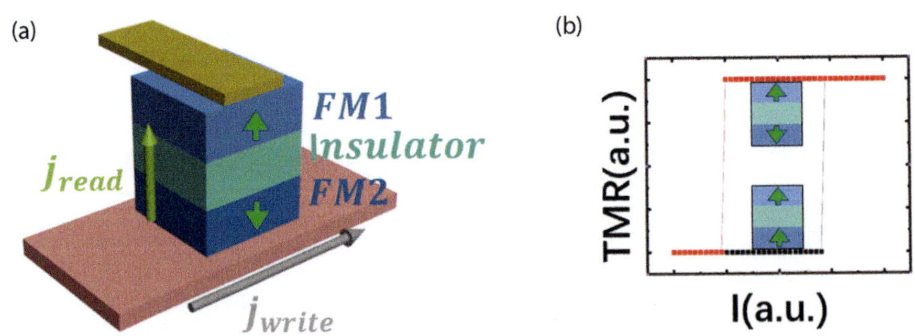

Figure 5.7 Schematics of SOT-MTJ. (a) The writing and reading process of the MTJ. Writing current is applied along the bottom HM layer and switches the free layer FM2. A low current is applied across the junction to read the state of the MTJ. (b) Schematic of the dependence of TMR on the states of SOT-MTJ. The current can switch the magnetization; as a result, the TMR changes with the current.

demonstrated by Huai et al. [145]. To replace the SRAM, 3D SOT MRAM is needed and Cao et al. [146] propose a two-terminal 3D cross-point SOT-MRAM based on perpendicular MTJs. The 1T1S1R configuration is shown in Figure 5.8, which consists of intercalated strong SOC layers with large spin Hall angles and interlayers with exchange bias or an exchange coupling field to drive field-free magnetization switching in the free ferromagnetic layer. Although the in-plane and out-of-plane electrical current are simultaneously generated under programmable bias voltage, SOT-induced magnetization switching is dominant.

5.4.2 In-memory Computing

The primary performance principle of modern computers is the von Neumann architecture. A typical computer consists of a control unit, an arithmetic/logic unit, a memory unit, inputs and an output device. In modern computers, the control unit and arithmetic/logic unit are incorporated within the central processing unit (CPU) [147]. Due to the speed mismatch between CPU and memory, a faster cache is introduced. The information is processed in the CPU first, and then in turn transported to the static random access memories, dynamic random access memories, and hard disk drive. But it does not solve the problem of the speed mismatch from the source. With the development of device fabrication technology, power consumption and even the time cost of data transfer are larger than the computation, which is the biggest bottleneck of the von Neumann architecture [148]. With the demand of big data processing in the information era, this "memory wall," or "von Neumann bottleneck,"

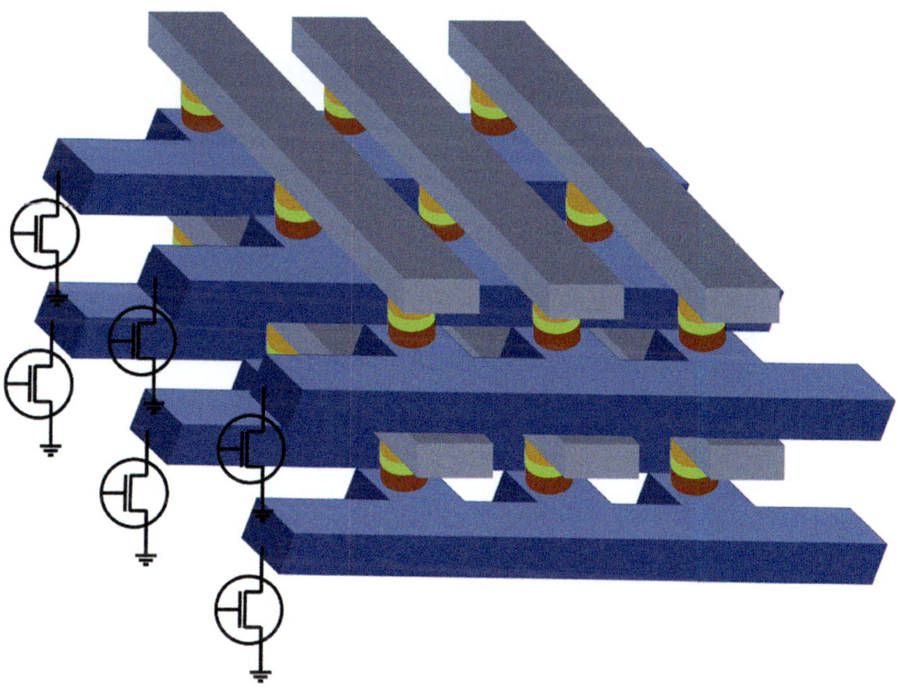

Figure 5.8 Schematic of the proposed 1T1S1R configured field-free 3D SOT-MTJ.

problem needs to be overcome. In-memory computing is one of the solutions proposed to mitigate speed mismatch. Information processing in nonvolatile memories enables greater speed and improved energy efficiency. The candidates for in-memory computing include phase change memory, a memristor, ferroelectric memory, STT and SOT-MRAM [149]. Although the phase change memory and memristor have the advantages of large dynamic range and the feasibility of fabrication, switching variability and resistance instability issues hinder their practical application. The advantages of SOT-MRAM make it a good candidate for in-memory computing. Compared with other candidates, the disadvantages of SOT-MRAM are its three-terminal architecture and dynamic range – the dynamic range of MRAM depends on the TMR. Although 604% TMR [150] has been demonstrated in MTJ, the dynamic range is still much smaller than other candidates. The three-terminal SOT-MTJ architecture requires a larger cell size.

5.4.3 SOT Artificial Intelligence Device

With the development of artificial intelligence (AI), neuromorphic computing has attracted a lot of attention because of its potential for solving AI tasks with much higher efficiency [151]. Several neuromorphic computing models, such as artificial/spiking/recurrent neuron networks, have been demonstrated [152]. Usually, the learning process can be divided into supervised learning with training data "labeled" and unsupervised learning. The primary issues for neuromorphic computing are building artificial synapses and neurons. Hardware which simulates synapses and neurons directly could make AI more efficient. To make an artificial synapse, the multilevel differentiated nonvolatile states are needed to store synaptic weight, which is difficult to achieve in a binary MTJ. However, the domain wall motion driven by SOT is a good candidate for artificial synapses. Recently, Yue et al. [153] reported a spiking neuromorphic system based on magnetic domain wall analog memristors. The TMR in the MTJ depends on the relative magnetization direction of the free layer and fixed layer. The position of the domain wall in the free layer could be continuously driven by current, so the TMR could also change continuously, which could be used to imitate the synapse. Cao et al. [154] demonstrated the synaptic functionality of spike-timing dependent plasticity in experiment, where the interlayer coupling between the in-plane ferromagnetic layer and the perpendicular ferromagnetic layer in a FM/HM/FM structure could induce the multistate magnetization set by SOT by tuning the domain pinning potential. The artificial synapse based on SOT devices has the potential to reduce circuit complexity and enable increased capabilities over an all-CMOS implementation [153].

SOT devices also have potential in probabilistic computing. Recently, quantum computing has drawn much attention due to the potential to solve some of the world's toughest challenges. Quantum computing using qubits to represent a superposition of 0 and 1 could perform inference, invertible logic, sampling and optimization tasks efficiently [155–157]. However, there are many challenges, such as short decoherence time and cryogenic temperature operation. Probabilistic computing is an intermediate approach between conventional computers and quantum computers, which is based on probabilistic bits (p-bits) that fluctuate between 0 and 1 [158, 159]. The probabilistic computing is a practical approach in solving the constraint satisfaction, optimization and sampling based on quantum computing-inspired algorithms [160]. Compared with traditional COMS implementations, probabilistic computing based on spintronic devices potentially offers a smaller size and higher efficiency. By utilizing a low-barrier nanomagnet or single domain nanomagnet, where the magnetization could reverse randomly, the probabilistic computing device and circuit could be realized. In 2019, Borders et al. [160] demonstrated integer factorization in stochastic STT-MTJs. The stochastic output of low barrier nanomagnet devices could also be tuned by SOT. Nanoscale SOT-MTJs with a low barrier showing stochastic behavior may be implemented as three-terminal p-bits. Tunable stochastic behavior, with correlated fluctuations, was also demonstrated in SOT-coupled perpendicular nanomagnets [161].

5.4.4 Internet of Things

The internet of things (IoT) is regarded as an important technology in future life and industry. The IoT not only connects computers and smartphones, but also the hardware between sensors communicating and the cloud. Multiple devices that are used in daily life send information to the cloud and are controlled by the cloud. The logic operation in SOT devices has potential for applications in the embedded nonvolatile memory market, which makes SOT devices good candidates in IoT technologies. The information processing and storage through edge computing could enhance the efficiency [162]. Magnetoresistance sensors have been widely used, and more application scenarios, such as wearable equipment based on flexible SOT devices, are envisioned.

5.5 Conclusion

SOT devices show the advantages of nonvolatility, high speed, high endurance and low power consumption. We have reviewed the development of SOT in mechanisms, materials and devices. SOT can be induced by interfacial and bulk spin orbit coupling, and has shown high efficiency in magnetization switching, domain wall and skyrmion motions. However, to promote SOT devices to practical application, many further efforts need to be made.

Although there are many proposed approaches to demonstrate SOT-induced magnetization switching in the absence of a magnetic field, reliable and integration-friendly field-free SOT-induced magnetization switching methods need to be further developed. With the demands of high density SOT-MRAM, it is important to scale down the device size while maintaining reliability, but also developing the 3D architecture.

To improve SOT efficiency and lower the energy consumption, many materials with a large spin Hall angle have been investigated. To further enhance the energy efficiency, the new materials or structures with high charge-to-spin conversion efficiency should be focused on, which should be industry production-friendly rather than only growing in the lab.

The SOT-MRAM is an excellent candidate for next-generation memory, which could also be used in information processing and storage. In addition, SOT devices have shown potential for applications in in-memory computing, neuromorphic computing, probabilistic computing, avionics and internet of things. However, these attempts are still elementary steps, and further investigations on these topics are extremely necessary. With the combined efforts of the academic and industry scientists, we believe that spin-orbitronics will begin to roll out to large-scale applications in the near future.

References

1 Martin, E.R. (1964). Wiley International Edition. New York.
2 Dresselhaus, G. (1955). *Phys. Rev.* 100: 580.
3 Manchon, A., Koo, H.C., Nitta, J., Frolov, S., and Duine, R. (2015). *Nat. Mater.* 14: 871.
4 Rashba, E. (1959). *Soviet Phys.-Solid State* 1: 368.
5 Soumyanarayanan, A., Reyren, N., Fert, A., and Panagopoulos, C. (2016). *Nature* 539: 509.
6 Dzyaloshinsky, I. (1958). *J. Phys. Chem. Solids* 4: 241.
7 Moriya, T. (1960). *Phys. Rev.* 120: 91.
8 Fert, A. and Levy, P.M. (1980). *Phys. Rev. Lett.* 44: 1538.
9 Bogdanov, A. and Hubert, A. (1994). *J. Magn. Magn. Mater.* 138: 255.
10 Hirsch, J. (1999). *Phys. Rev. Lett.* 83: 1834.
11 Dyakonov, M. and Perel, V. (1971). *Phys. Lett. A* 35: 459.
12 Kato, Y.K., Myers, R.C., Gossard, A.C., and Awschalom, D.D. (2004). *Science* 306: 1910.
13 Wunderlich, J., Kaestner, B., Sinova, J., and Jungwirth, T. (2005). *Phys. Rev. Lett.* 94: 047204.
14 Smit, J. (1958). *Physica* 24: 39.
15 Berger, L. (1964). *Physica* 30: 1141.
16 Miron, I.M., Moore, T., Szambolics, H., Buda-Prejbeanu, L.D., Auffret, S., Rodmacq, B., Pizzini, S., Vogel, J., Bonfim, M., and Schuhl, A. (2011). *Nat. Mater.* 10: 419.
17 Liu, L., Lee, O., Gudmundsen, T., Ralph, D., and Buhrman, R. (2012). *Phys. Rev. Lett.* 109: 096602.
18 Sinova, J., Valenzuela, S.O., Wunderlich, J., Back, C., and Jungwirth, T. (2015). *Rev. Mod. Phys.* 87: 1213.
19 Yang, M., Cai, K., Ju, H., Edmonds, K.W., Yang, G., Liu, S., Li, B., Zhang, B., Sheng, Y., and Wang, S. (2016). *Sci. Rep.* 6: 20778.
20 Amin, V. and Stiles, M. (2016). *Phys. Rev. B* 94: 104420.
21 Haney, P.M., Lee, H.-W., Lee, K.-J., Manchon, A., and Stiles, M.D. (2013). *Phys. Rev. B* 87: 174411.
22 Fukami, S., Anekawa, T., Zhang, C., and Ohno, H. (2016). *Nat. Nanotechnol.* 11: 621.
23 Sheng, Y., Li, Y., Ma, X., and Wang, K. (2018). arXiv preprint arXiv:1806.00964.
24 Manchon, A., Železný, J., Miron, I.M., Jungwirth, T., Sinova, J., Thiaville, A., Garello, K., and Gambardella, P. (2019). *Rev. Mod. Phys.* 91: 035004.
25 Pi, U.H., Won Kim, K., Bae, J.Y., Lee, S.C., Cho, Y.J., Kim, K.S., and Seo, S. (2010). *Appl. Phys. Lett.* 97: 162507.
26 Garello, K., Miron, I.M., Avci, C.O., Freimuth, F., Mokrousov, Y., Blügel, S., Auffret, S., Boulle, O., Gaudin, G., and Gambardella, P. (2013). *Nat. Nanotechnol.* 8: 587.
27 Hayashi, M., Kim, J., Yamanouchi, M., and Ohno, H. (2014). *Phys. Rev. B* 89: 144425.
28 Tulapurkar, A., Suzuki, Y., Fukushima, A., Kubota, H., Maehara, H., Tsunekawa, K., Djayaprawira, D., Watanabe, N., and Yuasa, S. (2005). *Nature* 438: 339.
29 Liu, L., Moriyama, T., Ralph, D., and Buhrman, R. (2011). *Phys. Rev. Lett.* 106: 036601.
30 Yu, G., Upadhyaya, P., Fan, Y., Alzate, J.G., Jiang, W., Wong, K.L., Takei, S., Bender, S.A., Chang, L.-T., and Jiang, Y. (2014). *Nat. Nanotechnol.* 9: 548.
31 You, L., Lee, O., Bhowmik, D., Labanowski, D., Hong, J., Bokor, J., and Salahuddin, S. (2015). *Proc. Natl. Acad. Sci.* 112: 10310.

32 Luo, Z., Dao, T.P., Hrabec, A., Vijayakumar, J., Kleibert, A., Baumgartner, M., Kirk, E., Cui, J., Savchenko, T., and Krishnaswamy, G. (2019). *Science* 363: 1435.

33 Oh, Y.-W., Baek, S.-H.C., Kim, Y., Lee, H.Y., Lee, K.-D., Yang, C.-G., Park, E.-S., Lee, K.-S., Kim, K.-W., and Go, G. (2016). *Nat. Nanotechnol.* 11: 878.

34 Wang, Z., Zhang, L., Wang, M., Wang, Z., Zhu, D., Zhang, Y., and Zhao, W. (2018). *IEEE Electron Device Lett.* 39: 343.

35 Cai, K., Yang, M., Ju, H., Wang, S., Ji, Y., Li, B., Edmonds, K.W., Sheng, Y., Zhang, B., and Zhang, N. (2017). *Nat. Mater.* 16: 712.

36 Yang, M., Deng, Y., Wu, Z., Cai, K., Edmonds, K.W., Li, Y., Sheng, Y., Wang, S., Cui, Y., and Luo, J. (2019). *IEEE Electron Device Lett.* 40: 1554.

37 Chen, S., Yu, J., Xie, Q., Zhang, X., Lin, W., Liu, L., Zhou, J., Shu, X., Guo, R., and Zhang, Z. (2019). *ACS Appl. Mater. Interfaces* 11: 30446.

38 Cao, Y., Sheng, Y., Edmonds, K.W., Ji, Y., Zheng, H., and Wang, K. (2020). *Adv. Mater.* 32: 1907929.

39 Zhang, N., Cao, Y., Li, Y., Rushforth, A.W., Ji, Y., Zheng, H., and Wang, K. (2020). arXiv preprint arXiv:2003.06563.

40 Seung-heon, C.B., Amin, V.P., Oh, Y.-W., Go, G., Lee, S.-J., Lee, G.-H., Kim, K.-J., Stiles, M.D., Park, B.-G., and Lee, K.-J. (2018). *Nat. Mater.* 17: 509.

41 Wang, X., Wan, C., Kong, W., Zhang, X., Xing, Y., Fang, C., Tao, B., Yang, W., Huang, L., and Wu, H. (2018). *Adv. Mater.* 30: 1801318.1.

42 Sheng, Y., Edmonds, K.W., Ma, X., Zheng, H., and Wang, K. (2018). *Adv. Electron. Mater.* 4: 1800224.

43 Kittel, C. (1946). *Phys. Rev.* 70: 965.

44 Kaplan, B. and Gehring, G. (1993). *J. Magn. Magn. Mater.* 128: 111.

45 Liu, X., Liu, W., Dai, Z., Li, S., Wang, T., Cui, W., Li, D., Komarek, A., Chang, C., and Zhang, Z. (2017). *Appl. Phys. Lett.* 111: 212401.

46 Parkin, S.S., Hayashi, M., and Thomas, L. (2008). *Science* 320: 190.

47 Hayashi, M., Thomas, L., Moriya, R., Rettner, C., and Parkin, S.S. (2008). *Science* 320: 209.

48 Allwood, D.A., Xiong, G., Faulkner, C., Atkinson, D., Petit, D., and Cowburn, R. (2005). *Science* 309: 1688.

49 Cao, A., Zhang, X., Koopmans, B., Peng, S., Zhang, Y., Wang, Z., Yan, S., Yang, H., and Zhao, W. (2018). *Nanoscale.* 10: 12062.

50 Thiaville, A., Rohart, S., Jué, É., Cros, V., and Fert, A. (2012). *EPL (Europhysics Letters)* 100: 57002.

51 Emori, S., Bauer, U., Ahn, S.-M., Martinez, E., and Beach, G.S. (2013). *Nat. Mater.* 12: 611.

52 Emori, S., Martinez, E., Lee, K.-J., Lee, H.-W., Bauer, U., Ahn, S.-M., Agrawal, P., Bono, D.C., and Beach, G.S. (2014). *Phys. Rev. B* 90: 184427.

53 Baumgartner, M., Garello, K., Mendil, J., Avci, C.O., Grimaldi, E., Murer, C., Feng, J., Gabureac, M., Stamm, C., and Acremann, Y. (2017). *Nat. Nanotechnol.* 12: 980.

54 Yu, X., Kanazawa, N., Zhang, W., Nagai, T., Hara, T., Kimoto, K., Matsui, Y., Onose, Y., and Tokura, Y. (2012). *Nat. Commun.* 3: 988.

55 Sampaio, J., Cros, V., Rohart, S., Thiaville, A., and Fert, A. (2013). *Nat. Nanotechnol.* 8: 839.

56 Jiang, W., Zhang, X., Yu, G., Zhang, W., Wang, X., Jungfleisch, M.B., Pearson, J.E., Cheng, X., Heinonen, O., and Wang, K.L. (2017). *Nat. Phys.* 13: 162.

57 Fert, A., Reyren, N., and Cros, V. (2017). *Nat. Rev. Mater.* 2: 17031.

58 Woo, S., Litzius, K., Krüger, B., Im, M.-Y., Caretta, L., Richter, K., Mann, M., Krone, A., Reeve, R.M., and Weigand, M. (2016). *Nat. Mater.* 15: 501.

59 Reichhardt, C. and Reichhardt, C.O. (2016). *New J. Phys.* 18: 095005.

60 Zhang, X., Zhou, Y., and Ezawa, M. (2016). *Sci. Rep.* 6: 24795.

61 Jin, C., Song, C., Wang, J., and Liu, Q. (2016). *Appl. Phys. Lett.* 109: 182404.

62 Barker, J. and Tretiakov, O.A. (2016). *Phys. Rev. Lett.* 116: 147203.

63 Yang, S.-H., Ryu, K.-S., and Parkin, S. (2015). *Nat. Nanotechnol.* 10: 221.

64 Bi, C., Almasi, H., Price, K., Newhouse-Illige, T., Xu, M., Allen, S.R., Fan, X., and Wang, W. (2017). *Phys. Rev. B* 95: 104434.

65 Shi, G., Wan, C., Chang, Y., Li, F., Zhou, X., Zhang, P., Cai, J., Han, X., Pan, F., and Song, C. (2017). *Phys. Rev. B* 95: 104435.

66 Železný, J., Gao, H., Výborný, K., Zemen, J., Mašek, J., Manchon, A., Wunderlich, J., Sinova, J., and Jungwirth, T. (2014). *Phys. Rev. Lett.* 113: 157201.

67 Wadley, P., Howells, B., Železný, J., Andrews, C., Hills, V., Campion, R.P., Novák, V., Olejník, K., Maccherozzi, F., and Dhesi, S. (2016). *Science* 351: 587.

68 Bodnar, S.Y., Šmejkal, L., Turek, I., Jungwirth, T., Gomonay, O., Sinova, J., Sapozhnik, A., Elmers, H.-J., Kläui, M., and Jourdan, M. (2018). *Nat. Commun.* 9: 348.

69 Zhou, X., Zhang, J., Li, F., Chen, X., Shi, G., Tan, Y., Gu, Y., Saleem, M., Wu, H., and Pan, F. (2018). *Phys. Rev. Appl.* 9: 054028.

70 Olejnik, K., Seifert, T., Kaspar, Z., Novak, V., Wadley, P., Campion, R., Baumgartner, M., Gambardella, P., Nemec, P., and Wunderlich, J. (2017). arXiv preprint arXiv:1711.08444.

71 Reichlová, H., Kriegner, D., Holý, V., Olejník, K., Novák, V., Yamada, M., Miura, K., Ogawa, S., Takahashi, H., and Jungwirth, T. (2015). *Phys. Rev. B* 92.

72 Chen, X., Zarzuela, R., Zhang, J., Song, C., Zhou, X., Shi, G., Li, F., Zhou, H., Jiang, W., and Pan, F. (2018). *Phys. Rev. Lett.* 120: 207204.

73 Tsai, H., Higo, T., Kondou, K., Nomoto, T., Sakai, A., Kobayashi, A., Nakano, T., Yakushiji, K., Arita, R., and Miwa, S. (2020). *Nature* 580: 608.

74 Lin, P.-H., Yang, B.-Y., Tsai, M.-H., Chen, P.-C., Huang, K.-F., Lin, -H.-H., and Lai, C.-H. (2019). *Nat. Mater.* 18: 335.

75 Liu, X., Edmonds, K., Zhou, Z., and Wang, K. (2020). *Phys. Rev. Appl.* 13: 014059.

76 Wang, H., Du, C., Pu, Y., Adur, R., Hammel, P.C., and Yang, F. (2014). *Phys. Rev. Lett.* 112: 197201.

77 Valenzuela, S.O. and Tinkham, M. (2006). *Nature* 442: 176.

78 Ando, K. and Saitoh, E. (2010). *J. Appl. Phys.* 108: 113925.

79 Pai, C.-F., Liu, L., Li, Y., Tseng, H., Ralph, D., and Buhrman, R. (2012). *Appl. Phys. Lett.* 101: 122404.

80 Akyol, M., Jiang, W., Yu, G., Fan, Y., Gunes, M., Ekicibil, A., Khalili Amiri, P., and Wang, K.L. (2016). *Appl. Phys. Lett.* 109: 022403.

81 Liu, J., Ohkubo, T., Mitani, S., Hono, K., and Hayashi, M. (2015). *Appl. Phys. Lett.* 107: 232408.

82 Niimi, Y., Morota, M., Wei, D., Deranlot, C., Basletic, M., Hamzic, A., Fert, A., and Otani, Y. (2011). *Phys. Rev. Lett.* 106: 126601.

83 Niimi, Y., Suzuki, H., Kawanishi, Y., Omori, Y., Valet, T., Fert, A., and Otani, Y. (2014). *Phys. Rev. B* 89: 054401.

84 Woo, S., Mann, M., Tan, A.J., Caretta, L., and Beach, G.S. (2014). *Appl. Phys. Lett.* 105: 212404.

85 Miao, B., Huang, S., Qu, D., and Chien, C. (2013). *Phys. Rev. Lett.* 111: 066602.

86 Du, C., Wang, H., Yang, F., and Hammel, P.C. (2014). *Phys. Rev. B* 90: 140407.

87 Zhang, W., Jungfleisch, M.B., and Jiang, W., et al. (2014) *Phys. Rev. Lett.* 113: 196602.

88 Fan, X., Celik, H., Wu, J., Ni, C., Lee, K.-J., Lorenz, V.O., and Xiao, J.Q. (2014). *Nat. Commun.* 5: 3042.

89 Fan, X., Wu, J., Chen, Y., Jerry, M.J., Zhang, H., and Xiao, J.Q. (2013). *Nat. Commun.* 4: 1799.

90 Nan, T., Emori, S., Boone, C.T., Wang, X., Oxholm, T.M., Jones, J.G., Howe, B.M., Brown, G.J., and Sun, N.X. (2015). *Phys. Rev. B* 91: 214416.

91 Nguyen, M.-H., Pai, C.-F., Nguyen, K.X., Muller, D.A., Ralph, D.C., and Buhrman, R.A. (2015). *Appl. Phys. Lett.* 106: 222402.

92 Qiu, X., Legrand, W., He, P., Wu, Y., Yu, J., Ramaswamy, R., Manchon, A., and Yang, H. (2016). *Phys. Rev. Lett.* 117: 217206.

93 Qiu, X., Narayanapillai, K., Wu, Y., Deorani, P., Yang, D.-H., Noh, W.-S., Park, J.-H., Lee, K.-J., Lee, H.-W., and Yang, H. (2015). *Nat. Nanotechnol.* 10: 333.

94 Emori, S., Bauer, U., Woo, S., and Beach, G.S. (2014). *Appl. Phys. Lett.* 105: 222401.

95 Zhang, H., Kang, W., Wang, L., Wang, K.L., and Zhao, W. (2017). *IEEE Trans. Electron Devices* 64: 4295.

96 Baek, S.-H.C., Park, K.-W., Kil, D.-S., Lee, K.-J., and Park, B.-G. (2017). arXiv preprint arXiv:1711.11172.

97 Wang, K. (2018). *Nat. Electron.* 1: 378.

98 Ohtomo, A. and Hwang, H.Y. (2004). *Nature* 427: 423.

99 Lesne, E., Fu, Y., Oyarzun, S., Rojas-Sánchez, J., Vaz, D., Naganuma, H., Sicoli, G., Attané, J.-P., Jamet, M., and Jacquet, E. (2016). *Nat. Mater.* 15: 1261.

100 Wang, Y., Ramaswamy, R., Motapothula, M., Narayanapillai, K., Zhu, D., Yu, J., Venkatesan, T., and Yang, H. (2017). *Nano Lett.* 17: 7659.

101 Liu, L., Qin, Q., Lin, W., Li, C., Xie, Q., He, S., Shu, X., Zhou, C., Lim, Z., and Yu, J. (2019). *Nat. Nanotechnol.* 14: 939.

102 Wang, X., Tang, J., Xia, X., He, C., Zhang, J., Liu, Y., Wan, C., Fang, C., Guo, C., and Yang, W. (2019). *Sci. Adv.* 5: eaaw8904.

103 Alghamdi, M., Lohmann, M., Li, J., Jothi, P.R., Shao, Q., Aldosary, M., Su, T., Fokwa, B.P., and Shi, J. (2019). *Nano Lett.* 19: 4400.

104 Ostwal, V., Shen, T., and Appenzeller, J. (2020). *Adv. Mater.* 32: 1906021.

105 MacNeill, D., Stiehl, G., Guimaraes, M., Buhrman, R., Park, J., and Ralph, D. (2017). *Nat. Phys.* 13: 300.

106 Shi, S., Liang, S., Zhu, Z., Cai, K., Pollard, S.D., Wang, Y., Wang, J., Wang, Q., He, P., and Yu, J. (2019). *Nat. Nanotechnol.* 14: 945.

107 Jamali, M., Lee, J.S., Jeong, J.S., Mahfouzi, F., Lv, Y., Zhao, Z., Nikolić, B.K., Mkhoyan, K.A., Samarth, N., and Wang, J.-P. (2015). *Nano Lett.* 15: 7126.

108 Moore, J.E. (2010). *Nature* 464: 194.

109 Zhang, H., Liu, C.-X., Qi, X.-L., Dai, X., Fang, Z., and Zhang, S.-C. (2009). *Nat. Phys.* 5: 438.

110 Hasan, M.Z. and Kane, C.L. (2010). *Rev. Mod. Phys.* 82: 3045.

111 Kondou, K., Yoshimi, R., Tsukazaki, A., Fukuma, Y., Matsuno, J., Takahashi, K., Kawasaki, M., Tokura, Y., and Otani, Y. (2016). *Nat. Phys.* 12: 1027.

112 Mellnik, A., Lee, J., Richardella, A., Grab, J., Mintun, P., Fischer, M.H., Vaezi, A., Manchon, A., Kim, E.-A., and Samarth, N. (2014). *Nature* 511: 449.

113 Fan, Y., Upadhyaya, P., Kou, X., Lang, M., Takei, S., Wang, Z., Tang, J., He, L., Chang, L.-T., and Montazeri, M. (2014). *Nat. Mater.* 13: 699.

114 Khang, N.H.D., Ueda, Y., and Hai, P.N. (2018). *Nat. Mater.* 1.

115 Mahendra, D., Grassi, R., Chen, J.-Y., Jamali, M., Hickey, D.R., Zhang, D., Zhao, Z., Li, H., Quarterman, P., and Lv, Y. (2018). *Nat. Mater.* 17: 800.

116 Valenzuela, S. and Tinkham, M. (2007). *J. Appl. Phys.* 101: 09B103.

117 Mihajlović, G., Pearson, J., Garcia, M., Bader, S., and Hoffmann, A. (2009). *Phys. Rev. Lett.* 103: 166601.

118 Seki, T., Hasegawa, Y., Mitani, S., Takahashi, S., Imamura, H., Maekawa, S., Nitta, J., and Takanashi, K. (2008). *Nat. Mater.* 7: 125.

119 Seki, T., Sugai, I., Hasegawa, Y., Mitani, S., and Takanashi, K. (2010). *Solid State Commun.* 150: 496.

120 Sugai, I., Mitani, S., and Takanashi, K. (2010). *IEEE Trans. Magn.* 46: 2559.

121 Mosendz, O., Pearson, J., Fradin, F., Bauer, G., Bader, S., and Hoffmann, A. (2010). *Phys. Rev. Lett.* 104: 046601.

122 Hung, H., Luo, G., Chiu, Y., Chang, P., Lee, W., Lin, J., Lee, S., Hong, M., and Kwo, J. (2013). *J. Appl. Phys.* 113: 17C507.

123 Obstbaum, M., Härtinger, M., Bauer, H., Meier, T., Swientek, F., Back, C., and Woltersdorf, G. (2014). *Phys. Rev. B* 89: 060407.

124 Gu, B., Sugai, I., Ziman, T., Guo, G., Nagaosa, N., Seki, T., Takanashi, K., and Maekawa, S. (2010). *Phys. Rev. Lett.* 105: 216401.

125 Laczkowski, P., Rojas-Sánchez, J.-C., Savero-Torres, W., Jaffrès, H., Reyren, N., Deranlot, C., Notin, L., Beigné, C., Marty, A., and Attané, J.-P. (2014). *Appl. Phys. Lett.* 104: 142403.

126 Hou, D., Qiu, Z., Harii, K., Kajiwara, Y., Uchida, K., Fujikawa, Y., Nakayama, H., Yoshino, T., An, T., and Ando, K. (2012). *Appl. Phys. Lett.* 101: 042403.

127 Morota, M., Niimi, Y., Ohnishi, K., Wei, D., Tanaka, T., Kontani, H., Kimura, T., and Otani, Y. (2011). *Phys. Rev. B* 83: 174405.

128 Vlaminck, V., Pearson, J.E., Bader, S.D., and Hoffmann, A. (2013). *Phys. Rev. B* 88: 064414.

129 Kondou, K., Sukegawa, H., Mitani, S., Tsukagoshi, K., and Kasai, S. (2012). *Appl. Phys. Express* 5: 073002.

130 Zhang, W., Vlaminck, V., Pearson, J.E., Divan, R., Bader, S.D., and Hoffmann, A. (2013). *Appl. Phys. Lett.* 103: 242414.

131 Rojas-Sánchez, J.-C., Reyren, N., Laczkowski, P., Savero, W., Attané, J.-P., Deranlot, C., Jamet, M., George, J.-M., Vila, L., and Jaffrès, H. (2014). *Phys. Rev. Lett.* 112: 106602.

132 Ganguly, A., Kondou, K., Sukegawa, H., Mitani, S., Kasai, S., Niimi, Y., Otani, Y., and Barman, A. (2014). *Appl. Phys. Lett.* 104: 072405.

133 Nakayama, H., Althammer, M., Chen, Y.-T., Uchida, K., Kajiwara, Y., Kikuchi, D., Ohtani, T., Geprägs, S., Opel, M., and Takahashi, S. (2013). *Phys. Rev. Lett.* 110: 206601.

134 Althammer, M., Meyer, S., Nakayama, H., Schreier, M., Altmannshofer, S., Weiler, M., Huebl, H., Geprägs, S., Opel, M., and Gross, R. (2013). *Phys. Rev. B* 87: 224401.

135 Choi, J.-G., Lee, J.W., and Park, B.-G. (2017). *Phys. Rev. B* 96: 174412.
136 Hahn, C., De Loubens, G., Klein, O., Viret, M., Naletov, V.V., and Youssef, J.B. (2013). *Phys. Rev. B* 87: 174417.
137 Gómez, J., Tedlla, B.Z., Álvarez, N., Alejandro, G., Goovaerts, E., and Butera, A. (2014). *Phys. Rev. B* 90: 184401.
138 Zhang, W., Jungfleisch, M.B., Jiang, W., Pearson, J.E., Hoffmann, A., Freimuth, F., and Mokrousov, Y. (2014). *Phys. Rev. Lett.* 113: 196602.
139 Soh, W.T., Yeow, Y., Zhong, X., and Ong, C. (2015). *J. Phys. D: Appl. Phys.* 48: 345002.
140 Zhao, B., Khokhriakov, D., Zhang, Y., Fu, H., Karpiak, B., Hoque, A.M., Xu, X., Jiang, Y., Yan, B., and Dash, S.P. (2020). *Phys. Rev. Res.* 2: 013286.
141 Benítez, L.A., Torres, W.S., Sierra, J.F., Timmermans, M., Garcia, J.H., Roche, S., Costache, M.V., and Valenzuela, S.O. (2020). *Nat. Mater.* 19: 170.
142 Xu, H., Wei, J., Zhou, H., Feng, J., Xu, T., Du, H., He, C., Huang, Y., Zhang, J., and Liu, Y. (2020). *Adv. Mater.* 32: 2000513.
143 Cubukcu, M., Boulle, O., Drouard, M., Garello, K., Onur Avci, C., Mihai Miron, I., Langer, J., Ocker, B., Gambardella, P., and Gaudin, G. (2014). *Appl. Phys. Lett.* 104: 042406.
144 Aggarwal, S., Almasi, H., DeHerrera, M., Hughes, B., Ikegawa, S., Janesky, J., Lee, H., Lu, H., Mancoff, F., and Nagel, K. (2019). Demonstration of a reliable 1 gb standalone spin-transfer torque mram for industrial applications. Presented at *2019 IEEE International Electron Devices Meeting (IEDM).*
145 Huai, Y., Yang, H., Hao, X., Wang, Z., Malmhall, R., Sato, K., Zhang, J., Jung, D.H., Wang, X., and Xu, P. (2018). High density 3D cross-point STT-MRAM. Presented at *2018 IEEE International Memory Workshop (IMW).*
146 Cao, Y., Xing, G., Lin, H., Zhang, N., Zheng, H., and Wang, K. (2020). *IScience.* 23: 12062.
147 Von Neumann, J. (1958). New Haven: Press.
148 Sebastian, A., Le Gallo, M., Khaddam-Aljameh, R., and Eleftheriou, E. (2020). *Nat. Nanotechnol.* 1.
149 Meena, J.S., Sze, S.M., Chand, U., and Tseng, T.-Y. (2014). *Nanoscale Res. Lett.* 9: 526.
150 Ikeda, S., Hayakawa, J., Ashizawa, Y., Lee, Y., Miura, K., Hasegawa, H., Tsunoda, M., Matsukura, F., and Ohno, H. (2008). *Appl. Phys. Lett.* 93: 082508.
151 Guo, R., Lin, W., Yan, X., Venkatesan, T., and Chen, J. (2020). *Appl. Phys. Rev.* 7: 011304.
152 Milo, V., Malavena, G., Monzio Compagnoni, C., and Ielmini, D. (2020). *Materials (Basel)* 13: 166.
153 Yue, K., Liu, Y., Lake, R.K., and Parker, A.C. (2019). *Sci. Adv.* 5: eaau8170.
154 Cao, Y., Rushforth, A., Sheng, Y., Zheng, H., and Wang, K. (2019). *Adv. Funct. Mater.* 29: 1808104.
155 Feynman, R.P. (1982). *Int. J. Theor. Phys.* 21.
156 Shor, P.W. (1999). *SIAM Rev.* 41: 303.
157 Vandersypen, L.M., Steffen, M., Breyta, G., Yannoni, C.S., Sherwood, M.H., and Chuang, I.L. (2001). *Nature* 414: 883.
158 Chowdhury, S., Datta, S., and Camsari, K. (2019). A probabilistic approach to quantum inspired algorithms. Presented at *2019 IEEE International Electron Devices Meeting (IEDM).*
159 Kirkpatrick, S., Gelatt, C.D., and Vecchi, M.P. (1983). *Science* 220: 671.

160 Borders, W.A., Pervaiz, A.Z., Fukami, S., Camsari, K.Y., Ohno, H., and Datta, S. (2019). *Nature* 573: 390.

161 Debashis, P., Faria, R., Camsari, K.Y., Datta, S., and Chen, Z. (2020). *Phys. Rev. B* 101: 094405.

162 Liu, X., Lam, K., Zhu, K., Zheng, C., Li, X., Du, Y., Liu, C., and Pong, P.W. (2019). *IEEE Trans. Magn.* 55: 1.

6

Spin Oscillators

Huayao Tu and Zhongming Zeng

Suzhou Institute of Nano-tech and Nano-bionics, Chinese Academy of Sciences, Suzhou, China

CHAPTER MENU

6.1 Introduction

Based on the manipulation of the magnetization dynamics of magnetic material, a variety of magnetic devices have been successfully invented and widely used [1, 2]. The most famous is the readout head in the hard disk drive, which significantly increases storage capacities. However, the conventional method to control magnetization in these devices relies on an external magnetic field, limiting their scalability and integration.

Spintronics: Materials, Devices and Applications, First Edition. Edited by Kaiyou Wang, Meiyin Yang, and Jun Luo.

In 1996, Slonczewski and Berger theoretically predicted that a spin-polarized current could exert a spin-transfer torque (STT) on the magnetization in magnetic multilayers [3, 4], which was later demonstrated experimentally in around 2000 [5–8]. The discovery of STT provides an innovative way to effectively manipulate the magnetization in magnetic systems through transferring angular momentum from spin-polarized current to a local magnetic moment. Combined with the giant magnetoresistance (GMR) effect [9] and tunnel magnetoresistance (TMR) effect [10], STT makes it possible to create a wide class of devices with novel functionalities. The most attractive applications are STT-based magnetic random-access memories (STT-MRAM, see excellent reviews, Ref. [11]) [11–13] and spin oscillators [14] (SOs), which use STT for data writing and microwave generation, respectively. In this chapter, we will put our attention only on issues related to SOs. (Readers interested in other spintronic devices can refer to other chapters.)

Compared with conventional microwave devices, SOs feature several advantages [15]. The first is the great tunability of frequency, which can be tuned between a few MHz and several tens of GHz. The second lies in their intrinsic nonlinear nature. The third is their excellent scalability and robustness. Finally, they can realize multiple functionalities, ranging from microwave generation to magnetic field sensing, to signal processing, even with the same device configuration. Such multifunctionality offers great opportunities for its implementation in the future.

The rest of this chapter is organized as follows: Section 2 gives a brief introduction of the fundamental physics of SOs, covering the basic concepts and the principle of magnetization dynamics. Section 3 then describes the device classification based on geometries, magnetic equilibrium states, and material structure. Section 4 offers a brief introduction of the emerging SOs based on vortices and skyrmions. Section 5 reviews the recent advances in functional properties, including frequency, output power, linewidth, and phase-locking. Section 6 remarks on the application of SOs. Finally, Section 7 provides a summary and an outlook for future research on SOs by discussing current challenges and the possible solutions.

6.2 Fundamental Physics

6.2.1 Spin Transfer Torque and Magnetization Dynamics

In general, STT refers to the change of magnetization direction of a magnetic material by a spin-polarized current [3, 4]. Usually, an electrical current is overall unpolarized with an equal number of spin-up and spin-down electrons. In the typical structure of STT, as shown in Figure 6.1(a), a nonmagnetic metal (NM) layer is sandwiched between two ferromagnetic (FM) layers, with one magnetic layer's magnetization fixed ($\mathbf{M}_{\text{fixed}}$) and the other's magnetization free to rotate ($\mathbf{M}_{\text{free}}$). Given the simplest scenario where an electrical current is injected into the magnetic stack from the free layer (electrons flow in the opposite direction), the electrons' spins will first align with the $\mathbf{M}_{\text{fixed}}$. If the magnetizations of the fixed layer and the free layer are not in the same direction, there will exist an exchange effect between the electrons' spins and the magnetization of the free layer when the spin-polarized current flows through. Because the angular momentum is conserved, electrons will lose their spin

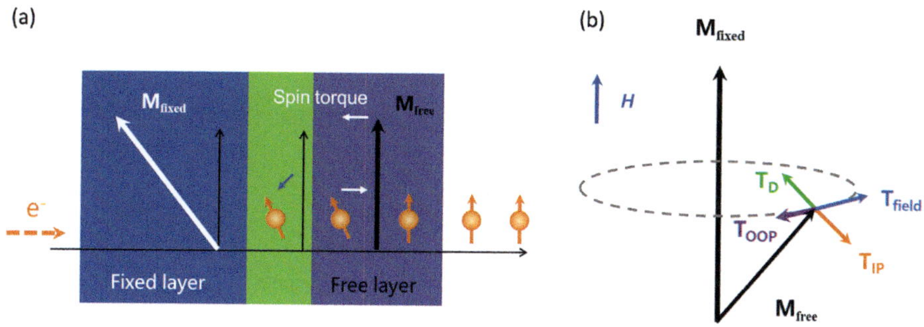

Figure 6.1 (a) Spin transfer torque principle in a simple FM/NM/FM trilayer. (b) The general principle of magnetization dynamics when the external field aligns with M_{fixed}.

component transverse to M_{free}, which is then transferred to the free layer. Depending on the direction of the injected current, the STT can cause the magnetization of the free layer to rotate either towards or away from the direction of the fixed layer magnetization. And the amplitude of STT can be fine-tuned by changing the driving current. With STT, no external magnetic field is needed to control magnetization dynamics, thus promising scalability and low power consumption.

Figure 6.1(b) illustrates the general principle of magnetization dynamics. The STT is divided into two components, in-plane, and out-of-plane torques (where the plane is defined by $\mathbf{M}_{fixed}$ and $\mathbf{M}_{free}$), which have different effects on the magnetization dynamics. The in-plane torque behaves as damping or anti-damping, while the out-of-plane torque acts like an effective external magnetic field. In the absence of spin-polarized current, if M_{free} is brought away from its equilibrium position, the effective-field torque will force it into precession around the effective field. However, due to the existence of the damping torque, it will finally be brought back to equilibrium. By contrast, in the presence of STT, because T_{IP} is aligned with the damping torque, the T_{IP} will reinforce or compensate the T_D depending on the injected direction of the spin current, stabilizing or destabilizing the magnetization, respectively. As for the T_{OOP}, it is aligned with the T_{field} and will change the energy landscape. (For a more in-depth introduction, turn to Chapter 4.) It should be noted that study has found T_{OOP} to be nearly zero in spin valves but to reach about 0.4 in magnetic tunnel junctions (MTJs) [16].

By using different materials and device structures or applying various injected currents, we can adjust the relative amplitude of T_{IP} and T_{OOP}, thereby controlling the output of STT-based devices and realizing multiple functions.

6.2.2 Spin Hall Effect (SHE) and Spin-Orbit Torque (SOT)

In addition to the spin-polarized electrical current, it was later discovered that STT can also be delivered by the pure spin current [17], which is known as spin-orbit torque (SOT). To distinguish these two mechanisms, STT is referred to as a torque generated from a spin-polarized electrical current in the remaining part of this chapter. And the corresponding spin oscillator (SO) based on these two mechanisms is referred to as the spin-Hall oscillator (SHO) and the spin-transfer oscillator (STO), respectively. The SOT originates from the charge-spin conversion in materials with strong spin-orbit coupling by means of the SHE

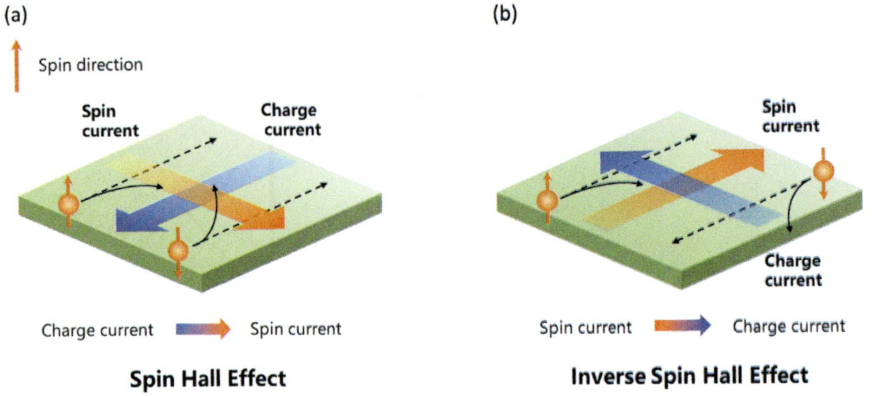

Figure 6.2 Principle of the Spin Hall Effect and the Inverse Spin Hall Effect.

[18, 19]. Figure 6.2 illustrates the basic principle of the SHE and the corresponding inverse SHE. (For a detailed introduction, please turn to Chapter 5.) An electrical current passing through a heavy metal (HM), such as Pt, will generate a transverse pure spin current and vice versa. Such conversion can be described as follows

$$j_s = j_c \times \frac{\theta_{SH}\mathbf{M}}{|\mathbf{M}|} \tag{6.1}$$

where j_s, j_c are the spin current and the charge current, respectively. θ_{SH} is the spin Hall angle. M is the magnetic moment of ferromagnetic materials. The SHE offers another means to effectively control the magnetization [20–22], which holds great promise in constructing efficient spintronic devices. Recently, in addition to the normal SHE, it has been found that the SOT can also be induced by means of the spin anomalous Hall effect in the CoFeB [23], providing the possibility to manipulate the orientation of the spin polarization by rotating the magnetization in the source ferromagnet. Furthermore, researchers have successfully combined the SOT and STT to drive the free layer into auto-oscillations in a three-terminal SO based on a nanopillar MTJ [24]. The combination of these two mechanisms excites the free layer into dynamic regimes beyond what can be achieved by each excitation mechanism individually, significantly enhancing output power.

6.2.3 Operation Principle of SO

As is shown in Figure 6.3(a), the operation principle of the STT-based spin torque oscillator with nanopillar geometry can be described as follows: when a spin current large enough to compensate the damping torque is injected into the magnetic nanostructure, it will force the magnetization of the free layer into sustained precession. Then, GMR or TMR, depending on the nonmagnetic spacer used, will translate this precession into a microwave voltage. When either a nonmagnetic metal or a thin dielectric is used as the spacer, the corresponding magnetic structure is called a magnetic spin valve or a magnetic tunnel junction (MTJ). Moreover, if the d.c. current is replaced by an a.c. one, the device will exhibit spin diode effect [25], leading to applications in microwave detection.

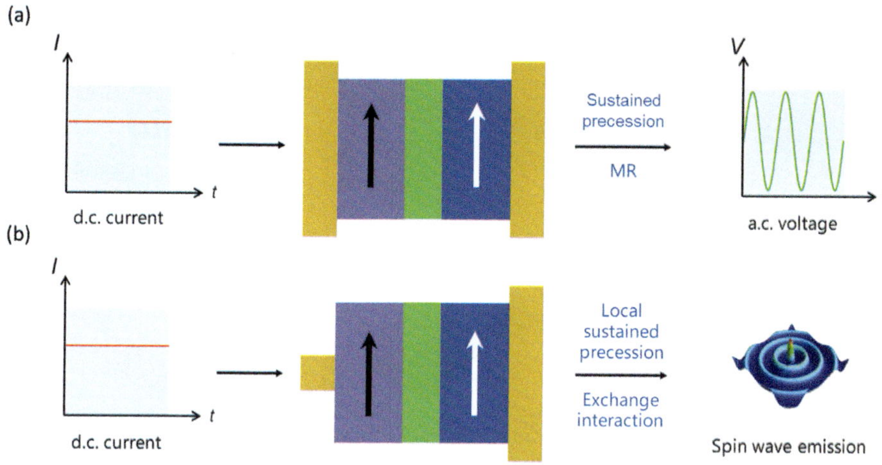

Figure 6.3 Operation principle of the spin oscillator. Injecting d.c. current leads to (a) microwave emission in a nanopillar geometry and (b) spin wave emission in a point contact/nanocontact geometry.

In contrast, as is illustrated in Figure 6.3(b), when a point contact/nanocontact geometry, where the nano-scale contact is placed on the extended free layer, is adopted, the confined high-density current will cause sustained precession of local magnetization and thereby lead to STT-driven spin wave emission and propagation outwards from the contact region [26, 27], making the devices spin wave emitters and useful in magnonics.

6.3 Device Classification

Since the initial theoretical prediction and the first experimental demonstration of microwave generation in magnetic multilayers [3, 4, 14], tremendous efforts have been devoted to the research on SOs. So far, many different types of devices have been explored, which can be classified into following types according to device geometries, the equilibrium magnetization orientation of magnetic layers, and the material structure.

6.3.1 Geometries

Figure 6.4 summarized the geometries used in SOs so far [28]. In the early days, the electrical contact widely used in STOs is mechanical point-contact, which consists of a highly sharp metal tip on an extended multilayer. And the contact area is on the order of 10^2 nm^2 [5, 29, 30]. Then, with the development of the nano-fabrication technique, point-contact has been replaced by a circular or elliptic electrical contact which is defined by electron beam lithography (EBL) or hole–mask colloidal lithography (HCL) process [31], called nanocontact [21, 32]. In these devices, the typical current densities for magnetization dynamics excitations are 10^7-10^8 A/cm^2. By comparison, the current density required to excite STT-induced magnetization switching or oscillations can be greatly lowered in SOs based on nanopillar geometry, with our previous work indicating that the value can be reduced to 10^6A/cm^2 [33]. This is because, in nanopillar structure, all of the magnetic layers are patterned into nano

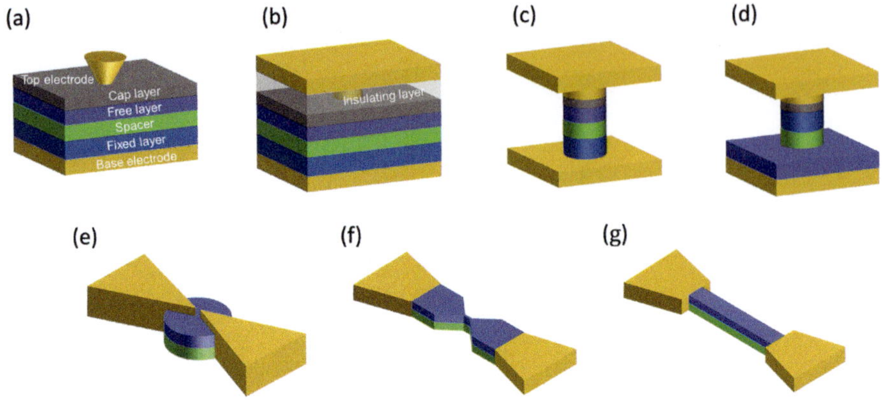

Figure 6.4 Device geometries [28]. (a) Point contact. (b) Nanocontact. (c) Nano-pillar. (d) Hybrid. (e) Nanogap. (f) Nanoconstriction. (g) Nanowire.

scale [7, 14] while only the current injection region is laterally confined in point-contact/nanocontact devices. However, devices with nanopillar geometry may suffer from a larger linewidth. Due to their smaller magnetic volume, they are more susceptible to thermally induced fluctuations [34]. Moreover, significant stray fields need to be taken into consideration in this structure when analyzing. As for the hybrid device architecture as shown in Figure 6.4(d) [26], it refers to a nanopillar grown on top of an extended film. It was in this structure that the first experimental observation and two-dimensional mapping of spin waves emitted by STOs was realized [26].

In 2011, Demokritov et al. studied the influence of pure spin current generated by SHE on the ferromagnetic resonance (FMR) of an adjacent magnetic layer in Py/Cu/Pt discs [35, 36]. They found that although the effective damping constant in Permalloy could be remarkably reduced by a factor of two, the uniform torques couldn't fully compensate the damping torque in these devices, leading to the failure of the generation of sustained oscillation [28]. Therefore, four main device geometries have been designed to generate nonuniform torques since then, including nanopillar, nanogap, nanoconstriction, and nanowire.

In the three-terminal nanopillar geometry [20, 37], the nanoscale magnetic free layer is put on the top of an extended plate of the SHE-generating normal metal. By using an MTJ for read-out, the output power can be enhanced while realizing the independent control of amplitude and frequency of SHOs at the same time. It is worth noting that the successful combination of STT and SOT mechanism was achieved in a device based on such geometry in 2019 [24]. For the nanogap geometry, two highly conductive electrodes with sharp points are placed on the top of the FM/HM bilayer with a 100-nm-wide gap [21]. In 2014, nanoconstriction geometry-based devices were demonstrated by Demidov et al. exhibiting a single-mode auto-oscillation and generating highly-coherent electronic microwave signals with high output power and a narrow linewidth about 6.2 MHz at room temperature [38]. Later, Åkerman et al. fabricated a 20 nm nanoconstriction device in Pt/NiFe bilayers [39]. The results showed that the dimension of SHO can be scaled down to truly nanoscopic dimensions, with the additional benefit of ultra-low operating currents and improved power conversion efficiency. Moreover, a Pt/Py bilayer with a 190-nm-wide nanowire geometry has been reported in which the SOT induced by SHE can excite self-oscillations of magnetization [40].

6.3.2 Magnetic Equilibrium States

Magnetic equilibrium state is another classification criterion, according to which the spin oscillators can be divided into four types, including all in-plane, all perpendicular, orthogonal, and tilted. Among them, the majority of SOs studied in the early stages were based on all in-plane configuration [14, 32, 41], where the magnetizations of both the free and fixed FM layers lie in the film plane when no external field is applied. However, to generate microwave output, devices utilizing such geometry require large external magnetic fields to overcome the demagnetization field to drive the magnetization of the free layer out-of-plane. The all perpendicular structure is very similar with both the free layer and polarizer layer perpendicular to the sample plane [42]. As for devices with orthogonal configuration, the equilibrium magnetization directions of the fixed and free layers are orthogonal to each other [43–45]. This design is very promising in producing a large output power [33, 44], even in the absence of an external magnetic field [46]. The last is the tilted configuration where magnetizations are all tilted, thus having both in-plane and out-of-plane components. Compared with conventional all in-plane or all perpendicular magnetic films, using tilted fixed layers can realize the optimization of microwave generation and enhanced spin-torque efficiency while maintaining high output powers at low or zero field [28, 47–50].

6.3.3 Material Structures

6.3.3.1 Spin Valves

The SOs based on spin valves have been extensively researched. And many excitation modes have been found to exist in the spin valves depending on the geometry, magnetic states, external field strength, and external field angles. For nanocontact spin valves, the modes are identified experimentally and theoretically as exchange-dominated cylindrical spin wave propagating modes [29, 51]. These modes showed the possibility of achieving phase-locking between different adjacent nanocontacts with two breakthrough results separately reported by Kaka et al. [52] and Mancoff et al. [53]. Later, nonlinear localized spin-wave modes, known as bullet modes, were identified and found to strongly depend on the angle of the applied magnetic field [54]. Especially, it was found that the Oersted field will lead to the existence of a critical angle, below which the bullet mode is excited, otherwise a Slonczewski mode will be excited [55]. As for spin valves with uniform current injection, the oscillation mode is largely determined by the size of the free FM layer and the strength of the applied magnetic field. Only when the external magnetic field is large enough, or the diameter or thickness of the free layer is sufficiently small, can uniform precessions be generated or the vortex state favored. Interestingly, Matsumoto et al. have recently reported the chaotic excitation state realized in a double free layer spin valve [56], which can be adjusted by changing the magnetization structure, anisotropy, and d.c. current bias. It is worth noting that the characteristics of the chaotic excitation state is similar to the neuron pattern in the brain and can be used for neuromorphic computing.

Particular efforts have been made to enhance spin valves' SO performance, especially the realization of a zero-field operation and the generation of large output power. Approaches proposed include using perpendicular/tilted magnetic layers [57, 58], introducing half-metallic materials with large spin polarization [59], using two in-plane-oriented free layers with perpendicularly oriented polarizers [60], or deploying device arrays [52, 53], etc.

6.3.3.2 Magnetic Tunnel Junctions

The first measurement of persistent magnetization dynamics in MTJ was reported by Fuchs et al. in 2003 [61]. The sample used was composed of Py 4/Cu 80/Ta 10/CoFeB 8/Al 0.65(O_x)/CoFeB 2/Cu 5/Pt 30 (thickness in nm) with AlO as the spacer. However, the temperature required is very low. It was not until the room-temperature TMR was realized with MgO tunnel barrier in 2004 [62, 63] that MTJ-based spin-torque oscillators began to develop rapidly. In 2005, Tulapurkar et al. made a notable discovery of a spin-torque diode effect in MTJs [25], opening a new way for a nanometer-scale radio-frequency detector. In 2006, an impressive work by Nazarov et al. was reported [64]. They observed a pronounced narrow peak at the frequency of around 7 GHz with a full width at half-maximum as low as 21 MHz in MTJs. Later, the distinct behaviors from two different groups of MTJs were studied [65], one with high R and high TMR ratio (100%), labeled HTMR, and the other with low R and low TMR ratio (30%), labeled LTMR. Significantly, it was found that LTMR could generate high power and achieve narrow linewidth at the same time, in agreement with other experimental results [66, 67].

An important progress for MTJ-based SOs was the introduction of interfacial perpendicular anisotropy (IPA) in Ta/CoFeB/MgO/CoFeB/Ta MTJs by Ikeda et al. [68], making it possible to define easy axis of either the fixed or free layer by controlling the relative magnitude between shape anisotropy and IPA. In particular, the performance of SOs has been greatly enhanced in this way [69]. For example, our group has reported the generation of microwave signals with maximum measured power of 0.28 μW and simultaneously narrow linewidth of 25 MHz in CoFeB-MgO-based MTJs [70]. A large emission power (0.55 μW), a high frequency (6.3 GHz), and a high Q factor (135) have also been simultaneously achieved by Kubota et al. [71]. Soon after they also reported the achievement of large emission power over 2μW with a high Q factor of 350 [72].

Another milestone is the realization of operation without the external magnetic field. By introducing IPA in MTJ-based SOs, zero-field magnetization oscillations have been successfully demonstrated [33, 46] and giant spin-torque diode sensitivity of 75,400 $mVmW^{-1}$ in the absence of a biased magnetic field has also been achieved [73]. Recently, Rajapakse et al. reported a high-frequency microwave emission under zero external magnetic field by adopting an MTJ with three active magnetic layers [74]. An additional perpendicular layer is added as the third layer on top of an otherwise conventional stack to enhance the IPA, resulting in emission frequency reaching as high as 6 GHz.

6.3.3.3 Bilayer

The FM/HM bilayer structure is widely adopted in SHOs. When a charge current flows through the bilayer structure, the portion in the HM layer will compromise most of the total current due to its low resistivity. This electrical current will generate a pure spin current flowing into the FM layer through SHE. Then, the pure spin current injected into the FM layer exerts a SOT on its magnetization, leading to magnetization dynamics.

Since the first experimental observation of microwave auto-oscillation in a Py/Pt bilayer [21], many types of materials have been explored. For the material used to generate a pure spin current,

Pt is the most widely used one while other materials have also been researched [38, 39, 75–78]. A giant SHE in β-Ta, that generates spin currents sufficient to induce efficient spin-torque switching of ferromagnets at room temperature, was reported by Buhrman in 2012 [20]. Later, a thin film of the topological insulator bismuth selenide (Bi_2Se_3) was demonstrated to deliver a strong SOT on an adjacent ferromagnetic Py thin film [79]. Recently, large spin Hall angles of up to approximately −0.5 have been achieved in a $Co_{40}Fe_{40}B_{20}$/W bilayer by incorporating oxygen into W [80]. As for the FM materials, $Co_{40}Fe_{40}B_{20}$ [78], Co/Ni multilayers with PMA (perpendicular magnetic anisotropy) [81–84], NiMnSb [85], and the ferrimagnetic insulator yttrium iron garnet (YIG) [77, 78, 86–88] have been studied in addition to NiFe [39, 89–92].

6.3.3.4 Single Layer

Different from the conventional trilayer structure for STOs, theoretical work predicts that a sizable spin accumulation in a single layer at each interface can be achieved [93, 94]. However, these interfaces are required to be asymmetric so that an imbalance of spin accumulation can lead to a sufficient STT to drive magnetization precession. Early experiments on nanopillars with a single Co layer confirmed that the spin-wave excitations, and even the switching of magnetization, could be realized [95, 96]. The work in 2013 on NiMnSb-based single layer devices has been in the nanocontact geometry, where evidence of a periodic vortex/antivortex creation/annihilation mode has been found [97]. In the latest work by Dürrenfeld et al. [98], they show that nanocontact STOs based on single layers of the epitaxially grown half-metal NiMnSb feature wide tunable frequency, low threshold current density, and small linewidth.

For the SHOs, theories have found that SOT can exist at the interfaces of magnetic material [99, 100]. In particular, interfaces to oxides can cause a non-uniform spin distribution over the film thickness [101, 102]. In 2019, a theoretical work showed that charge current through a single antiferromagnetic layer can excite and control self-oscillations [103]. In the same year, as depicted in Figure 6.5, Åkerman et al. experimentally demonstrated a SOT nano-oscillator based on a single permalloy layer with thickness in the 15–20 nm range [104], grown on an Al_2O_3 substrate and capped with SiO_2.

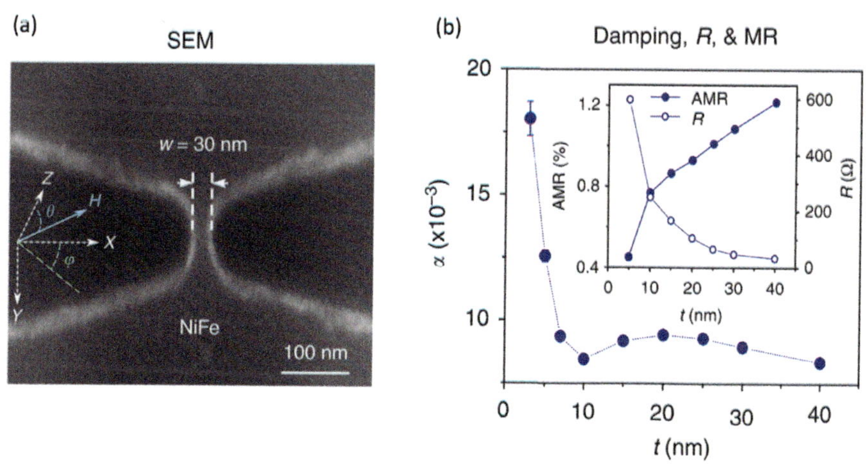

Figure 6.5 A single-layer SHO. (a) Scanning electron microscope image of the device. (b) Gilbert damping of blanket permalloy films as a function of film thickness. *Source:* Reproduced from [104] under CC BY 4.0.

6.4 Emerging Spin-torque Oscillators Based on Magnetic Solitons

A soliton is a self-reinforcing wave packet that can be observed in material with a non-linear and dispersive nature. To date, many kinds of magnetic solitons have been discovered and studied in magnetism, including (i) vortices [105–107], (ii) bubbles [108, 109], (iii) skyrmions [81, 110], etc. These magnetic solitons are valuable and hopeful for applications in many fields with their great stability and tunability. Particularly, magnetic vortex- and skyrmion-based oscillators are attractive candidates for future high-efficiency spin oscillators.

6.4.1 Vortex

A magnetic vortex is a curling magnetization distribution whose magnetization orientation points perpendicular to the film plane within a nano-scaled vortex core [111], which is characterized by two Boolean parameters: chirality and core polarity. The dynamics of the magnetic vortex has been extensively studied since its discovery [112–115]. Particularly, it's found that SOs based on a vortex can exhibit comparatively small linewidth, a wide frequency tuning range, and high output power [116, 117], making them a hot topic in the academic community.

Vortex dynamics has been studied in both nanopillar and nanocontact geometries. In the nanopillar-based SOs, the demagnetization field and the spin transfer torque determine the dynamic behavior of the vortex due to the boundary confinement. Increasing the current density, damped motion, stationary vortex rotation, and the switching of the vortex core will be successively observed [113]. As for nanocontact structure [117, 118], the dynamic behavior is determined by the STT and the current-induced Oersted field. As shown in Figure 6.6, under the action of the Oersted field, a vortex will be produced and stabilized in the free layer. When the spin-polarized electrons carrying a magnetic moment opposite to the core polarity are injected into the device, the vortex will then be driven away from the nanocontact area. Otherwise, if the vortex and the current have the same chirality, the Oersted field will attract the vortex towards the contact. Thus, a steady gyrotropic precession of the vortex around its equilibrium position is sustained by the STT exerted by the current on the vortex tail under the contact area. Similar to normal SO, this type of oscillation will result in a changing voltage response.

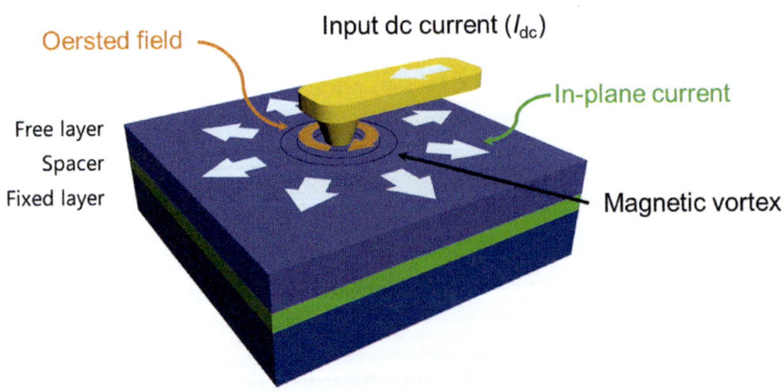

Figure 6.6 A schematic of a nanocontact vortex spin valve oscillator.

The sub-gigahertz, narrow linewidth auto-oscillations of SOs were first observed in both nanocontact and nanopillar spin valves [119, 120], resulting from the spin-torque-driven gyrotropic motion of a magnetic vortex. Several methods have been explored to improve the performance, including tailoring the device structure and using different material [107, 120]. An important breakthrough toward increasing the output power to a practical level was made in 2010. In a nanopillar MTJ the microwave emission generated by the persistent motion of magnetic vortices features a stronger emitted power and similar spectral quality compared to those based on spin valves [121]. Then, in 2016, by inserting a thin $Co_{70}Fe_{30}$ layer between the tunnel layer and the free layer, Kubota et al. successfully made microwave emission power exceed 10 μW, which is the largest value reported so far. The same year, G. Siracusano et al. first reported a novel magnetic configuration named a radial vortex, different from the traditional circular vortex, in the presence of an interfacial Dzyaloshinskii–Moriya interaction (DMI) with a much lower threshold current density smaller than 10^6 A/cm^2 [122, 123]. It's predicted that this new class of vortex is able to greatly enhance the microwave power and reduce the energy dissipation [124].

In addition, the phase-locking and synchronization of vortex-based SOs have attracted much attention. In 2009 it was shown that vortices can interact with each other via antivortices in a zero field [106], leading to synchronization when they are closely spaced in a nanocontact spin valve. The microwave power and the corresponding linewidth could be simultaneously improved. Experimentally, Locatelli et al. demonstrated the coupled auto-oscillations of two vortices, each within a separate NiFe layer in two spin-valve nano-pillars, and were able to exhibit enhanced microwave coherence [125]. A later theoretical work found that two nanocontact oscillators interacting through an intermediate antivortex can achieve phase-locking over a wide range of nanocontact currents [126]. Recently, it's been put forward numerically that the strength of phase locking between two dipolarly coupled vortex-based SOs can be tuned by an intermediate oscillator between them [127].

6.4.2 Skyrmion

A magnetic skyrmion is a kind of topological spin texture found in chiral magnets with a strong DMI induced by a lack or breaking of inversion symmetry [128, 129] (readers interested in skyrmions can refer to ref. [130]). Due to topological protection, it is very difficult to destroy a skyrmion, thus making it promising to create skyrmion-based devices. Similar to vortices, magnetic skyrmions can also be excited into sustained oscillations by current-induced STT [131–136], as shown in Figure 6.7, thereby holding potential in SO devices. What's more, skyrmions are localized spin texture and are usually smaller than magnetic vortices; therefore, multiple skyrmions can be put in one SO to implement synchronization and improve device performance [131, 137].

A skyrmion-based oscillator was first proposed by Senfu Zhang et al. (2015) using micromagnetic simulations with attractive features of nanoscale size, small linewidth, and broad working frequency [131]. However, the skyrmion may be destroyed at the nanodisk edge when the driving current density is too large due to the skyrmion Hall effect [132], which limits oscillation frequency and ultrahigh-frequency applications based on a skyrmion. A modified nanodisk structure overcoming this frequency limit is then presented by Y. Feng et al. [138]. Their simulation results demonstrate that the upper threshold of the skyrmion

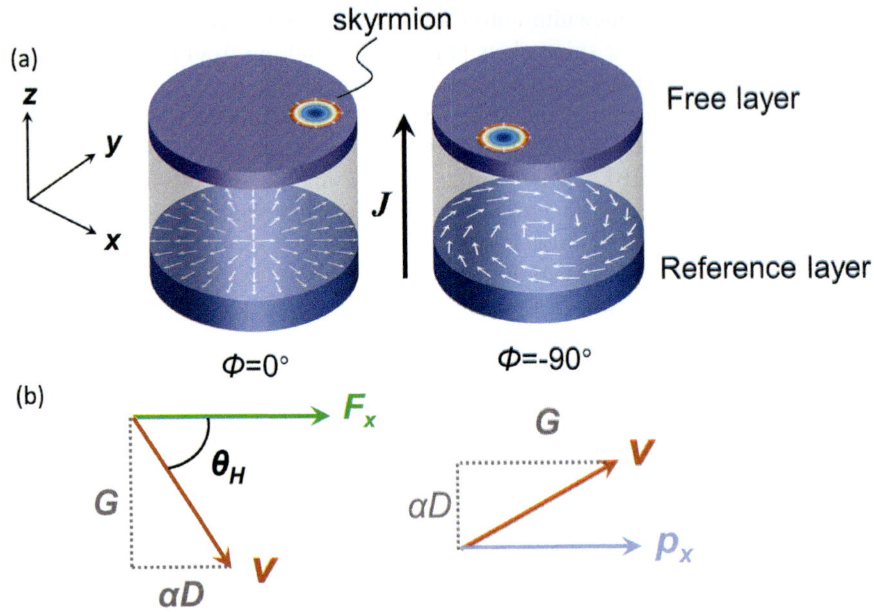

Figure 6.7 (a) Circular nanopillar with a gyrating skyrmion in the free layer and two possible vortex-like magnetic configurations j in the reference layer. (b) Skyrmion velocity *v* in response to a uniform force F_x and spin torques due to a uniform spin polarization P_x.

oscillation frequency can be increased by 75% compared to the original structure without the enhanced edge. Other approaches include using a FM/spacer/FM/HM structure [139] or adopting an antiferromagnetically exchange-coupled bilayer system [140], etc. Most recently, a new concept of spin-Hall-effect-modulation skyrmion oscillator has been predicted numerically to be able to reach a frequency of up to 15 Ghz [141]. (For an in-depth review of skyrmions, see ref [142].)

6.5 Functional Properties

6.5.1 Frequency

The microwave frequency of SO devices covers a wide tunable range, from a few hundred MHz to tens of GHz, and can be regulated by current and external magnetic fields. According to Kittel's equation, the precession frequency increases with the square root of the applied field at low fields and linearly at high fields, and decreases with the input d.c. current in line with Slavin's non-linear oscillator theory [143]. Figure 6.8 depicts the typical dependence of microwave frequencies on magnetic fields and current.

In 2003, the first microwave oscillator based on a Co/Cu/Co spin valve was reported to emit a 12~25 GHz microwave signal under a 2 KOe magnetic field. The field tunability of the frequency is 4.5 GHz/KOe [14]. Subsequently, Rippard et al. measured the microwave emission in a CoFe/NiFe nano-contact spin valve under in-plane and out-of-plane fields, observing frequencies tunable from below 5 to above 40 GHz with maximum field tunability

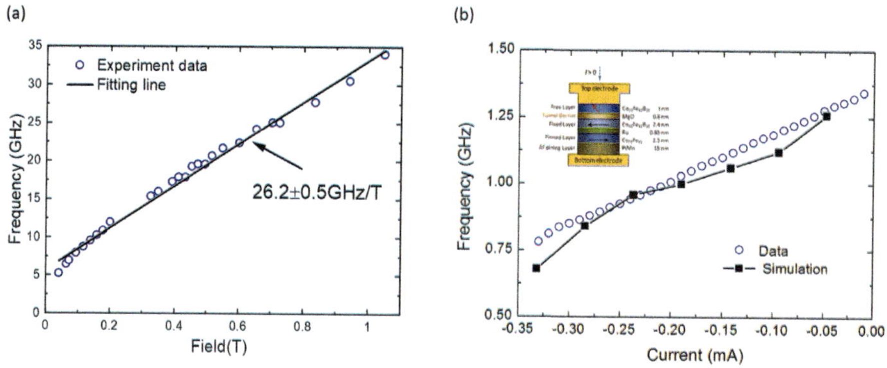

Figure 6.8 (a) Magnetic field dependence of frequency for a spin valve. Reproduced with permission [41]. Copyright 2004, American Physical Society. (b) Dependence of microwave frequencies on current for an MTJ having a planar polarizer and a perpendicular free layer. *Source:* Reproduced with permission [33]. Copyright 2013, Springer Nature.

of 32 GHz/T [41]. In 2009, by applying fields normal to the film plane, a 46 GHz high frequency microwave signal in nanocontact spin valve device was observed [144], with the maximum expected operating frequency predicted to reach beyond 65 GHz under in-plane magnetic fields. Although SOs have several advantages over conventional oscillators, large phase noise made it embarrassing in applications. There was finally a breakthrough in 2015 by Tamaru et al[145]. They introduced a phase-locking loop with a SO as a voltage-controlled oscillator that generated a 7.34 GHz microwave signal stabilized by a 153 MHz reference signal and achieved a linewidth as narrow as 1 Hz.

Although the microwave emission frequency of SO devices is commonly higher in high magnetic fields, it is not suitable for practical application. A variety of researches are focused on improving the emission frequency of the device under low or zero field by enhancing PMA.

In 2007, the CoFe/Cu/CoFe spin valve with the magnetization of the reference layer being pinned perpendicularly to the easy axis of the free layer achieved microwave emission without an external field for the first time [146]. Most recently, Rajapakse et al. published an experimental study on zero-field microwave emission in a SO consisting of three active magnetic layers [74]. In this device, the microwave emission frequency can reach as high as 6 GHz in the absence of an applied magnetic field. This could be ascribed to enhanced PMA due to an extra CoFeB/MgO interface [147].

Additionally, it was proposed that the operation frequency of SOs can be increased to much higher THz by replacing ferromagnets with antiferromagnets (AFs) because the eigenfrequencies of typical AFs lie in the THz range [148, 149]. And similar to ferromagnets, they can also be driven into sustained oscillations by STT [150–152]. In 2016, Cheng and Xiao theoretically put forward a THz antiferromagnetic SHO in an antiferromagnet/HM heterostructure using a dynamic feedback mechanism derived from the combined effect of the SHE and its reverse process [153]. The THz oscillation arises when the amplitude of the spontaneous motion induced by the STT is balanced to the magnetic damping. In addition, a kind of antiferromagnetic tunnel junction is also theoretically designed which uses an antiferromagnetic pinned layer and a Pt mental layer separated by a thin MgO constructing a three-layer sandwich structure [154, 155]. The strong antiferromagnetic oscillator signal is driven by a spin current originating from SHE in the precious metal Pt.

6.5.1.1 Modulation Properties

Modulation refers to the process of changing the frequency, phase, and amplitude of the high-frequency periodic carrier signal according to the low-frequency signal information. SOs' modulation properties are essential for their potential in modern wireless communication.

The typical modulation method of SOs is the frequency shift keying (FSK) modulation by controlling the injected current. In 2005, Pufall found that the input current could make the frequency of the microwave signal increase or decrease [156]. In 2010, Muduli et al. discovered three principally different cases of frequency nonlinearity when studying the current-controlled modulation of a nanocontact spin-torque oscillator [157]. Later, the FSK was first reported using vortex-based SOs whose frequency lies in the 0.1–2 GHz range by modulating the current between two values at data rates of 0–10 Mbit/s [158]. Then, Purbawati et al. proposed a binary keying frequency shift based on SO devices. When the injected current value changes between 1.06 mA and 1.2 mA, a frequency shift of 380 MHz is observed, and the maximum data rate reaches 20 Mbit/s [159]. Especially, FSK up to data rates of 400 Mbps has been achieved by Ruiz-Calaforra et al. [160], with Figure 6.9 showing the current modulation of the device. In addition to direct current injection, the microwave Oersted field can also be used to modulate the SO frequency [161]. In 2011, Nagasawa studied the MTJs with an additional magnetic field line and realized a transition time down to 2 ns [162]. Recently, Sknowronski et al. prepared an independent microwave line on top of the SO device. The Oersted field generated by the microwave line enables field-modulation at a rate up to 3 GHz [163].

6.5.2 Output Power

Output power is the most important parameter of SO in practical applications. Therefore, increasing the power of the device is a key issue. According to the operation principle of SOs, the direct current flowing through the magnetic stack will excite the magnetization

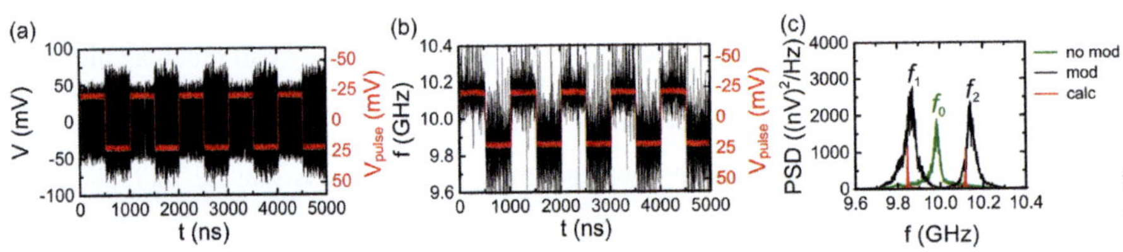

Figure 6.9 Current modulation of a STO [160]. (a) Snapshot from the amplified output voltage of the modulated STO with $f_m = 1$ MHz (black) together with the injected current (red). (b) Corresponding instantaneous frequency obtained from the Hilbert transformation. The voltage axis of the pulses is inverted for better comparison with the frequency shift due to the negative df/dI. (c) Corresponding power spectral density (PSD) (black) together with the PSD of the free-running STO (green). The red lines mark the calculated position of the peaks of the modulated signal. *Source:* Reproduced with permission [160]. Copyright 2017, AIP Publishing.

oscillation in the free layer [14, 29, 164], causing the resistance to change with time due to the *MR* (magnetoresistance) effect. The time-dependent resistance of SO is given by

$$R(t) = R + \frac{\Delta R}{2} \times \cos(\omega t) \tag{6.2}$$

in which ΔR is the oscillation amplitude of resistance induced by the spin current, R and ω are the average resistance and the precession frequency, respectively. The resistance generates a microwave voltage $V(t) = I \times R(t)$. In the load impedance of R_L, the output power of SO can be described by [165]

$$P_{out} = \frac{V_{out}^2}{2R} = \frac{I^2}{8} \frac{\Delta R^2 R_L}{(R + R_L)^2}. \tag{6.3}$$

According to the equation, the microwave output power in SO depends on the ΔR, which is proportional to the magnetoresistance of the device. To maximize the output of SO, the most efficient way is to enlarge the magnetoresistance variation ΔR, which is named the resistance area product and is defined by $\Delta R = A_m \times MR$, where A_m is related to the amplitude of magnetization oscillation and MR is the magnetoresistance of the device. $MR = (R_{AP}\text{-}R_P)/R_P$, where R_{AP} and R_P are the resistances when the magnetic moments between the fixed and free layer are aligned anti-parallel and parallel, respectively. Thus, the output power is approximately proportional to the *MR* effect, the oscillation amplitude of magnetization and the impedance match ratio R/R_L.

The first effective way to enhance the output power is to increase the magnetoresistance value of the device. Usually, the magnetoresistance of tunnel junction devices is generally higher than that of spin valves. The magnetoresistance value of the first microwave SO is less than 1.1% [14], such a microwave emission power is too small (<pW) for practical application. In 2008, the first MTJ-based SO achieved a microwave output power of 43 nW [65]. The tunneling magnetoresistance of the device is only 30%. Although it is much higher than the GMR of the spin valve, it is still not large enough to generate high-power microwave signals (~μW). Compared with traditional ferromagnetic alloys, Heusler alloy has low saturation magnetization, low damping, high Curie temperature, and high spin polarizability. In 2013, Seki et al. used semi-metallic Co_2MnSi to prepare a spin valve (CMS/Ag/CMS). Thc device has a giant magnetoresistance of 36% and a microwave output power of 1 nW[166]. Subsequently, in their similar work in 2014, they introduced PMA and adjusted the new materials in $Co_2Fe_{0.4}Mn_{0.6}Si$ device (CFMS/Ag/CFMS), and then realized an enhanced 0.3 μW microwave output power and 48% *MR* with high *Q* factor (1124) [167]. One year later, Maehara et al. fabricated a sombrero-shaped nanocontact SO with high output power of 2 μW and high stability($Q = 350$) [72]. By reducing the shunt current passing through the free layer, the device displayed a 46% *MR*.

Secondly, according to the nonlinear coupling theory [143] and micromagnetic simulation, the high input current can increase the precession angle of the free layer magnetic moment, thereby enhancing the oscillation amplitude and the output power. Slavin and Tiberkevich proposed that the amplitude of the magnetic moment oscillation increases as the ratio of the input current to the threshold current increases (I/I_c) [143]. Therefore, higher input current or lower device threshold current can enlarge the output power of the device. Figure 6.10 demonstrates

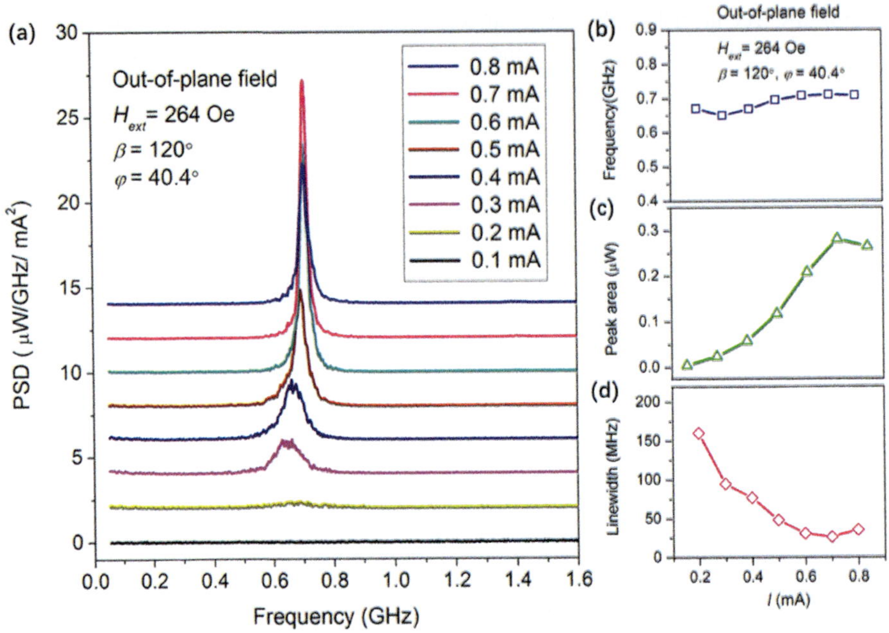

Figure 6.10 (a) Microwave emission spectra under 264 Oe out-of-plane field. Here β is the angle between the external magnetic field H_{ext} and the reference layer in the x-y plane, and φ is the angle with respect to the x–y plane; (b), (c) and (d) are the plot of frequency, power, and linewidth as a function of I_{dc}. *Source:* Adapted with permission [70]. Copyright 2012, American Chemical Society.

the typical microwave emission spectra under different input currents. Particularly, for MTJ SO devices, because they are based on the principle of quantum tunneling, a large bias voltage will cause device breakdown. Therefore, in these devices, methods such as adjusting the thickness of the non-magnetic layer MgO, increasing the device perpendicular magnetic anisotropy, and reducing the thickness of the free layer are usually adopted.

Apart from improving the *MR* value and the magnitude of the oscillator, some researchers found that output power can be enhanced by tailoring the magnetization structure [43, 168-170]. Ikeda et al. first demonstrated a perpendicular magnetized MTJ cell by adjusting the thickness of CoFeB layers, suggesting that it could be used to fabricate a perpendicular magnetized SO device based on CoFeB/MgO/CoFeB [68]. Due to the large PMA in CoFeB/MgO interface counteracting the demagnetizing field, the threshold current density decreases to only 2.2×10^6 A/cm^2. Then in 2013, Kubota et al. prepared a nanopillar SO with a perpendicular magnetized FeCo free layer. The device exhibited a 0.55μW emission power under a high vertical magnetic field (3 KOe) [71]. Fang et al. has recently achieved 3.2 μW microwave emission power in SO with canted polarized free layer [171], which can not only be used for microwave emission, but also can be used as a microwave energy source to provide energy for other microwave devices. Besides, vortex-type SO with optimized structures has shown larger microwave emission power [171], as is shown in Figure 6.11. The optimized SO's magnetoresistance ratio is up to 190% with a resistance area value of 4.1 Ω·μm^2. And a 10.1 μW output power has been obtained, which is the highest value from a single SO to date, with excellent frequency stability ($f_{SO}/\Delta f = 2000$ with the timing jitter of 12 ps).

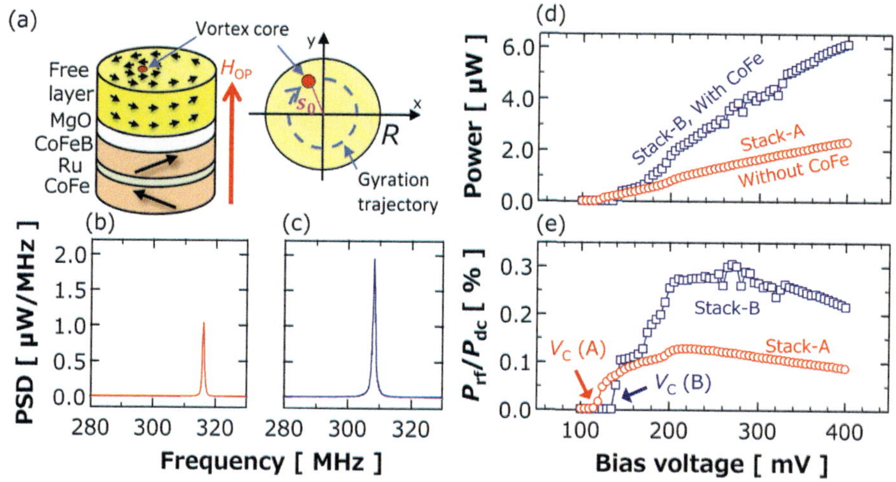

Figure 6.11 (a) Schematics of the stacking structure with magnetization configuration (left) and a top view of vortex gyration (right) of the SO. Typical power spectral density (PSD) spectra for (b) stack-A (without CoFe insertion) and (c) stack-B (with CoFe insertion) measured at $H_{OP} = 4.3$ kOe and $V_{dc} = 240$ mV. (d) Bias voltage (V_{dc}) dependence of emission power for stack-A and stack-B at $H_{OP} = 4.3$ kOe. (e) V_{dc} dependence of conversion efficiency (P_{rf}/P_{dc}) for stack-A and stack-B. Arrows indicate the critical voltages V_C for both stacks. *Source:* Reproduced with permission[172]. Copyright 2016, AIP Publishing.

Moreover, the output power of SO can be enhanced by phase-locking or constructing a synchronized SO array. To achieve high microwave power, Shehzaad Kaka et al. fabricated a coupled spin valve device with 500 nm space interval. Through phase-locking between the two oscillators they obtained an enhanced 9.9 pW output power [52]. Then in 2009 Fert et al. demonstrated a similar four nanocontact spin valve coupled by antivortices, with the integrated power of 83 nW/A^2 [106]. Because synchronization between spatially separated MTJ or spin valve through microwave signal or spin wave was usually small. Researchers have adopted a phase-lock loop and self-locking technique to optimize microwave emissions based on SOs. Yuasa and Fukushima's group made a breakthrough that brings the phase-lock loop into SO, and thus obtained a 0.1 μW microwave output power [145].

6.5.3 Linewidth

The linewidth (Δf, full width at half maximum of the power spectra), as defined in Figure 6.12, is another important parameter to evaluate the performance of the SOs. In general, the linewidth is dependent on the device geometry, material structure, and operation conditions. Nanocontact oscillators usually show a narrow linewidth at varied currents and external magnetic field (1 ~ 100 MHz) [173-175], while it is ranging from tens of MHz to GHz level for the nanopillar device [176–180]. One critical goal of SO design is to achieve narrow linewidth while maintaining a large output power at the same time. Because of the dependence of the effective field on magnetization, there existed an intrinsic coupling between magnetic moment oscillator amplitude and output signal phase [143, 181]. Such nonlinear coupling leads to an external contribution to the phase noise coming from a renormalization of the power noise via the non-linear frequency shift (N).

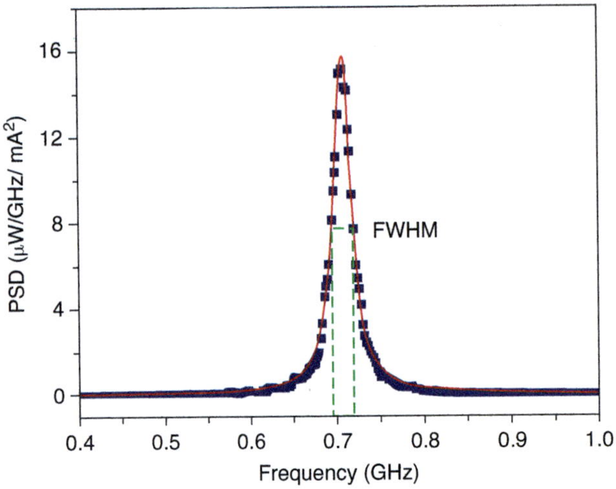

Figure 6.12 A typical MTJ-based SO with a narrow linewidth. *Source:* Reproduced with permission [182]. Copyright 2013, Royal Society of Chemistry.

There are many factors that have effects on the linewidth broadening of the device. On the one hand, it's found theoretically [183, 184] and experimentally [185, 186] that the linewidth shows a strong dependence on temperature. In the ideal case, the magnetic moment precession angle is unchanged under specific temperature. But the thermal fluctuations (ΔT) will strongly affect the precession angle by changing the equilibrium trajectory of the magnetic dynamics and lead to transition between different dynamic modes. Thus, a linewidth broadening and frequency shift is induced [187–189]. On the other hand, materials defects and unregular device boundaries would result in redundant non-uniform oscillation modes, thereby worsening device linewidth. Such intrinsic linewidth broadening in the fabricating process is usually inevitable and difficult to eliminate.

Extensive efforts have been devoted to reducing the nonlinear frequency shift (N) to as low as possible to narrow the linewidth [98, 177, 190–192]. Experimental results in recent years have indicated that the nonlinear factor is the excitation of a vortex mode [121, 126, 158], where the linewidth can be adjusted by the strength and direction of the magnetic field [193], and the vertical anisotropy of the interface in the free layer of the SO device itself. Jiang S. et al. prepared a nanocontact [Co/Pd]/Cu/[Co/Ni] spin valve device [191], and adjusted the vertical anisotropy of the interface in the Co/Ni multilayer film through He ion diffusion. As the vertical anisotropy field at the interface decreases from –0.68 T to 0.03 T, the nonlinear factor is reduced by two orders of magnitude.

Another way to reduce the linewidth of SO was to utilize the vortex mode. Recent research shows that SO devices based on the vortex nucleus have a smaller linewidth than the uniformly precession SO devices [121, 126, 158]. In spin vortex-oscillators, sub-MHz linewidth with a sub-GHz oscillator frequency has been demonstrated [121, 194, 195], with Figure 6.13 giving one example. In addition, the phase-locked injection and synchronization of SO devices can not only increase the output power of the device, as mentioned earlier, but also reduce the linewidth of the device.

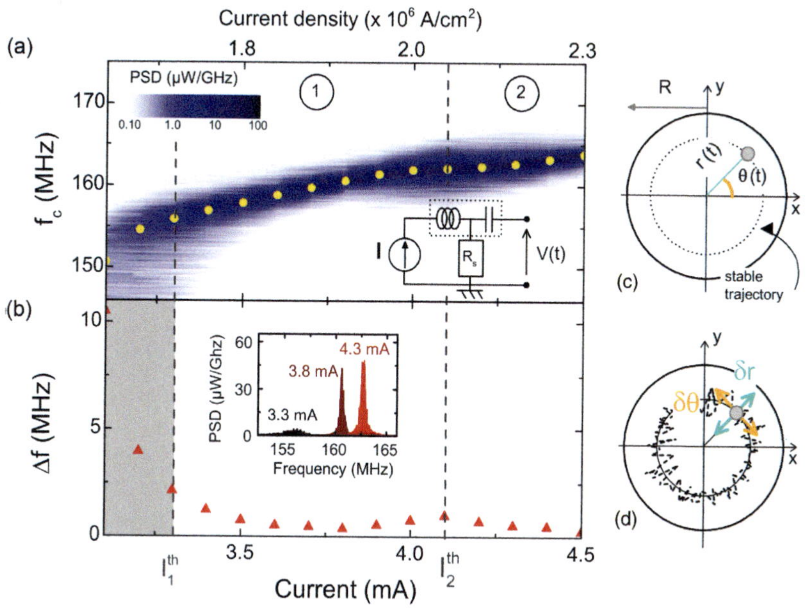

Figure 6.13 Rf characteristics of a vortex-based STNO having narrow linewidth measured for $H\perp = 4.4$ kOe. (a) Color scale of the PSD vs current and (yellow dots) carrier frequency *fc* extracted from fit. Inset: Schematic of the measurement set-up. The *Rs* resistance represents the STNO resistance, and *V*(t) is the input measurement voltage. (b) *f* vs current. Inset: Voltage PSD measured for 3.3, 3.8 and 4.3 mA. *Source:* Reproduced with permission [194]. Copyright 2014, American Physical Society.

6.5.4 Phase-locking and Synchronization

Similar to the oscillator synchronization in the IC circuit, the realization of phase lock or synchronization between SOs is an important means to reduce the microwave output linewidth and increase output power [52, 196, 197]. In 2005, Kaka et al. realized the current-driven STO phase-locked synchronization in two nanocontact spin valves separated by 400 nm [52]. The current regulates the microwave emission frequencies f_a and f_b of the two spin valves, respectively. When the two frequencies are close to each other, there are microwave signal spikes on the PSD graph, with the power density of the phase-locking signal ten times higher than that of a single signal. Subsequently, Rippard et al. realized the phase lock of AC current injection in a single STO device and proved that the signal phase changed by ± 90° within the phase lock range [196]. Experiments show that the output linewidth of the oscillating-coupled STO array is at least one order of magnitude smaller than the linewidth of a single SO [52, 53, 106].

At present, the main methods to realize SO synchronization include phase-locked loop circuit [198], current coupling [199–201], spin wave[173], etc. Experimental progress on this issue remains modest and will be a very hot research topic shortly soon [106]. In 2015, some researchers proposed that the propagating spin waves can be utilized to synchronize the multiple nanocontact spin-torque oscillators [202]. The experimental results show that the separation range of robust synchronization between two oscillators exceeds 1 μm. One year later, Awad and his colleagues demonstrated mutual synchronization of up to nine individual SHOs, each separated by 300 nm [90]. Through further tailoring of the connection regions, the researchers extend the synchronization range to 4 μm. However, the short-range feature of the magnetic coupling has largely hampered control of the synchronization process. In 2017, as is shown in Figure 6.14, Lebrun et al. demonstrated a mutual synchronization of two spin-torque oscillators with a large separation distance through their long-range self-emitted microwave currents [203], which led to a great improvement of both the emitted power and the linewidth. Most recently, Zahedinejad et al. achieved a robust mutual synchronization of two-dimensional SHO arrays ranging from 22 to 88 nano-constrictions [204], where the auto-oscillation linewidth Δf is governed by white noise and the signal quality factor Q. The Q increases linearly with the number of mutually synchronized nano-constrictions (N), reaching 170000 in the largest arrays. The same synchronization maps can be used for neuromorphic vowel recognition, which will be discussed in the next section.

Although there have been a lot of experimental studies on the phase-locking of SO devices, there is still a long way to deploy the large-scale phase-locking arrays. There are some challenges: SO devices based on MTJs are composed of multilayer films with a few nanometers thick, especially the thickness of the middle layer MgO is only 1 nm. Large-scale SO arrays are difficult to prepare, and adjacent SO devices usually have different microwave emission performances, which increase the difficulty of phase-locked synchronization. To further improve synchronization, the method of combining vortex core-based SO and spin-torque diode (STD) devices to form an array to achieve broadband microwave detection and neuromorphic calculations has been proposed [205, 206]. Moreover, the structure and processing technology of SHO are simpler compared with STO. Although there is little research on its array phase locking at present, it holds great potential in implementing large scale arrays in the future.

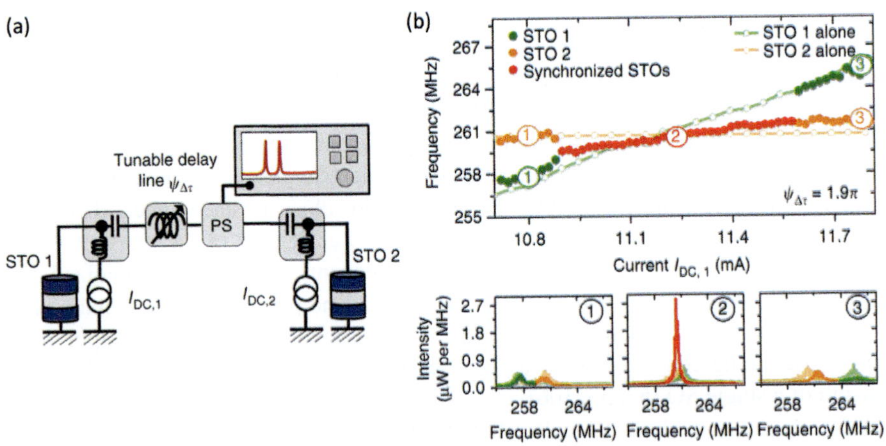

Figure 6.14 Scheme of phase-locking circuit. (a) Frequency varies as a function of input d.c. current (b), and (c) mutual synchronization of two SOs by adjusting their d.c. bias current. *Source:* Reproduced from [203] under CC BY 4.0.

6.6 Applications

As mentioned in the introduction, one particular advantage of SOs is their ability to realize multiple functions, from microwave source to magnetic sensors to signal processing, even with the same device. In this section, we will introduce a variety of applications of SOs, including microwave emission, spin wave emitter, microwave detector, magnetic field detectors, energy harvester, and neuromorphic computing.

6.6.1 Microwave Source

In comparison to traditional microwave oscillators based on semiconductor devices, SO holds several advantages, including wide tunable frequency range (from several megahertz to tens of gigahertz) [74, 144, 210, 211], high switching speed [212, 213] (0.2~0.8 ns), and low power consumption [14, 29, 52, 212, 213]. Here we focus on the relevant history and advances.

Following the first experimental microwave emission in 2003, the microwave emission approximation and synchronization theory based on free layer/interlayer/pinned layer sandwich structure devices was put forward [14, 41, 196]. In 2006, Nazarov et al. prepared a SO device with a 250 nm diameter, achieving microwave emission from 100 MHz to 18 GHz, and the peak of the microwave signal appeared at 7 GHz with a 21 MHz linewidth [64]. However, these microwave emitters need an external magnetic field of 200/440 Oe in the horizontal and vertical directions, limiting their practical use. Subsequently, a lot of research concentrated on reducing the linewidth of the microwave signal and increasing the output power under a zero magnetic field [143, 214–216]. Experiments have confirmed that the narrowest linewidth of SOs based on MTJ can reach 3.4 MHz [217] and the output power can reach over 10.1 μW [72, 172]. Besides, by using a tilted magnetization free layer, an SO based on CoFeB/MgO/CoFeB MTJ can generate a microwave emission without the applied magnetic field, which gives more opportunity to MTJ–SO-based applications. Rajapakse et al. also took this approach to realize 6 GHz microwave emission at zero field. With the development of mutual synchronization techniques, the linewidth and output power can be improved conspicuously while the phase noise has been reduced sharply [218, 219]. These improvements have made great contributions to the early realization of SO as a microwave source.

Table 6.1 summarizes the performance parameters of numerous microwave sources reported in recent years. As can be seen from the table, spin valve devices can be used as oscillators but usually need a magnetic field, while a single STO based on an MTJ device has realized small linewidth and operation under zero or weak magnetic field. Notably, the spin vortex oscillator had narrow linewidth compared to uniform precession mode STO. Besides, the phase-locking loop or self-locking also could lead to higher output power and smaller linewidth down to several MHz. These techniques had been key factors in ~μW scale microwave emission. SHO also could be utilized as a microwave source with a simpler structure compared with STO, but up to now there are no SHO devices able to obtain a tens-of-microwatts output, and more attention should be paid on the performance enhancement of a SHO device.

Table 6.1 Device parameters summary.

Time	Device configuration	Frequency (GHz)	Output power (dBm)	Linewidth (MHz)	Magnetic field (KOe)	Reference
2004	spin valve	9.7–34.4	–74	3.58–1.8	73	[207]
2008	spin valve phase-locking	1.88	–22	17	2.65	[208]
2009	nanocontact spin valve	10–46	-	4.5	138	[144]
2010	STO	8	–26	98	0.3	[67]
2012	STO	0.7	–35	25	0.26	[70]
2013	nanocontact STO	3.92–4.75	–27	<20	0.21–0.3	[72]
2015	STO phase-locking	7.34	>–40	4.1	-	[145]
2016	STO phase-locking	2.565–2.567	–60	10	0.092	[209]
2016	STO	0.234	–20	0.116	-	[172]

6.6.2 Spin Wave Emitter

The use of spin waves as a post-CMOS paradigm to process digital and analog signals is an important application of magnonics [220, 221]. How to generate and utilize spin waves is the most important subject of magnonics. Using the SO as a spin wave emitter is one of the most promising methods [26, 222–224].

Zhu X. et al. predicted the spin wave in the spin valve device with a vertical injection of current using the micromagnetic model [222]. And then in 2005, Slavin et al. discovered that there are different spin wave modes in point-contact SO devices [223], which ignited the research on spin wave modes [178, 225–227]. Notably, Demidov et al. has directly observed the propagation of spin waves caused by STO and confirmed the propagation distance to be several micrometers for the first time by means of the microfocus Brillouin light scattering (BLS) spectroscopy in 2010 [26]. The study found that the spin wave mode is related to experimental conditions such as the structure of the oscillator, the direction angle of the external magnetic field, the polarity of the current, and temperature.

In addition, particular attention has been paid to phase-locking of SO devices driven by spin waves. However, since the early seminal experiments [52, 53], which demonstrated spin wave (SW)-based mutual synchronization of two nanocontact SOs [228, 229], progress has been relatively slow in terms of synchronizing more oscillators. It was not until 2013 that SW-based mutual synchronization of three high-frequency nanocontact SOs was demonstrated. Then, in 2016 Åkerman's group reported a new record of synchronization of up to five oscillators [202].

6.6.3 Microwave Detector and Energy Harvester

As well as a microwave source, SO can be used as a microwave detector and energy harvester based on the spin-torque diode effect [25, 230, 231]. The principle can be described as follows [25]: When an alternating current whose frequency approaches the natural FMR frequency of the MTJ's free layer is injected into the MTJ, it will excite magnetic precession via STT. When the sign of the current is positive, the magnetization is rotated towards the pinned-layer magnetization, resulting in a low resistance state (we assume that this direction is positive). During the next half of the alternating current, the magnetic moment is forced to rotate in the opposite direction, which corresponds to a high-resistance state. The microwave signal injected into SO is then transferred to a kind of varied d.c. voltage, resulting in the so-called spin transfer diode. A radio-frequency current then passes through the device, generating a measurable d.c.-current. This effect has been widely used in microwave detectors.

Originally in 2005, Tulapurkar et al. displayed a CoFeB/MgO/CoFeB spin diode [25], as is illustrated in Figure 6.15. Over the next few years, a lot of attempts were made to enhance

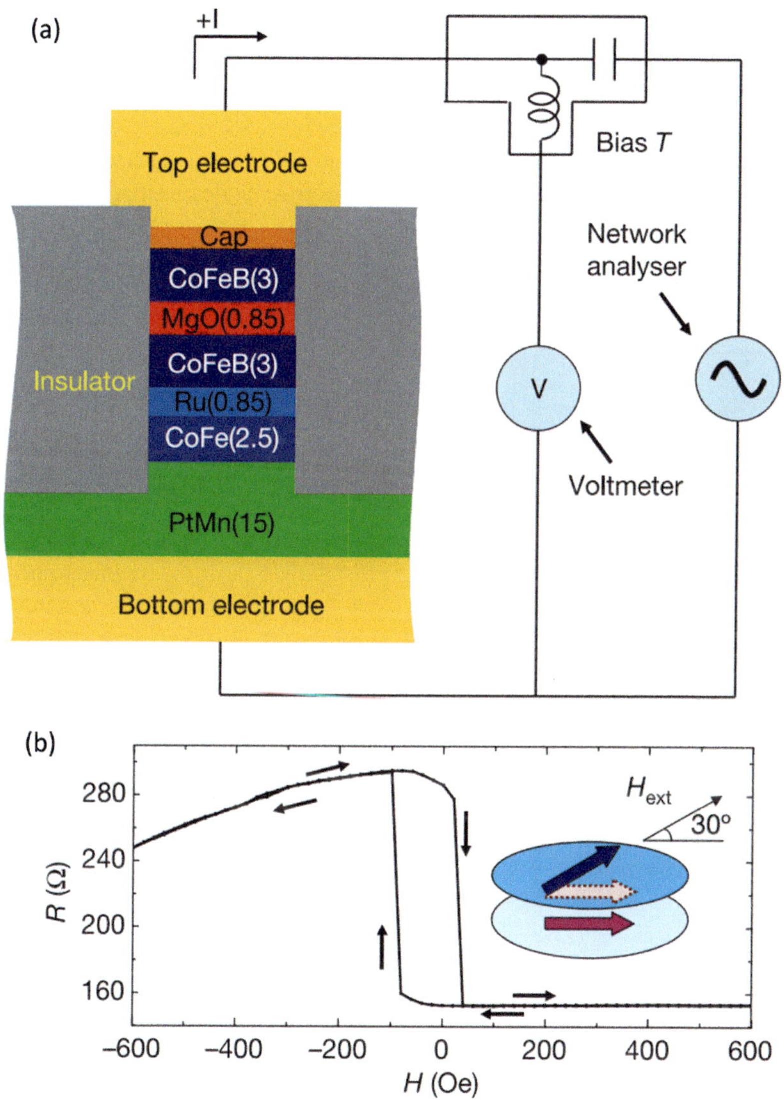

Figure 6.15 Microwave detector of a magnetic tunnel junction. (a) Schematic illustrations of the experiment's setup and (b) a magnetoresistance curve varies with the external magnetic field in both directions. *Source:* Reproduced with permission [25]. Copyright 2005, Springer Nature.

the detector sensitivity of the spin diode; for example, using some new tunnel junction structures and materials [232–235]. Then, in 2010, Ishibashi et al. reported a large diode sensitivity (170 mV/mW) CoFeB/MgO/CoFeB MTJ under a perpendicular magnetic field [236]. By adjusting the optimal direction of the external magnetic field and using an Fe-rich CoFeB free layer, a large voltage (180 μV) was obtained when the RF signal power was –25 dBm. The enriched Fe in the free layer leads to a lower critical current and avoids the *MR* reduction, thus increasing the output voltage. In 2014, d.c. bias current to MTJ, along with precise control, reached a detection sensitivity of 12 kV/W, which needs a very large external magnetic field applied in a particular direction [237]. In 2016, Fang et al. demonstrated nanoscale MTJ microwave detectors, featuring high-detection sensitivity of 75400 mV/mW at a low input microwave power of 10 nW based on an injection locking mechanism without an external magnetic field [73]. Figure 6.16 illustrates the device structure and experimental setup. After some optimizations, an ultrahigh detection sensitivity of 210 kV/W was demonstrated at room temperature [238]. In addition, a new type of spintronics-based RF detector generating a d.c. voltage and based on the expulsion of the vortex core in MTJ, with a detection sensitivity of 80 kV/W, was proposed [73, 239, 240]. And the spin-torque diode sensitivity in MTJ

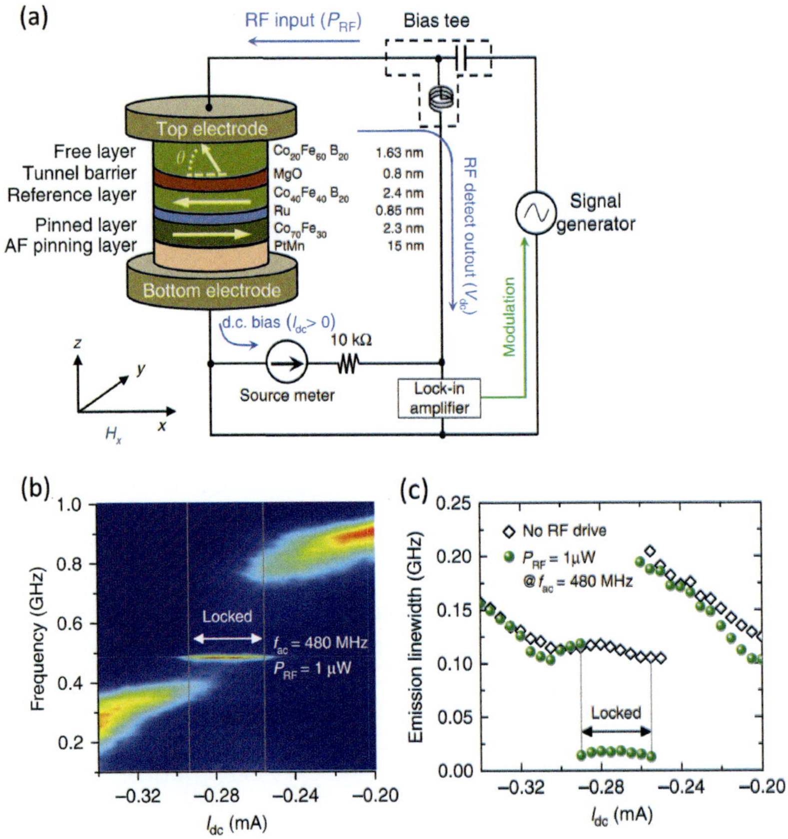

Figure 6.16 (a) Schematic illustration of the experimental setup. (b) Plot of oscillator frequency as a function of input d.c. current. (c) Same as (b), but linewidth as a function of d.c. current. There is a sharp decrease when phase-locking happened. *Source:* Reproduced from [73] under CC BY 4.0.

can be further improved by using synchronization [241, 242]. An ultra-wide RF detection bandwidth of over 3 GHz in MTJ has been experimentally investigated by tuning the free layer of PMA. In 2019, a nonlinear regime broadband detection for nanoscale spintronic diodes, where the rectification properties are independent of the input microwave frequency, was demonstrated[171]. Also, a wide RF detection bandwidth (about 1.8 GHz) can be implemented, even without any external bias field [243].

The concept of an energy harvester based on a broadband STD was first theoretically predicted in 2012 by Prokopenko et al. [244]. Later, broadband STD rectification was demonstrated experimentally with zero d.c. bias by Fang et al.[171]. A nonlinear regime broadband detection of STD was achieved. In this regime, the efficiency of rectification does not depend on RF in a wide range (the widest bandwidth approaching 1.2 GHz). This mode was demonstrated in MTJ with a tilted magnetization of the free layer due to perpendicular magnetic anisotropy and can be operated even without an external field. It was shown that the STD considered maintains a high enough d.c. voltage to supply the low power nanodevice (black phosphorus photosensor). Zhang et al. reported that the bandwidth of broadband STD detection was extended up to 3 GHz by means of free layer interface PMA engineering [243]. Moreover, with a free layer thickness of about 1.65 nm a wide RF rectification bandwidth (up to 1.8 GHz) was demonstrated without any external magnetic field. This result shows that SO microwave energy harvesters can be used as building blocks for the wireless powering of IoT nanodevices, medical implanted sensors, and portable and wearable electronic devices.

6.6.4 Magnetic Field Detector

SOs can not only be used for microwave emission but also for magnetic field detection [245–247]. Compared with traditional macro-sized magnetic field detectors, SOs' small size, high sensitivity and excellent anti-interference performance makes them especially promising candidates for miniature magnetic field detectors.

The principle of magnetic field detection based on STO is given as follows [248]: The d.c. bias current flowing through the device causes the precession of the magnetic moment of the free layer. When the external magnetic field is zero, the precession of the magnetic moment is not affected by the external magnetic field torque. When an external magnetic field is in the same direction as the free layer magnetic moment, the free layer magnetic moment is affected by the damping-like spin torque and tends to stabilize in the direction of the external magnetic field, causing the precession angle to decrease. Therefore, the frequency of the output microwave signal will increase by Δf. Similarly, when the direction of the external magnetic field is opposite to the direction of the free layer magnetic moment, it tends to move away from the precession axis due to the damping-like spin torque, the precession angle increases, and the output microwave signal frequency decreases by Δf. Thus, the frequency response of the SO device to different magnetic fields can be used to measure the strength of the external magnetic field. What's more, it doesn't require the detector to have a large signal-to-noise ratio.

In 2010, Braganca et al. prepared a CoFe/Ru/CoFe/Cu/NiFe in-plane magnetized spin-valve magnetic field detector with an 80 nm diameter [248] and achieved high sensitivity up to 180 GHz/T. In 2016, Ramaswamy et al. used an MTJ SO to detect the tiny magnetic field generated by the current in the micro loop [249]. They injected a current of 300–700 μA into the wire micro coil and used a SO device with a distance of 6.5 mm to detect the magnetic field generated by the current in the micro-coil. It's noted that the power of a single SO device is very low, and the power generated is only –70 dB even when the distance from the current micro coil is 1 mm. To achieve higher detection sensitivity and detection range, it is necessary to build a SO array [183, 215, 216].

6.6.5 Neuromorphic Computing

As the application of artificial intelligence booms, the demand for greater computing capabilities has soared. However, classic computers based on the Von Neumann architecture cannot meet the needs of the fast processing of big data and parallel computing. Therefore, it's urgent to create new efficient neuromorphic devices. SOs, with their nonlinear nature and ability to synchronize, are one of the most promising candidates to fabricate novel neuromorphic devices.

So far, many neural network models based on SO have been proposed [15, 250–256]. Much experimental progress has also been made. In 2017, Grollier's group in France used a single SO device with a diameter of 325 nm to achieve reservoir computing, and realized a spiking neural network in a single SO through time multiplexing technology [257]. Figure 6.17 gives the schematic of the device and transport performance. This new network is used to distinguish the sine/square wave component in the input signal and achieves a 99.6% recognition rate. In 2018, they took a different approach and realized the neural network recognition of 12 vowel signals with the phase-locking of four SOs [258], as shown in Figure 6.18. The output of the SO was adjusted by d.c. bias and the vowel signals were encoded into a different

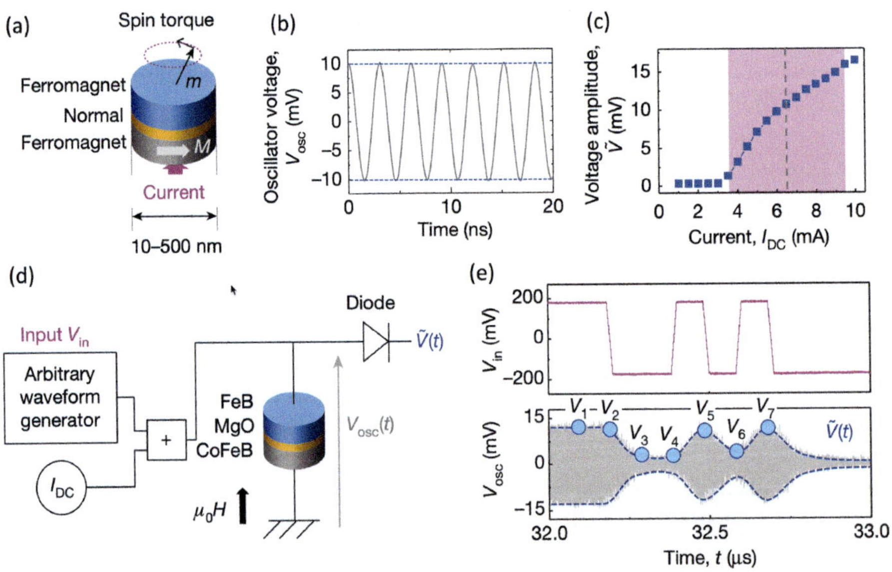

Figure 6.17 Schematic of a SO and transport performance. (a) Schematic illustration of a single SO device. (b) Oscillator voltage varies similarly as a sine function with time. (c) Voltage amplitude varies as current increases from zero to 10 milliampere. (d) Schematic illustration of the measurement circuit. (e) Oscillator voltage varies with the input voltage. *Source:* Reproduced with permission [257]. Copyright 2017, Springer Nature.

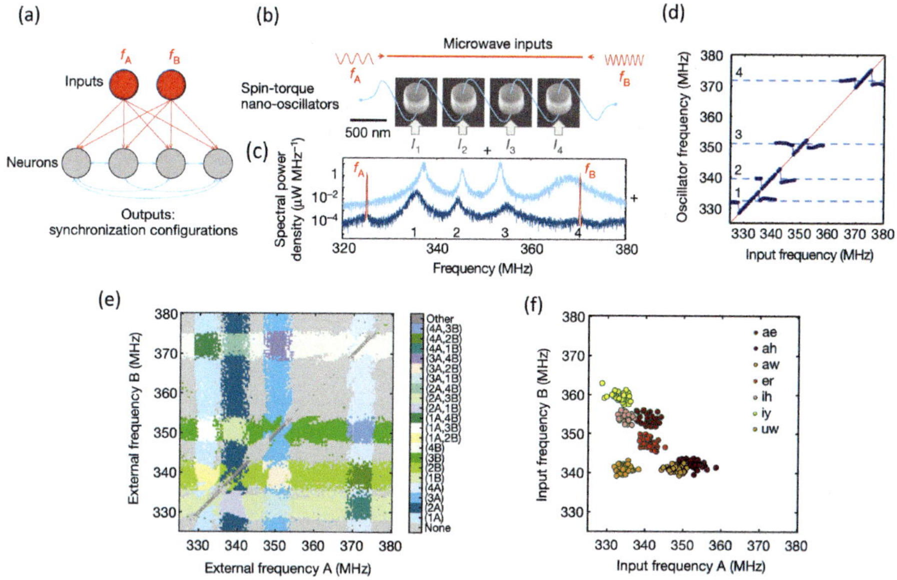

Figure 6.18 Phase-locking of four SOs and the corresponding neural network for vowel recognition. (a) Schematic illustration of the full connected neural network and (b) phase-locking of four SO devices by adjusting d.c. bias current I. (c) Microwave output emitted by the network of four oscillators without (light blue) and with (dark blue) the two microwave signals applied to the system. (d) Oscillator frequencies change as the frequency of external source A is swept from 330 MHz to 380 MHz. (e) Experimental synchronization map as a function of the frequencies of the external signals f_A and f_B. (f) Inputs applied to the system, represented in the (f_A, f_B) plane. *Source:* Reproduced with permission [258]. Copyright 2018, Springer Nature.

frequency signal. This SO array realizes speech recognition based on spintronics for the first time. However, it should be noted that this method, based on current-controlled phase-locking, has strict requirements on the performance of the single device and the preparation process.

Taking inspiration from these studies, Cai et al. studied sparse neuron simulation based on SO, and constructed a neural network based on SO using the time-division multiplexing technology in a single device [259], with Figure 6.19 showing the experimental setup and results. With the nonlinear function relationship between the input microwave signal frequency of the SO and the rectified output voltage as the neuron activation function, the MNIST(Mixed National Institute of Standards and Technology) handwritten digit recognition rate of the neural network reached 92.7%, which is slightly better than the artificial activation function constructed by a ReLu (Rectified linear unit) with an 85.9% recognition rate. The SO sparse neural network provides a new direction for the hardware development of sparse neuromorphic computing.

Recently, the Grollier group committed to a wireless connection between SO and STD devices to realize a device array and proposed an electrically synchronized microwave transmission and neural network hardware in theory [260]. Inspired by Grollier's work, Åkerman et al. fabricated a 4×4 SHO array neuromorphic chip, realizing vowel recognition [204]. This device has a 60 KHz linewidth with a 10 GHz high operating frequency, leading to a Q value of 170000. By adding two independent microwave currents to the current, it could achieve 20 phase-locking states corresponding to different vowel frequencies.

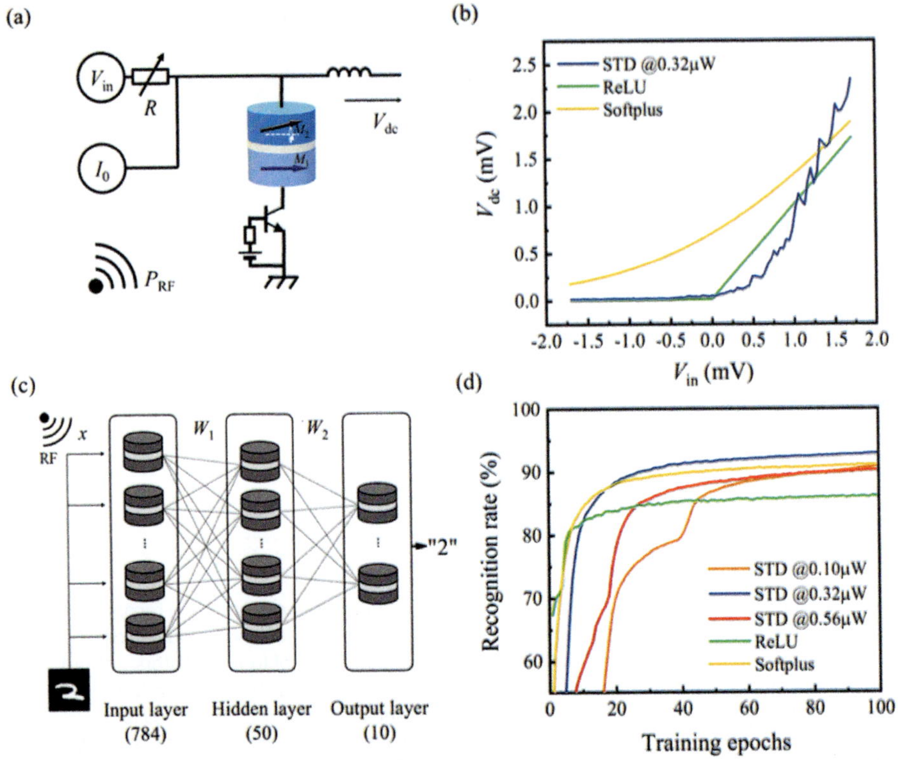

Figure 6.19 Sparse neuromorphic computing based on spin-torque diodes. (a) A schematic illustration of a single STD neuron. (b) The activation function of STD neurons (blue line) at the microwave frequency $f_{RF} = 1.92$ GHz and $P_{RF} = 0.32$ μW, compared to the Softplus (yellow line) and ReLU (green line) activation functions. (c) The structure of the ANN constructed to recognize the handwriting digit in the MNIST database. (d) The transition in the recognition rate obtained from the neural network constructed with STDs compared to a conventional neural network with a ReLU activation function. *Source:* Reproduced with permission [259]. Copyright 2019, AIP Publishing.

6.7 Summary and Outlook

This chapter has presented an overview of fundamental concepts such as spin valves, MTJs, and STT first. Then we briefly reviewed the structure and performance of the spin nano-oscillator including frequency, linewidth, output power and phase-locking. Finally, we summarized recent advances in application of SOs.

In the past decade, extensive theoretical and experimental works have been devoted to understanding and exploiting the novel physical phenomena of SOs. However, the difficulties and challenges remain on both the device and system level. In the first place, microwave emitters or microwave detectors based on SOs usually need an external strong magnetic field, which is impractical for application, even though the microwave emission can be obtained in a perpendicular device structure without the need for a magnetic field. The lower output power and large linewidth seriously limit their further application. Besides, the difference between SO devices, for example, their intrinsic oscillating frequency and device size, makes it difficult to realize synchronization with each other.

The adoption of new materials and new structures should be helpful. Recently, some researchers utilized piezoelectric materials combined with MTJ to control the interface perpendicular anisotropy, reducing threshold current density. Shigemi's group et al. demonstrated a $Co_3Mn/MgO/Co_3Mn$-based MTJ showing a 200%~250% *MR* at room temperature [261]. A SO device having two perpendicular magnetization free layers and a fixed layer has been shown to feature the maximum output frequency (6 GHz) at zero magnetic fields [74]. The combination of STT and SOT is also promising in producing large output power. These advances open up new ways of developing high output power, low linewidth, and low power consumption for SO-based devices.

References

1 Žutić, I., Fabian, J. et al. (2004). Spintronics: Fundamentals and applications. Rev. Mod. Phys. 76: 323–410.
2 Bader, S.D. and Parkin, S.S.P. (2010). Spintronics. Annu. Rev. Condens. Matter Phys. 1: 71–88.
3 Berger, L. (1996). Emission of spin waves by a magnetic multilayer traversed by a current. *Phys. Rev. B* 54: 9353–9358.
4 Slonczewski, J.C. (1996). Current-driven excitation of magnetic multilayers. *J. Magn. Magn. Mater.* 159: L1–L7.
5 Tsoi, M., Jansen, A.G.M. et al. (1998). Excitation of a Magnetic Multilayer by an Electric Current. *Phys. Rev. Lett.* 80: 4281–4284.
6 Myers, E.B., Ralph, D.C. et al. (1999). Current-induced switching of domains in magnetic multilayer devices. *Science* 285: 867–8701.
7 Katine, J.A. and Buhrman, R.A. (2000). Current-Driven Magnetization Reversal and Spin-Wave Excitations in Co. *Phys. Rev. Lett.* 84: 3149–3152.
8 Grollier, J., Cros, V. et al. (2001). Spin-polarized current induced switching in Co/Cu/Co pillars. *Appl. Phys. Lett.* 78: 3663–3665.
9 Baibich, M.N., Broto, J.M. et al. (1988). Giant Magnetoresistance of (001)Fe/(001)Cr Magnetic Superlattices. *Phys. Rev. Lett.* 61: 2472–2475.
10 Julliere, M. (1975). Tunneling between ferromagnetic films. *Phys. Lett. A* 54: 225–226.
11 Bhatti, S., Sbiaa, R. et al. (2017). Spintronics based random access memory: a review. *Mater. Today* 20: 530–548.
12 Khvalkovskiy, A.V., Apalkov, D. et al. (2013). Basic principles of STT-MRAM cell operation in memory arrays. *J. Phys. D: Appl. Phys.* 46: 074001.
13 Sbiaa, R. and Piramanayagam, S.N. (2017). Recent Developments in Spin Transfer Torque MRAM. *Physica Status Solidi (RRL) – Rapid Research Letters* 11: 1700163.
14 Kiselev, S.I., Sankey, J.C. et al. (2003). Microwave oscillations of a nanomagnet driven by a spin-polarized current. *Nature* 425: 380–383.
15 Locatelli, N., Cros, V. et al. (2014). Spin-torque building blocks. *Nat. Mater.* 13: 11–20.
16 Matsumoto, R., Chanthbouala, A. et al. (2011). Spin-Torque Diode Measurements of MgO-Based Magnetic Tunnel Junctions with Asymmetric Electrodes. *Appl. Phys. Express* 4: 063001.
17 Dyakonov, M. and Perel, V. (1971). Possibility of Orienting Electron Spins with Current. (1971) Soviet Journal of Experimental and Theoretical Physics Letters. *Soviet J. Exp. Theor. Phys.* 13: 467.

18 Sinova, J., Valenzuela, S.O. et al. (2015). Spin Hall effects. *Rev. Mod. Phys.* 87: 1213–1260.

19 Jungwirth, T., Wunderlich, J. et al. (2012). Spin Hall effect devices. *Nat. Mater.* 11: 382–390.

20 Liu, L., Pai, C.F. et al. (2012). Spin-torque switching with the giant spin Hall effect of tantalum. *Science* 336: 555–558.

21 Demidov, V.E., Urazhdin, S. et al. (2012). Magnetic nano-oscillator driven by pure spin current. *Nat. Mater.* 11: 1028–1031.

22 Liu, L., Lee, O.J. et al. (2012). Current-Induced Switching of Perpendicularly Magnetized Magnetic Layers Using Spin Torque from the Spin Hall Effect. *Phys. Rev. Lett.* 109.

23 Iihama, S., Taniguchi, T. et al. (2018). Spin-transfer torque induced by the spin anomalous Hall effect. *Nat. Electron.* 1: 120–123.

24 Tarequzzaman, M., Böhnert, T. et al. (2019). Spin torque nano-oscillator driven by combined spin injection from tunneling and spin Hall current. *Commun. Phys.* 2: 20.

25 Tulapurkar, A.A., Suzuki, Y. et al. (2005). Spin-torque diode effect in magnetic tunnel junctions. *Nature* 438: 339–342.

26 Demidov, V.E., Urazhdin, S. et al. (2010). Direct observation and mapping of spin waves emitted by spin-torque nano-oscillators. *Nat. Mater.* 9: 984–988.

27 M.M. (2011). Direct observation of a propagating spin wave induced by spin-transfer torque. *Nat. Nanotechnol.* 6: 635–638.

28 Chen, T., Dumas, R.K. et al. (2016). Spin-Torque and Spin-Hall Nano-Oscillators. *Proc. IEEE* 104: 1919–1945.

29 Tsoi, M., Jansen, A.G.M. et al. (2000). Generation and detection of phase-coherent current-driven magnons in magnetic multilayers. *Nature* 406: 46–48.

30 Sun, J.Z. (1999). Current-driven magnetic switching in manganite trilayer junctions. *J. Magn. Magn. Mater.* 202: 157–162.

31 Sani, S., Persson, J. et al. (2013). Mutually synchronized bottom-up multi-nanocontact spin-torque oscillators. *Nat. Commun.* 4: 2731.

32 Rippard, W.H., Pufall, M.R. et al. (2003). Quantitative studies of spin-momentum-transfer-induced excitations in Co/Cu multilayer films using point-contact spectroscopy. *Appl. Phys. Lett.* 82: 1260–1262.

33 Zeng, Z., Finocchio, G. et al. (2013). Ultralow-current-density and bias-field-free spin-transfer nano-oscillator. *Sci. Rep.* 3: 1426.

34 Rippard, W.H., Pufall, M.R. et al. (2006). Comparison of frequency, linewidth, and output power in measurements of spin-transfer nanocontact oscillators. *Phys. Rev. B* 74: 224409.

35 Demidov, V.E., Urazhdin, S. et al. (2011). Control of Magnetic Fluctuations by Spin Current. *Phys. Rev. Lett.* 107.

36 Demidov, V.E., Urazhdin, S. et al. (2011). Wide-range control of ferromagnetic resonance by spin Hall effect. *Appl. Phys. Lett.* 99: 172501.

37 Liu, L., Pai, C.-F. et al. (2012). Magnetic Oscillations Driven by the Spin Hall Effect in 3-Terminal Magnetic Tunnel Junction Devices. *Phys. Rev. Lett.* 109: 186602.

38 Demidov, V.E., Urazhdin, S. et al. (2014). Nanoconstriction-based spin-Hall nano-oscillator. *Appl. Phys. Lett.* 105: 172410.

39 Durrenfeld, P., Awad, A.A. et al. (2017). A 20 nm spin Hall nano-oscillator. *Nanoscale* 9: 1285–1291.

40 Duan, Z., Smith, A. et al. (2014). Nanowire spin torque oscillator driven by spin orbit torques. *Nat. Commun.* 5: 5616.

41 Rippard, W., Pufall, M. et al. (2004). Direct-Current Induced Dynamics in Co90Fe10/Ni80Fe20 Point Contacts. *Phys. Rev. Lett.* 92: 027201.

42 Mangin, S., Ravelosona, D. et al. Current-induced magnetization reversal in nanopillars with perpendicular anisotropy. *INTERMAG 2006 – IEEE International Magnetics Conference*, IEEE.

43 Houssameddine, D., Ebels, U. et al. (2007). Spin-torque oscillator using a perpendicular polarizer and a planar free layer. *Nat. Mater.* 6: 441–447.

44 Rippard, W.H., Deac, A.M. et al. (2010). Spin-transfer dynamics in spin valves with out-of-plane magnetized CoNi free layers. *Phys. Rev. B* 81: 014426.

45 Mohseni, S.M., Sani, S.R. et al. (2014). Magnetic droplet solitons in orthogonal nano-contact spin torque oscillators. *Phys. B Condens. Matter* 435: 84–87.

46 Fang, B., Feng, J. et al. (2016). Zero-field spin transfer oscillators based on magnetic tunnel junction having perpendicular polarizer and planar free layer. *AIP Advances* 6: 125305.

47 Chen, T.Y., Huang, S.X. et al. (2006). Enhanced magnetoresistance induced by spin transfer torque in granular films with a magnetic field. *Phys. Rev. Lett.* 96: 207203.

48 Zhou, Y., Zha, C.L. et al. (2008). Spin-torque oscillator with tilted fixed layer magnetization. *Appl. Phys. Lett.* 92: 262508.

49 Zhou, Y., Bonetti, S. et al. (2009). Zero-field precession and hysteretic threshold currents in a spin torque nano device with tilted polarizer. *New J. Phys.* 11: 103028.

50 He, P.-B., Li, Z.-D. et al. (2009). Current-driven ferromagnetic resonance in magnetic trilayers with a tilted spin polarizer. *J. Appl. Phys.* 105: 043908.

51 Slonczewski, J.C. (1999). Excitation of spin waves by an electric current. *J. Magn. Magn. Mater.* 195: L261–L268.

52 Kaka, S., Pufall, M.R. et al. (2005). Mutual phase-locking of microwave spin torque nano-oscillators. *Nature* 437: 389–392.

53 Mancoff, F.B., Rizzo, N.D. et al. (2005). Phase-locking in double-point-contact spin-transfer devices. *Nature* 437: 393–395.

54 Consolo, G., Azzerboni, B. et al. (2007). Excitation of self-localized spin-wave bullets by spin-polarized current in in-plane magnetized magnetic nanocontacts: A micromagnetic study. *Phys. Rev. B* 76.

55 Bonetti, S., Tiberkevich, V. et al. (2010). Experimental Evidence of Self-Localized and Propagating Spin Wave Modes in Obliquely Magnetized Current-Driven Nanocontacts. *Phys. Rev. Lett.* 105: 217204.

56 Matsumoto, R., Lequeux, S. et al. (2019). Chaos and Relaxation Oscillations in Spin-Torque Windmill Spiking Oscillators. *Phys. Rev. Appl.* 11: 044093.

57 Okura, R., Sakuraba, Y. et al. (2011). *Appl. Phys. Lett.* 99: 052510.

58 Huang, H.B., Ma, X.Q. et al. (2013). Micromagnetic simulation of high-power spin-torque oscillator in half-metallic Heusler alloy spin valve nanopillar. *AIP Advances* 3: 032132.

59 Sinha, J., Hayashi, M. et al. (2011). Large amplitude microwave emission and reduced nonlinear phase noise in Co2Fe(Ge0.5Ga0.5) Heusler alloy based pseudo spin valve nanopillars. *Appl. Phys. Lett.* 99: 162508.

60 Moriyama, T., Finocchio, G. et al. (2012). Phase locking and frequency doubling in spin-transfer-torque oscillators with two coupled free layers. *Phys. Rev. B* 86: 060411.

61 Fuchs, G.D., Emley, N.C. et al. (2004). Spin-transfer effects in nanoscale magnetic tunnel junctions. *Appl. Phys. Lett.* 85: 1205–1207.

62 Parkin, S.S., Kaiser, C. et al. (2004). Giant tunnelling magnetoresistance at room temperature with MgO (100) tunnel barriers. *Nat. Mater.* 3: 862–867.

63 Yuasa, S., Nagahama, T. et al. (2004). Giant room-temperature magnetoresistance in single-crystal Fe/MgO/Fe magnetic tunnel junctions. *Nat. Mater.* 3: 868–871.

64 Nazarov, A.V., Olson, H.M. et al. (2006). Spin transfer stimulated microwave emission in MgO magnetic tunnel junctions. *Appl. Phys. Lett.* 88: 162504.

65 Houssameddine, D., Florez, S.H. et al. (2008). Spin transfer induced coherent microwave emission with large power from nanoscale MgO tunnel junctions. *Appl. Phys. Lett.* 93: 022505.

66 Deac, A.M., Fukushima, A. et al. (2008). Bias-driven high-power microwave emission from MgO-based tunnel magnetoresistance devices. *Nat. Phys.* 4: 803–809.

67 Zeng, Z.M., Upadhyaya, P. et al. (2011). Enhancement of microwave emission in magnetic tunnel junction oscillators through in-plane field orientation. *Appl. Phys. Lett.* 99: 032503.

68 Ikeda, S., Miura, K. et al. (2010). A perpendicular-anisotropy CoFeB-MgO magnetic tunnel junction. *Nat. Mater.* 9: 721–724.

69 Taniguchi, T. (2013). Critical Field of Spin Torque Oscillator with Perpendicularly Magnetized Free Layer. *Applied Physics Express* 6: 123003.

70 Zeng, Z., Amiri, P.K. et al. (2012). High-Power Coherent Microwave Emission from Magnetic Tunnel Junction Nano-oscillators with Perpendicular Anisotropy. *ACS Nano* 6: 6115–6121.

71 Kubota, H., Yakushiji, K. et al. (2013). Spin-Torque Oscillator Based on Magnetic Tunnel Junction with a Perpendicularly Magnetized Free Layer and In-Plane Magnetized Polarizer. *Appl. Phys. Express* 6: 103003.

72 Maehara, H., Kubota, H. et al. (2013). Large Emission Power over 2 μW with HighQFactor Obtained from Nanocontact Magnetic-Tunnel-Junction-Based Spin Torque Oscillator. *Appl. Phys. Express* 6: 113005.

73 Fang, B., Carpentieri, M. et al. (2016). Giant spin-torque diode sensitivity in the absence of bias magnetic field. *Nat. Commun.* 7: 11259.

74 Rajapakse, R.N.S., Zeng, Z.M. et al. (2020). High-Frequency Microwave Emission of a Trilayer Magnetic Tunnel Junction in the Absence of External Magnetic Bias Field. *Phys. Rev. Appl.* 14: 014040.

75 Demidov, V.E., Urazhdin, S. et al. (2014). Spin Hall controlled magnonic microwaveguides. *Appl. Phys. Lett.* 104: 152402.

76 Ranjbar, M., Durrenfeld, P. et al. (2014). CoFeB-Based Spin Hall Nano-Oscillators. *IEEE Magn. Lett.* 5: 3000504.

77 Safranski, C., Barsukov, I. et al. (2017). Spin caloritronic nano-oscillator. *Nat. Commun.* 8: 117.

78 Collet, M., de Milly, X. et al. (2016). Generation of coherent spin-wave modes in yttrium iron garnet microdiscs by spin-orbit torque. *Nat. Commun.* 7: 10377.

79 Mellnik, A.R., Lee, J.S. et al. (2014). Spin-transfer torque generated by a topological insulator. *Nature* 511: 449–451.

80 Pai, C.F., Liu, L.Q. et al. (2012). Spin transfer torque devices utilizing the giant spin Hall effect of tungsten. *Appl. Phys. Lett.* 101: 122404.

81 Liu, R.H., Lim, W.L. et al. (2015). Dynamical skyrmion state in a spin current nano-oscillator with perpendicular magnetic anisotropy. *Phys. Rev. Lett.* 114: 137201.

82 Divinskiy, B., Demidov, V.E. et al. (2017). Nanoconstriction spin-Hall oscillator with perpendicular magnetic anisotropy. *Appl. Phys. Lett.* 111: 032405.

83 Divinskiy, B., Urazhdin, S. et al. (2019). Controlled nonlinear magnetic damping in spin-Hall nano-devices. *Nat. Commun.* 10: 5211.

84 Chen, L., Urazhdin, S. et al. (2020). Magnetic Droplet Mode in a Vertical Nanocontact-Based Spin Hall Nano-Oscillator at Oblique Fields. *Phys. Rev. Appl.* 13: 024034.

85 Durrenfeld, P., Gerhard, F. et al. (2015). Spin Hall effect-controlled magnetization dynamics in NiMnSb. *J. Appl. Phys.* 117: 17E103.

86 Padrón-Hernández, E., Azevedo, A. et al. (2011). Amplification of spin waves in yttrium iron garnet films through the spin Hall effect. *Appl. Phys. Lett.* 99.

87 Jungfleisch, M.B., Zhang, W. et al. (2016). Large Spin-Wave Bullet in a Ferrimagnetic Insulator Driven by the Spin Hall Effect. *Phys. Rev. Lett.* 116: 057601.

88 Evelt, M., Safranski, C. et al. (2018). Spin Hall-induced auto-oscillations in ultrathin YIG grown on Pt. *Sci. Rep.* 8: 1269.

89 Liu, L., Moriyama, T. et al. (2011). *Spin-torque ferromagnetic resonance induced by the spin Hall effect. Phys. Rev. Lett.* 106: 036601.

90 Awad, A.A., Dürrenfeld, P. et al. (2016). Long-range mutual synchronization of spin Hall nano-oscillators. *Nat. Phys.* 13: 292–299.

91 Mazraati, H., Chung, S. et al. (2016). Low operational current spin Hall nano-oscillators based on NiFe/W bilayers. *Appl. Phys. Lett.* 109: 242402.

92 Mazraati, H., Etesami, S.R. et al. (2018). Auto-oscillating Spin-Wave Modes of Constriction-Based Spin Hall Nano-oscillators in Weak In-Plane Fields. *Phys. Rev. Appl.* 10: 054017.

93 Stiles, M.D., Xiao, J. et al. (2004). Phenomenological theory of current-induced magnetization precession. *Phys. Rev. B* 69: 054408.

94 Polianski, M.L. and Brouwer, P.W. (2004). Current-Induced Transverse Spin-Wave Instability in a Thin Nanomagnet. *Phys. Rev. Lett.* 92: 026602.

95 Özyilmaz, B., Kent, A.D. et al. (2004). Current-Induced Excitations in Single Cobalt Ferromagnetic Layer Nanopillars. *Phys. Rev. Lett.* 93.

96 Özyilmaz, B. and Kent, A.D. (2006). Current-induced switching in single ferromagenetic layer nanopillar junctions. *Appl. Phys. Lett.* 88: 162506.

97 Sani, S.R. (2013). Microwave Signal Generation in Single-Layer Nano-Contact Spin Torque Oscillators. *IEEE Trans. Magn.* 49: 4331–4334.

98 Dürrenfeld, P., Gerhard, F. et al. (2016). Low-current, narrow-linewidth microwave signal generation in NiMnSb based single-layer nanocontact spin-torque oscillators. *Appl. Phys. Lett.* 109: 222403.

99 Kurebayashi, H., Sinova, J. et al. (2014). An antidamping spin–orbit torque originating from the Berry curvature. *Nat. Nanotechnol.* 9: 211–217.

100 Safranski, C., Montoya, E.A. et al. (2019). Spin–orbit torque driven by a planar Hall current. *Nat. Nanotechnol.* 14: 27–30.

101 An, H., Kageyama, Y. et al. (2016). Spin–torque generator engineered by natural oxidation of Cu. *Nat. Commun.* 7: 13069.

102 Gao, T., Qaiumzadeh, A. et al. (2018). Intrinsic Spin-Orbit Torque Arising from the Berry Curvature in a Metallic-Magnet/Cu-Oxide Interface. *Phys. Rev. Lett.* 121: 017202.

103 Troncoso, R.E., Rode, K. et al. (2019). Antiferromagnetic single-layer spin-orbit torque oscillators. *Phys. Rev. B* 99.

104 Haidar, M., Awad, A.A. et al. (2019). A single layer spin-orbit torque nano-oscillator. *Nat. Commun.* 10.

105 Cowburn, R.P., Koltsov, D.K. et al. (1999). Single-domain circular nanomagnets. *Phys. Rev. Lett.* 83: 1042–1045.

106 Ruotolo, A., Cros, V. et al. (2009). Phase-locking of magnetic vortices mediated by antivortices. *Nat. Nanotechnol.* 4: 528–532.

107 Pribiag, V.S., Krivorotov, I.N. et al. (2007). Magnetic vortex oscillator driven by d.c. spin-polarized current. *Nat. Phys.* 3: 498–503.

108 Vukadinovic, N. and Boust, F. (2007). Three-dimensional micromagnetic simulations of magnetic excitations in cylindrical nanodots with perpendicular anisotropy. *Physical Review B* 75.

109 Liu, T., Puliafito, V. et al. (2016). *Reproducible formation of single magnetic bubbles in an array of patterned dots*. *J. Phys. D: Appl. Phys.* 49: 245002.

110 Zhang, X., Zhou, Y. et al. (2016). Magnetic bilayer-skyrmions without skyrmion Hall effect. *Nat. Commun.* 7: 10293.

111 Hrkac, G., Keatley, P.S. et al. (2015). Magnetic vortex oscillators. *J. Phys. D: Appl. Phys.* 48: 453001.

112 Moriya, R., Thomas, L. et al. (2008). Probing vortex-core dynamics using current-induced resonant excitation of a trapped domain wall. *Nat. Phys.* 4: 368–372.

113 Khvalkovskiy, A., Grollier, J. et al. (2009). Vortex oscillations induced by spin-polarized current in a magnetic nanopillar: Analytical versus micromagnetic calculations. *Phys. Rev. B* 80: 140401.

114 Carpentieri, M., Martinez, E. et al. (2011). High frequency spin-torque-oscillators with reduced perpendicular torque effect based on asymmetric vortex polarizer. *J. Appl. Phys.* 110: 093911.

115 Dussaux, A., Khvalkovskiy, A.V. et al. (2012). Field dependence of spin-transfer-induced vortex dynamics in the nonlinear regime. *Phys. Rev. B* 86: 014402.

116 Lehndorff, R., Bürgler, D. et al. (2009). Magnetization dynamics in spin torque nano-oscillators: Vortex state versus uniform state. *Phys. Rev. B* 80: 054412.

117 Manfrini, M., Devolder, T. et al. (2009). Agility of vortex-based nanocontact spin torque oscillators. *Appl. Phys. Lett.* 95: 192507.

118 Jaromirska, E., Lopez-Diaz, L. et al. (2011). Influence of geometry on current-driven vortex oscillations in nanocontact devices. *Phys. Rev. B* 83: 094419.

119 Pufall, M.R., Rippard, W.H. et al. (2007). Low-field current-hysteretic oscillations in spin-transfer nanocontacts. *Phys. Rev. B* 75: 140404.

120 Mistral, Q., van Kampen, M. et al. (2008). Current-driven vortex oscillations in metallic nanocontacts. *Phys. Rev. Lett.* 100: 257201.

121 Dussaux, A., Georges, B. et al. (2010). Large microwave generation from current-driven magnetic vortex oscillators in magnetic tunnel junctions. *Nat. Commun.* 1: 8.

122 Siracusano, G., Tomasello, R. et al. (2016). Magnetic Radial Vortex Stabilization and Efficient Manipulation Driven by the Dzyaloshinskii-Moriya Interaction and Spin-Transfer Torque. *Phys. Rev. Lett.* 117: 087204.

123 Karakas, V., Gokce, A. et al. (2018). Observation of Magnetic Radial Vortex Nucleation in a Multilayer Stack with Tunable Anisotropy. *Sci. Rep.* 8.

124 Li, C., Wang, S. et al. (2020). Spin-torque nano-oscillators based on radial vortex in the presence of interface Dzyaloshinskii-Moriya interaction. *J. Magn. Magn. Mater.* 498: 166155.

125 Locatelli, N., Hamadeh, A. et al. (2015). Efficient Synchronization of Dipolarly Coupled Vortex-Based Spin Transfer Nano-Oscillators. *Sci. Rep.* 5: 17039.

126 Zaspel, C.E. (2013). Phase locking of vortex-based spin-torque nanocontact oscillators by antivortices. *Appl. Phys. Lett.* 102: 052403.

127 Araujo, F.A. and Grollier, J. (2016). Controlling the synchronization properties of two dipolarly coupled vortex based spin-torque nano-oscillators by the intermediate of a third one. *J. Appl. Phys.* 120: 103903.

128 Jonietz, F., Muhlbauer, S. et al. (2010). Spin transfer torques in MnSi at ultralow current densities. *Science* 330: 1648–1651.

129 Yu, X.Z., Kanazawa, N. et al. (2012). Skyrmion flow near room temperature in an ultralow current density. *Nat. Commun.* 3: 988.

130 Everschor-Sitte, K., Masell, J. et al. (2018). Perspective: Magnetic skyrmions—Overview of recent progress in an active research field. *J. Appl. Phys.* 124: 240901.

131 Zhang, S., Wang, J. et al. (2015). Current-induced magnetic skyrmions oscillator. *New J. Phys.* 17: 023061.

132 Garcia-Sanchez, F., Sampaio, J. et al. (2016). A skyrmion-based spin-torque nano-oscillator. *New J. Phys.* 18: 075011.

133 Shen, L., Xia, J. et al. (2019). Spin torque nano-oscillators based on antiferromagnetic skyrmions. *Appl. Phys. Lett.* 114: 042402.

134 Dai, Y., Wang, H. et al. (2015). Flower-like dynamics of coupled Skyrmions with dual resonant modes by a single-frequency microwave magnetic field. *Sci. Rep.* 4: 6153.

135 Wang, H., Dai, Y. et al. (2014). Dual-frequency microwave-driven resonant excitations of skyrmions in nanoscale magnets. *Rsc Advances* 4: 62179–62185.

136 Jiang, W., Xia, J. et al. (2018). Array of Synchronized Nano-Oscillators Based on Repulsion between Domain Wall and Skyrmion. *IEEE Magn. Lett.* 9: 1–5.

137 Jin, C.D., Wang, J.B. et al. (2018). Array of Synchronized Nano-Oscillators Based on Repulsion between Domain Wall and Skyrmion. *Phys. Rev. Appl.* 9: 044007.

138 Feng, Y., Xia, J. et al. (2019). A skyrmion-based spin-torque nano-oscillator with enhanced edge. *J. Magn. Magn. Mater.* 491: 165610.

139 Guo, J.H., Xia, J. et al. (2020). A ferromagnetic skyrmion-based nano-oscillator with modified profile of Dzyaloshinskii-Moriya interaction. *J. Magn. Magn. Mater.* 496: 165912.

140 Xia, J., Zhang, X. et al. (2018). Current-driven dynamics of frustrated skyrmions in a synthetic antiferromagnetic bilayer. *Physical Review Applied* 11: 044046.

141 Whang, H.-S. and Choe, S.-B. (2020). Spin-Hall-effect-modulation skyrmion oscillator. *Sci. Rep.* 10.

142 Das, D., Muralidharan, B. et al. (2019). Skyrmion based spin-torque nano-oscillator. *J. Magn. Magn. Mater.* 491: 165608.

143 Slavin, A. and Tiberkevich, V. (2009). Nonlinear Auto-Oscillator Theory of Microwave Generation by Spin-Polarized Current. *IEEE Trans. Magn.* 45: 1875–1918.

144 Bonetti, S., Muduli, P. et al. (2009). Spin torque oscillator frequency versus magnetic field angle: The prospect of operation beyond 65 GHz. *Appl. Phys. Lett.* 94: 102507.

145 Tamaru, S., Kubota, H. et al. (2015). Extremely Coherent Microwave Emission from Spin Torque Oscillator Stabilized by Phase Locked Loop. *Sci. Rep.* 5: 18134.

146 Devolder, T., Meftah, A. et al. (2007). Spin transfer oscillators emitting microwave in zero applied magnetic field. *J. Appl. Phys.* 101.

147 Monteblanco, E. (2013). Redshift and Blueshift Regimes in Spin-Transfer-Torque Nano-Oscillator Based on Synthetic Antiferromagnetic Layer. *IEEE Magn. Lett.* 4: 3500204.

148 Sievers, A.J. and Tinkham, M. (1963). Far Infrared Antiferromagnetic Resonance in MnO and NiO. *Phys. Rev.* 129: 1566–1571.

149 Keffer, F. and Kittel, C. (1952). Theory of Antiferromagnetic Resonance. *Phys. Rev.* 85: 329–337.

150 Cheng, R., Xiao, J. et al. (2014). Spin Pumping and Spin-Transfer Torques in Antiferromagnets. *Phys. Rev. Lett.* 113.

151 Daniels, M.W., Guo, W. et al. (2015). Spin-transfer torque induced spin waves in antiferromagnetic insulators. *New J. Phys.* 17: 103039.

152 Gomonay, H.V. and Loktev, V.M. (2010). Spin transfer and current-induced switching in antiferromagnets. *Phys. Rev. B* 81.

153 Cheng, R., Xiao, D. et al. (2016). Terahertz Antiferromagnetic Spin Hall Nano-Oscillator. *Phys. Rev. Lett.* 116: 207603.

154 Sulymenko, O.R., Prokopenko, O.V. et al. (2017). Terahertz-Frequency Spin Hall Auto-oscillator Based on a Canted Antiferromagnet. *Phys. Rev. Appl.* 8: 064007.

155 Sulymenko, O.R., Prokopenko, O.V. et al. (2018). Terahertz-Frequency Signal Source Based on an Antiferromagnetic Tunnel Junction. *IEEE Magn. Lett.* 9: 1–5.
156 Pufall, M.R., Rippard, W.H. et al. (2005). Frequency modulation of spin-transfer oscillators. *Appl. Phys. Lett.* 86: 082506.
157 Muduli, P.K., Pogoryelov, Y. et al. (2010). Nonlinear frequency and amplitude modulation of a nanocontact-based spin-torque oscillator. *Phys. Rev. B* 81: 140408.
158 Manfrini, M., Devolder, T. et al. (2011). Frequency shift keying in vortex-based spin torque oscillators. *J. Appl. Phys.* 109: 083940.
159 Ma, R., Purbawati, A. et al. (2017). Spin torque oscillator based BFSK modulation. *13th Conference on Ph.D. Research in Microelectronics and Electronics* 1-4.
160 Ruiz-Calaforra, A., Purbawati, A. et al. (2017). Frequency shift keying by current modulation in a MTJ-based STNO with high data rate. *Appl. Phys. Lett.* 111: 082401.
161 Purbawati, A., Garcia-Sanchez, F. et al. (2016). Enhanced modulation rates via field modulation in spin torque nano-oscillators. *Appl. Phys. Lett.* 108: 122402.
162 Nagasawa, T., Suto, H. et al. (2011). *Frequency transition of spin-torque oscillator under the magnetic-field pulse in nanosecond range. J. Appl. Phys.* 109: 07C907.
163 Skowronski, W., Checinski, J. et al. (2019). Microwave magnetic field modulation of spin torque oscillator based on perpendicular magnetic tunnel junctions. *Sci. Rep.* 9: 19091.
164 Lee, K.J., Deac, A. et al. (2004). Excitations of incoherent spin-waves due to spin-transfer torque. *Nat. Mater.* 3: 877–881.
165 Sattler, K. (2010). Handbook of Nanophysics: Functional Nanomaterials. CRC Press.
166 Seki, T., Sakuraba, Y. et al. (2014). High power radio frequency oscillation by spin transfer torque in a Co2MnSi layer: Experiment and macrospin simulation. *Appl. Phys. Lett.* 105: 092406.
167 Arai, H. and Imamura, H. (2017). Stochastic Phase Synchronization of Perpendicularly Magnetized Spin-Torque Oscillators With the Second-Order Uniaxial Anisotropy. *IEEE Trans. Magn.* 53: 6100705.
168 Lubarda, M.V., Kuteifan, M. et al. (2020). Stochastic Phase Synchronization of Perpendicularly Magnetized Spin-Torque Oscillators With the Second-Order Uniaxial Anisotropy. *Phys. Rev. B* 102: 014405.
169 Yuan, X., Lu, Z. et al. (2020). Current-driven transverse domain wall oscillations in perpendicular spin-valve structures. *Appl. Phys. Lett.* 116: 222405.
170 Fang, B., Carpentieri, M. et al. (2019). Lateral domain wall oscillations in IMA/PMA bilayered nano-strips driven by a perpendicular current: A type of domain wall based oscillators. *Phys. Rev. Appl.* 11: 014022.
171 Tsunegi, S., Yakushiji, K. et al. (2016). Experimental Demonstration of Spintronic Broadband Microwave Detectors and Their Capability for Powering Nanodevices. *Appl. Phys. Lett.* 109: 252402.
172 Kendziorczyk, T. and Kuhn, T. (2016). Microwave emission power exceeding 10 μW in spin torque vortex oscillator. *Phys. Rev. B* 93: 134413.
173 Zahedinejad, M. (2019). Mutual synchronization of nanoconstriction-based spin Hall nano-oscillators through evanescent and propagating spin waves. *Göteborgs universitet. Naturvetenskapliga fakulteten.*
174 Létang, J., Petit-Watelot, S. et al. (2019). Modulation and phase-locking in nanocontact vortex oscillators. *Phys. Rev. B* 100: 144414.
175 Bhoomeeswaran, H., Bakyalakshmi, R. et al. (2020). Frequency tunability via spin hall angle through spin hall spin torque nano oscillator. *AIP Conf. Proc.* 2220: 110037.
176 Iacocca, E., Heinonen, O. et al. (2014). Generation linewidth of mode-hopping spin torque oscillators. *Phys. Rev. B* 89: 054402.

177 Hirayama, E., Kanai, S. et al. (2015). Electric-field induced nonlinear ferromagnetic resonance in a CoFeB/MgO magnetic tunnel junction. *Appl. Phys. Lett.* 107: 132404.

178 Pauselli, M. and Carlotti, G. (2015). Spin wave eigenmodes excited by spin transfer torque in circular nanopillars: influence of lateral size and Oersted field studied by micromagnetic simulations. *J. Phys. D: Appl. Phys.* 48: 415001.

179 Zhang, J. (2017). *University of California, Irvine* 164.

180 Iwakiri, S., Sugimoto, S. et al. (2020). Generation of multipeak spectrum of spin torque oscillator in non-linear regime. *Appl. Phys. Lett.* 117: 022406.

181 Carpentieri, M. and Torres, L. (2010). *Micromagnetic simulations of linewidths and nonlinear frequency shift coefficient in spin torque nano-oscillators. J. Appl. Phys.* 107: 073907.

182 Zeng, Z., Finocchio, G. et al. (2013). Spin transfer nano-oscillators. *Nanoscale* 5: 2219.

183 Kim, J.-V., Tiberkevich, V. et al. (2008). Generation Linewidth of an Auto-Oscillator with a Nonlinear Frequency Shift: Spin-Torque Nano-Oscillator. *Phys. Rev. Lett.* 100: 017207.

184 Finocchio, G., Siracusano, G. et al. (2010). Time-domain study of frequency-power correlation in spin-torque oscillators. *Phys. Rev. B* 81: 184411.

185 Sankey, J.C., Krivorotov, I.N. et al. (2005). Mechanisms limiting the coherence time of spontaneous magnetic oscillations driven by dc spin-polarized currents. *Phys. Rev. B* 72: 224427.

186 Schneider, M.L., Rippard, W.H. et al. (2009). Temperature dependence of spin-torque-driven self-oscillations. *Phys. Rev. B* 80: 144412.

187 Houssameddine, D., Ebels, U. et al. (2009). Temporal Coherence of MgO Based Magnetic Tunnel Junction Spin Torque Oscillators. *Phys. Rev. Lett.* 102: 257202.

188 Kudo, K., Nagasawa, T. et al. (2009). Measurement of nonlinear frequency shift coefficient in spin-torque oscillators based on MgO tunnel junctions. *Appl. Phys. Lett.* 95: 022507.

189 Georges, B., Grollier, J. et al. (2009). Origin of the spectral linewidth in nonlinear spin-transfer oscillators based on MgO tunnel junctions. *Phys. Rev. B* 80: 060404.

190 Sierra, J.F., Quinsat, M. et al. (2012). Influence of thermal fluctuations on the emission linewidth in MgO-based spin transfer oscillators. *Appl. Phys. Lett.* 101: 062407.

191 Jiang, S., Khymyn, R. et al. (2020). Reduced spin torque nano-oscillator linewidth using He + irradiation. *Appl. Phys. Lett.* 116: 072403.

192 Xu, J.-W. and Kent, A.D. (2020). Charge-To-Spin Conversion Efficiency in Ferromagnetic Nanowires by Spin Torque Ferromagnetic Resonance: Reconciling Lineshape and Linewidth Analysis Methods. *Phys. Rev. Appl.* 14: 014012.

193 Eklund, A., Bonetti, S. et al. (2014). Dependence of the colored frequency noise in spin torque oscillators on current and magnetic field. *Appl. Phys. Lett.* 104: 092405.

194 Grimaldi, E., Dussaux, A. et al. (2014). Response to noise of a vortex based spin transfer nano-oscillator. *Phys. Rev. B* 89: 104404.

195 Sankey, J.C., Cui, Y.-T. et al. (2007). Measurement of the spin-transfer-torque vector in magnetic tunnel junctions. *Nat. Phys.* 4: 67–71.

196 Rippard, W., Pufall, M. et al. (2005). Injection Locking and Phase Control of Spin Transfer Nano-oscillators. *Phys. Rev. Lett.* 95: 067203.

197 Grollier, J., Cros, V. et al. (2006). Synchronization of spin-transfer oscillators driven by stimulated microwave currents. *Phys. Rev. B* 73: 060409.

198 Mishagin, K.G. and Shalfeev, V.D. (2010). Synchronization of spin-transfer nanooscillator using phase-locking loop. *Tech. Phys. Lett.* 36: 1049–1052.

199 Chang, J.-H., Chen, -H.-H. et al. (2011). Phase locking of dynamical modes in a nanomagnetic oscillator with a circularly spin-polarized current. *Phys. Rev. B* 84: 054457.

200 Carpentieri, M., Finocchio, G. et al. (2012). Micromagnetic Study of Synchronization of Nonlinear Spin-Torque Oscillators to Microwave Current and Field. *Adv. Condens. Matter Phys.* 2012: 1–5.

201 Giordano, A. and Pompei, M. (2017). Phase locking of current-driven spin-Hall oscillators: A micromagnetic study. *IEEE 3rd International Forum on Research and Technologies for Society and Industry (RTSI)* 1–4.

202 Houshang, A., Iacocca, E. et al. (2016). Spin-wave-beam driven synchronization of nanocontact spin-torque oscillators. *Nat. Nanotechnol.* 11: 280–286.

203 Lebrun, R., Tsunegi, S. et al. (2017). *Mutual synchronization of spin torque nano-oscillators through a long-range and tunable electrical coupling scheme. Nat. Commun.* 8: 15825.

204 Zahedinejad, M., Awad, A.A. et al. (2020). Two-dimensional mutually synchronized spin Hall nano-oscillator arrays for neuromorphic computing. *Nat. Nanotechnol.* 15: 47–52.

205 Marković, D., Leroux, N. et al. (2020). *Detection of the Microwave Emission from a Spin-Torque Oscillator by a Spin Diode. Phys. Rev. Appl.* 13: 044050.

206 Li, Y., de Milly, X. et al. (2018). Selective control of vortex polarities by microwave field in two robustly synchronized spin-torque nano-oscillators. *Appl. Phys. Lett.* 112: 022405.

207 Rippard, W., Pufall, M. et al. (2004). Current-driven microwave dynamics in magnetic point contacts as a function of applied field angle. *Phys. Rev. B* 70: 100406.

208 Georges, B., Grollier, J. et al. (2008). Coupling efficiency for phase locking of a spin transfer nano-oscillator to a microwave current. *Phys. Rev. Lett.* 101: 017201.

209 Kumar, D., Konishi, K. et al. (2016). Coherent microwave generation by spintronic feedback oscillator. *Sci. Rep.* 6: 30747.

210 Georges, B., Grollier, J. et al. (2009). Frequency Converter Based on Nanoscale MgO Magnetic Tunnel Junctions. *Appl. Phys. Express* 2: 123003.

211 Sim, C.H., Moneck, M. et al. (2012). Frequency-tunable perpendicular spin torque oscillator. *J. Appl. Phys.* 111: 07C914.

212 Nikonov, D.E., Bourianoff, G.I. et al. (2010). Strategies and tolerances of spin transfer torque switching. *J. Appl. Phys.* 107: 113910.

213 Safranski, C. and Sun, J.Z. (2019). Interface moment dynamics and its contribution to spin-transfer torque switching process in magnetic tunnel junctions. *Phys. Rev. B* 100: 014435.

214 Tiberkevich, V., Slavin, A. et al. (2009). Phase-locking and frustration in an array of nonlinear spin-torque nano-oscillators. *Appl. Phys. Lett.* 95: 262505.

215 Pogoryelov, Y., Muduli, P.K. et al. (2011). Spin-torque oscillator linewidth narrowing under current modulation. *Appl. Phys. Lett.* 98: 192506.

216 Bonetti, S., Puliafito, V. et al. (2012). Power and linewidth of propagating and localized modes in nanocontact spin-torque oscillators. *Phys. Rev. B* 85: 174427.

217 Maehara, H., Kubota, H. et al. (2014). High Q factor over 3000 due to out-of-plane precession in nano-contact spin-torque oscillator based on magnetic tunnel junctions. *Appl. Phys. Express* 7: 023003.

218 Yang, L., Verba, R. et al. (2015). Reduction of phase noise in nanowire spin orbit torque oscillators. *Sci. Rep.* 5: 16942.

219 Tamaru, S., Kubota, H. et al. (2016). Analysis of phase noise in a spin torque oscillator stabilized by phase locked loop. *Appl. Phys. Express* 9: 053005.

220 Demidov, V.E., Urazhdin, S. et al. (2015). Dipolar field-induced spin-wave waveguides for spin-torque magnonics. *Appl. Phys. Lett.* 106: 022403.

221 Grundler, D. (2016). Spintronics: Nanomagnonics around the corner. *Nat. Nanotechnol.* 11: 407–408.

222 Zhu, X., Zhu, J.-G. et al. (2004). Spin transfer excited regular and chaotic spin waves in current perpendicular to plane spin valves. *J. Appl. Phys.* 95: 6630–6632.

223 Slavin, A. and Tiberkevich, V. (2005). Spin wave mode excited by spin-polarized current in a magnetic nanocontact is a standing self-localized wave bullet. *Phys. Rev. Lett.* 95: 237201.

224 Xi, H., Yang, Y. et al. (2007). Spin waves excited by dc currents injected into single ferromagnetic thin films. *Phys. Rev. B* 75: 174411.

225 Rezende, S.M., de Aguiar, F.M. et al. (2007). Mode locking of spin waves excited by direct currents in microwave nano-oscillators. *Phys. Rev. Lett.* 98: 087202.

226 Yamaguchi, A., Motoi, K. et al. (2011). High-order standing spin wave modes in Fe19Ni81micron wire observed by homodyne method. *J. Phys. Conf. Ser.* 266: 012113.

227 Mancilla-Almonacid, D., Arias, R.E. et al. (2018). Spin wave modes of two magnetostatic coupled spin transfer torque nano-oscillators. *J. Appl. Phys.* 124: 162102.

228 Consolo, G., Lopez-Diaz, L. et al. (2013). Excitation of spin waves by a current-driven magnetic nanocontact in a perpendicularly magnetized waveguide. *Phys. Rev. B* 88: 014417.

229 Slavin, A. and Tiberkevich, V. (2008). Excitation of spin waves by spin-polarized current in magnetic nano-structures. *IEEE Trans. Magn.* 44: 1916–1927.

230 Ishibashi, S., Ando, K. et al. (2011). High Spin-Torque Diode Sensitivity in CoFeB/MgO/CoFeB Magnetic Tunnel Junctions Under DC Bias Currents. *IEEE Trans. Magn.* 47: 3373–3376.

231 Akinobu Yamaguchi, A.F. (2012). Spin Torque Diode Spectroscopy of Quantized Spin Wave in MTJ. *IEEE Trans. Magn.* 48: 2816–2819.

232 Saito, H., Yuasa, S. et al. (2006). Spin-polarized tunneling in metal-insulator-semiconductor Fe/ZnSe/Ga1-xMnxAs magnetic tunnel diodes. *Appl. Phys. Lett.* 89: 3.

233 Iovan, A., Andersson, S. et al. (2008). Spin Diode Based on Fe/MgO Double Tunnel Junction. *Nano. Lett.* 8: 805–809.

234 Saito, H., Yamamoto, A. et al. (2008). High tunneling magnetoresistance in Fe/GaOx/Ga1-xMnxAs with metal/insulator/semiconductor structure. *Appl. Phys. Lett.* 93: 3.

235 Wang, C., Cui, Y.T. et al. (2009). Sensitivity of spin-torque diodes for frequency-tunable resonant microwave detection. *J. Appl. Phys.* 106: 053905.

236 Ishibashi, S., Seki, T. et al. (2010). Large Diode Sensitivity of CoFeB/MgO/CoFeB Magnetic Tunnel Junctions. *Appl. Phys. Express* 3: 073001.

237 Miwa, S., Ishibashi, S. et al. (2014). Highly sensitive nanoscale spin-torque diode. *Nat. Mater.* 13: 50–56.

238 Zhang, L., Fang, B. et al. (2018). Ultrahigh detection sensitivity exceeding 105 V/W in spin-torque diode. *Appl. Phys. Lett.* 113: 102401.

239 Tsunegi, S., Taniguchi, T. et al. (2018). Achievement of high diode sensitivity via spin torque-induced resonant expulsion in vortex magnetic tunnel junction. *Appl. Phys. Express* 11: 053001.

240 Jenkins, A.S., Lebrun, R. et al. (2016). Spin-torque resonant expulsion of the vortex core for an efficient radiofrequency detection scheme. *Nat. Nanotechnol.* 11: 360–364.

241 Auerbach, E., Berkov, D. et al. (2019). Injection Locking at Fractional Frequencies of Magnetic-Tunnel-Junction-Based Read Sensors' Ferromagnetic Resonance Modes. *Phys. Rev. Appl.* 12: 054022.

242 Skirdkov, P.N. and Zvezdin, K.A. (2020). Spin-Torque Diodes: From Fundamental Research to Applications. *Annalen der Physik* 532: 1900460.

243 Zhang, L., Fang, B. et al. (2019). Enhanced Broad-band Radio Frequency Detection in Nanoscale Magnetic Tunnel Junction by Interface Engineering. *ACS Appl. Mater. Interfaces* 11: 29382–29387.

244 Prokopenko, O.V., Krivorotov, I.N. et al. (2012). Spin-torque microwave detector with out-of-plane precessing magnetic moment. *J. Appl. Phys.* 111: 123904.

245 Fu, L., Cao, Z.X. et al. (2012). Microwave reflection imaging using a magnetic tunnel junction based spintronic microwave sensor. *Appl. Phys. Lett.* 101: 232406.

246 Skowroński, W., Wiśniowski, P. et al. (2012). Magnetic field sensor with voltage-tunable sensing properties. *Appl. Phys. Lett.* 101: 192401.

247 Saha, R., Wu, K. et al. (2020). Spin current nano-oscillator (SCNO) as a potential frequency-based, ultra-sensitive magnetic biosensor: a simulation study. *Nanotechnology* 31: 375501.

248 Braganca, P.M., Gurney, B.A. et al. (2010). Nanoscale magnetic field detection using a spin torque oscillator. *Nanotechnology* 21: 235202.

249 Ramaswamy, B., Algarin, J.M. et al. (2016). Wireless current sensing by near field induction from a spin transfer torque nano-oscillator. *Appl. Phys. Lett.* 108: 242403.

250 Appeltant, L., Soriano, M.C. et al. (2011). Information processing using a single dynamical node as complex system. *Nat. Commun.* 2: 468.

251 Sarkar, A., Srinivasan, S. et al. (2011). Modeling all spin logic: multi-magnet networks interacting via spin currents. Ieee International Electron Devices Meeting.

252 Csaba, G. (2013). Computational Study of Spin-Torque Oscillator Interactions for Non-Boolean Computing Applications. *IEEE Trans. Magn.* 49: 4447–4451.

253 Sharad, M., Fan, D. et al. (2013). Ultra-low power neuromorphic computing with spin-torque devices. 2013 Third Berkeley Symposium on Energy Efficient Electronic Systems.

254 Sharad, M., Fan, D.L. et al. (2013). Spin-neurons: A possible path to energy-efficient neuromorphic computers. *J. Appl. Phys.* 114: 234906.

255 Locatelli, N., Vincent, A.F. et al. (2015). Spintronic devices as key elements for energy-efficient neuroinspired architectures. *Design, Automation & Test in Europe Conference & Exhibition (DATE)*, 994–999.

256 Zhang, D., Zeng, L. et al. (2015). Energy-efficient neuromorphic computation based on compound spin synapse with stochastic learning. *IEEE International Symposium on Circuits and Systems (ISCAS)*, 1538–1541.

257 Riou, M., Araujo, F.A. et al. (2017). Neuromorphic Computing through Time-Multiplexing with a Spin-Torque Nano-Oscillator. *IEEE Trans. Electron Devices* 2017: 8268505.

258 Romera, M., Talatchian, P. et al. (2018). Vowel recognition with four coupled spin-torque nano-oscillators. *Nature* 563: 230–234.

259 Cai, J.L., Zhang, L.K. et al. (2019). Sparse neuromorphic computing based on spin-torque diodes. *Appl. Phys. Lett.* 114: 192402.

260 Talatchian, P., Romera, M. et al. (2020). Designing Large Arrays of Interacting Spin-Torque Nano-Oscillators for Microwave Information Processing. *Phys. Rev. Appl.* 13: 024073.

261 Kunimatsu, K., Tsuchiya, T. et al. (2020). Magnetic tunnel junctions with metastable bcc Co3Mn electrodes. *Appl. Phys. Express* 13: 083007.

7

Magnetic Tunnel Junctions for Artificial Neural Network

Meiyin Yang, Tengzhi Yang, and Jun Luo

Key Laboratory of Microelectronic Devices and Integrated Technology, Institute of Microelectronics, Chinese Academy of Sciences, Beijing, China

CHAPTER MENU

7.1 Introduction of Neural Computing

In human brains, a neural network is composed of tens of billions of neural cells as the basic biological unit. Neural cells usually include neurons, synapses, and contacts, etc. Neurons are connected to each other through synapses to form a huge network. In this network each neuron can even connect with up to 106 other neurons. This complex connection system makes the computation of human brains highly parallel. Neurons are responsible for receiving signals from other neurons through synapses on dendrites. They continue to propagate downwards, finally reaching the target area. In order to maintain signal integrity, the signal on the neuron is a binary pulse signal so that it can be more tolerant to noise and loss. At the same time, in order to reduce energy consumption, the neuron adopts the strategy of "interval potential," that is, the input accumulates to a certain extent before it stimulates the neuron to discharge. Because of these mysterious and interesting features, although the human brain is not as fast as computers, it can handle many extremely complex tasks, such as complex scene recognition, emotion recognition, and so on. This drives people to study the human brain and try to design ANNs.

ANNs are a popular area of research over the past decade in computer science and neuroscience; they are a neurological form of computation that abstracts a human neural network from an information-processing perspective. Nerve morphology calculation was presented in 1990 by Carver Mead [1]. He imitated the learning method of the human brain by using a

Spintronics: Materials, Devices and Applications, First Edition. Edited by Kaiyou Wang, Meiyin Yang, and Jun Luo.

large number of perceptrons to form a neural network through the interconnection of nerve synapses, thus realizing a self-adaptive perception system that can complete complex tasks such as adaptive pattern recognition. Although conventional approaches have been proposed for solving a series of problems, such as prediction and pattern recognition, they are constrained in a certain environment which is not flexible enough to be generalized to other domains [2]. ANNs, by contrast, provide a much better generalization ability and flexibility in many applications.

The neural network has gone through three stages of development. The first generation of neural networks, namely single-layer perceptrons [3, 4], was proposed in the 1950s. The single-layer perceptron has had an enlightening effect on the neural network architecture. However, Minsky and Papert propose that the linear structure that the perceptrons adopted could not solve linear inseparable problems such as XOR operations [5], which is easy to realize by non-linear mapping in traditional methods. After a downturn for the first generation of artificial intelligence, multi-layer perceptrons were proposed in the 1980s. This second-generation neural network has greatly improved the learning and reasoning capabilities of complex problems at the cost of increasing the number of layers and parameters required by neurons. Typical representatives are the convolutional neural network (CNN) [6, 7], the recurrent neural network (RNN) [8], long short-term memory (LSTM) [9] etc, which are widely used neural network architectures nowadays. Compared with the first-generation neural network, the second generation contains multiple hidden layers between the input layer and the output layer, and introduces a nonlinear activation function, which can handle more complex nonlinear problems. In addition, because the neurons are continuously interconnected, the second-generation neural network generally supports the back-propagation learning method, which is another huge improvement in learning efficiency. However, follow-up studies have found that, as the number of layers increases, the gradient tends to disappear, and in the training process it is more inclined to a local optimal solution [10]; that is, overfitting. This has affected the inference results of the neural network. With the subsequent training algorithms and deep neural networks, this problem has been improved to a certain extent [11]. In recent years, with the development of biological neuroscience, the academic community will further focus on time coding and spatial accumulation. As early as 1995, Hopfield [12] and Sejnowski [13] predicted that time-coding would be an important direction. The third-generation neural network, also known as the spiking neural network, refers to biological neurology. On the basis of the neural network, the coding resources in the time domain are fully considered as a novel method to form a new type of neural network composed of neurons and synapses. The input and output of this kind of network are no longer level signals but a series of encoded pulse signals or square wave signals, with different intervals and pulse widths, which are processed by nerve synapses and input into neurons to accumulate and fire. This method is closer to the information transmission process of the biological nervous system, and is capable of processing higher density information. It is also more suitable for processing sound and other time-coded analog signals and discrete digital signals. But relatively speaking, the structure of SNN is more complicated and its processing signal is pulse-based.

The distinctive feature of neuromorphic computing architecture is highly interconnected and parallel, which requires a large number of storage and processing units to cooperate. In the 21st century, as the rapid development of general purpose chips, such as CPU and GPU, the scale of neural networks has also shown exponential growth. However, the characteristics of the neural network are contradictory to the von Neumann architecture. The memory and processing units are interconnected by a bus, and its bandwidth limits the processing speed of the neural network. The huge number of parameters and complex training processes pose severe challenges to storage capacity and computing speed. To solve this problem, researchers have proposed and designed a variety of Complementary Metal-Oxide-Silicon (CMOS)-based artificial intelligence acceleration chips [14, 15] to accelerate the training process through hardware. However, its huge power consumption still limits its use in large-scale networks and the internet of things (IoT), which is highly sensitive to power consumption.

The MTJ has the characteristics of extremely low power writing consumption and non-volatile storage. In addition, random output can be achieved under certain conditions. Therefore, it is very suitable for realizing low-power artificial intelligence acceleration chips. There are two main applications for MTJ in neural networks: 1. Mass parameter storage, 2. Synapses, neurons, and other neural devices. As mentioned above, a neural network needs thousands of parameters to finish training and interference. For example, the classic CNN model, VGG-16, has more than 526 Mbits weights, which was proposed in 2014 [17]. Nowadays, a mature model will have more than 1 billion parameters, which is an unimaginable workload for storage, reading, and writing. So we can use MRAM as parameter storage. Works about this part are related to MRAM application on memory and will not be introduced here. For the neuromorphic computation, the MTJ can be used as the hardware neurons and synapses in the neural network structure. Figure 7.1 shows [16] the schematics of neural networks with synapses and neurons, in which the synapses received an input signal and weighted it, and then all the weighted signals accumulated onto the output neuron. The synapse here is similar to the weighted accumulator or convolution kernel in the neural network, and the neuron is similar to the activation function. This structure mimics the working mechanism of neurons in the brain and is more suitable for applications in the field such as spike neural networks (SNN) [18, 19].

The random output of MTJs has always been a severe challenge in the memory field. Due to the influence of heat on the potential barrier, the magnetization of a free layer (FL) in a MTJ is likely to flip randomly. This flip is truly random, which is fatal to device reliability. In order to solve these error problems, people have also thought of many methods, such as designing MTJs with higher stability from materials and structure [20–22], or using error correction codes in software to reduce the effects of errors [23]. However, for neural networks, this random flip is exactly the same as the mechanism of biological nerves, and artificial neural networks are inherently robust to this random error, so the reliability

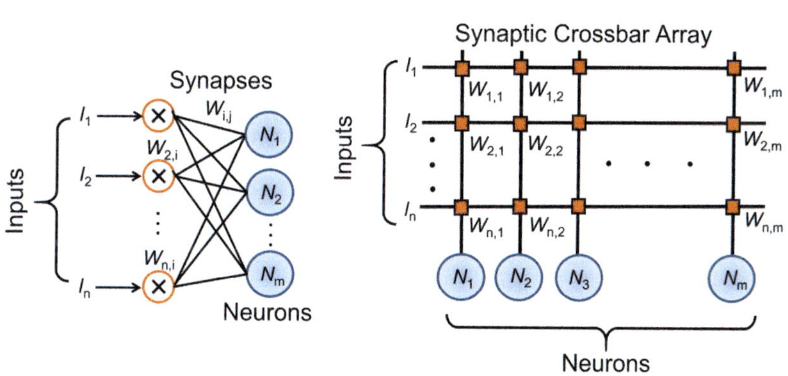

Figure 7.1 Synapses and neurons of crossbar arrays. *Source:* Based on Sengupta et al [16].

requirements for MTJs are not high [24, 25]. For some networks, such as spike neural networks, these writing errors can even be used to train the network [26, 27]. Spike neural networks need to use a large number of pulses to trigger low-probability changes, which is very different from traditional neural networks such as CNN. Because this learning strategy needs to generate random numbers to achieve probability flipping, a MTJ with a true random bit error rate can effectively implement this learning strategy.

7.2 Hardware Requirements for an Artificial Intelligence Neural Network

In recent years, a large number of neural network computing models have been proposed. Brain-like computing architectures have been implemented on hardware based on these models [28–33]. For these brain-like hardware computing structures, their basic units are a large amount of neurons interconnected by programmable nerve synapses [28]. The operation required by the neuron is shown in Figure 7.2(a). The neuron receives N input signals from the input nerve synapse and performs weighted accumulation, and then outputs the result after nonlinear calculation. The output follows equation (7.1)

$$Y = \phi\left(\sum W_i \bullet X_i - \theta\right) \tag{7.1}$$

where Y is the output of the neuron, W_i is the weight of the i-th synapse, X_i is the i-th input signal, θ is the biased signal of the neuron, and ϕ is the nonlinear transfer function. There are two important calculation processes involved here, one is multiplication and accumulation ($\sum W_i \bullet X_i - \theta$), and the other is the nonlinear transfer function (ϕ, also called an activation function). The multiplying and accumulating calculation is the most computationally intensive operation of all neural networks. Especially for hardware, the optimization of the amounts of multiplying and accumulating, and their speed, determine the overall power consumption and the design area. Multiply-accumulate operations include a lot of multiplication: floating-point multipliers especially, and multipliers to implement in hardware are very

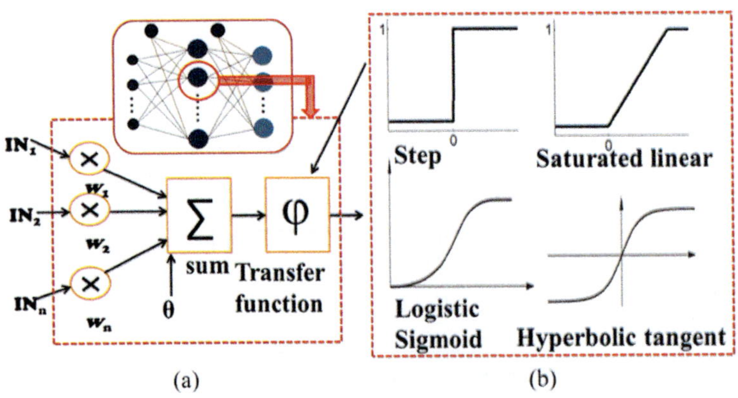

Figure 7.2 (a) Neuron operation process, (b) four different activation functions [38].

complex. In order to save area on the chip, the hardware implementation generally takes the form of fixed-point multipliers [34, 35]. So, parameters need to be stored as limited-bit data structures. For weighted accumulation, the weight of *Wi* needs to be continuously modified during the learning process, which is called synaptic plasticity; so, for neural networks, a memory that can achieve multi-state is needed. The MTJ device can realize the characteristics of multi-state storage by manipulating the movement of the magnetic domain wall, and it has been used in the weight storage of this neural network [36, 37].

The non-linear transfer function is also an important neural network unit. Various non-linear transfer functions have been proposed. Figure 7.2(b) depicts output *vs.* input loops of step function, a saturated linear function, a logistics sigmoid function and the hyperbolic tangent function [38]. The step function output is binary, so it is called a hard limit transfer function. The latter three function outputs are continuous signals, which are also called soft limit transfer functions. For binary-based digital integrated circuits, the realization of the soft limit transfer function will face the contradiction between computational complexity and accuracy. However, many mainstream models now use the soft limit transfer function as the activation function due to higher accuracy and the ability to solve nonlinear problems. The output of the MTJ is an analog signal. The output's dependence on the input curve, during the writing process, is a non-linear function which is close to the performance of the step function, and is a suitable neuron hardware device.

The input and output signals of non-volatile MTJs are current signals. According to Kirchhoff's law, it is easy to realize the multiplication and accumulation operation, and it has ultra-low computing power consumption and very small area occupation. The MTJ can be used for the realization of synapse and neuron functions in the artificial intelligence neural network. It has high speed and low energy consumption, and thus possesses huge advantages in the field of artificial intelligence hardware for the future.

The proposal for MTJ as the neuron hardware has gone through several iterations [39], as shown in Figure 7.3. Early modeling of neuron perception was approximated as a step function. Such neuron functions can be achieved by STT-MTJ or the spin-orbit torque MTJ (SOT-MTJ) directly [40, 41]. In addition, the structure using a lateral spin valve (LSV) and current-induced domain wall motion (DWM) methods can also be adopted or combined to realize neuron devices [42, 43], which can significantly reduce path resistance, thereby reducing power consumption. With the deepening of neuron modeling, non-step neuron devices with multiple states were proposed. To achieve multiple steady-state or even continuous-state outputs, the DWM mechanism can be used in the FL with a multi-domain FM (FM) layer, and the magnetic domain wall motion is driven by the underlying spin-orbit torque [44, 45]. Sengupta proposed a neuron device based on integrate-fire (IF) spiking [45], which is different from the continuous output caused by magnetic domain wall motion. The potential in an IF spiking-based neuron device steps up continuously under pulse input, which is compatible with the pulse-based coding method of the SNN. In addition, single-domain MTJ probability spiking neuron device designs have been proposed [46].

7.3 Introduction to Magnetic Tunnel Junction Devices

A MTJ basic structure is one FM layer and another FM layer sandwiched by a tunneling barrier layer [29–31]. One of the FM layers can be flipped freely in the directions of the two easy magnetization axes – the FL. The direction of the other FM layer is fixed, whose

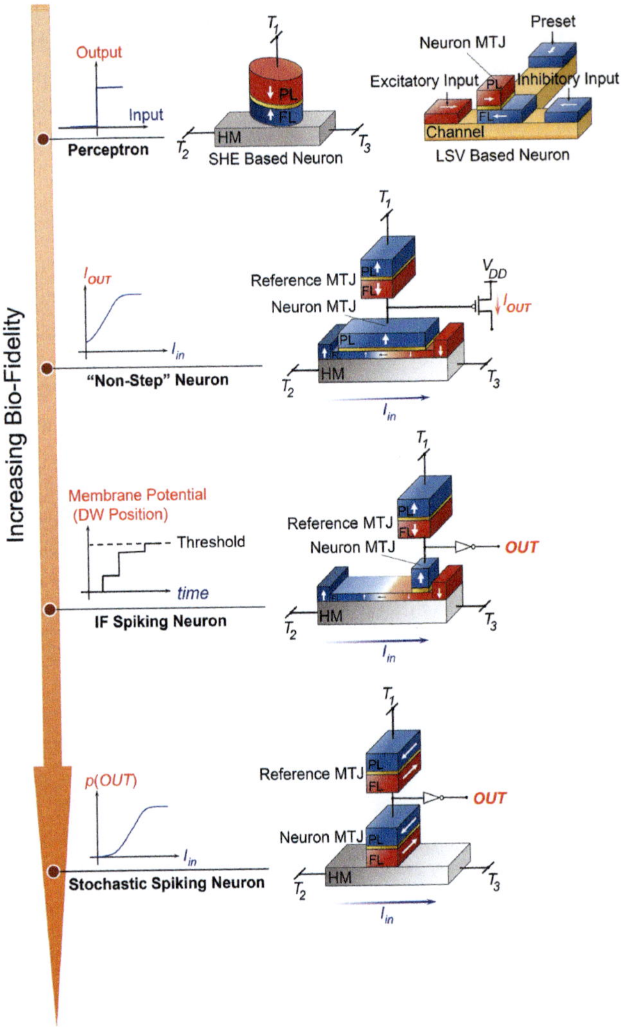

Figure 7.3 Artificial neurons with several different magnetic tunnel structures [39].

direction is pinned by a pinning layer such as an anti-FM layer. When the magnetization directions of the FL and the reference layer are the same, the MTJ is in a high resistance state, and vice versa. This is the core mechanism of the MTJ to achieve state storage. This phenomenon is called tunneling magnetoresistance (TMR).

TMR is defined as

$$\mathrm{TMR} = \frac{R_{AP} - R_P}{R_P} = \frac{G_{AP}^{-1} - G_P^{-1}}{G_P^{-1}} = \frac{2P_1P_2}{1 - P_1P_2},$$

where R_P and R_{AP} are the resistances when the magnetization directions of the two FM layers are parallel and antiparallel, respectively. P are the spin polarizabilities of the two FM layers, respectively. From this formula, it can be seen that the TMR depends on the spin polarizability of the FM layer; that is, the larger the spin polarizability, the larger the TMR value. Generally speaking, the higher the requirement of the MTJ for the TMR value, the better to ensure a high switching ratio and read/write resolution.

The MTJ has a much larger impedance because the intermediate layer is an insulating layer, which results in a larger readout voltage, and is easier to match with the MOSFET (metal–oxide–semiconductor field-effect transistor) and has better compatibility. This is extremely important for applications in brain-like computing and other fields.

7.4 Magnetic Tunnel Junction for Neuron Hardware

7.4.1 Introduction of STT-MTJ and SOT-MTJ

According to the writing path of MTJ devices, we can divide them into two major categories; that is, the spin transfer torque (STT-MTJ), and the spin-orbit-torque MTJ (SOT-MTJ). We will describe the STT-MTJ-based neurons first. The STT effect was proposed first by Berger in 1978 when studying magnetic domain wall motion [47]. It has been predicted by John Slonczewski [48] and confirmed by experiments [49]. The basic principle of the STT effect, as shown in Figure 7.4, is that spin-polarized electrons flow into the magnetic metal, and the spin angular momentum is transferred to the local electron spin of the metal, changing the direction of the metal magnetic moment. A basic STT-MTJ device consists of two FM layers, one of which is magnetically "pinned" and used as a reference layer, and the other layer that can be flipped freely is the FL. The two magnets are separated by the isolation layer, the isolation layer is typically Al_2O_3 [50] or MgO [51, 52]. When the magnetization directions of the FL and the reference layer are the same, the MTJ is in a low resistance state, and when the FL and the reference layer have opposite directions, the MTJ is in a high resistance state. When the current passes through the MTJ isolation layer due to the scattering mechanism of generating a spin current, the spin current exerts a spin-transfer torque on the FL magnetization. Spin flipping occurs when the current

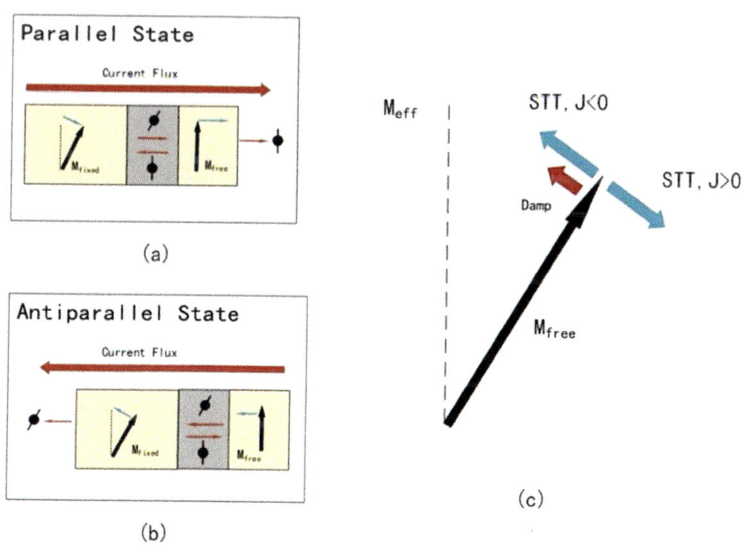

Figure 7.4 The physical mechanism of the STT effect: (a) parallel state; (b) anti-parallel state; (c) the moment of the free layer magnetic moment.

exceeds a value to overcome the damping torque. The output during the spin flipping process is similar to the step function.

SOT-MTJ is another MTJ device structure that has received widespread attention in recent years. Its storage principle, as well as STT-MTJ's, relies on the TMR effect, and the main difference lies in the different writing mechanism. SOT-MTJ does not need to rely on the scattering of the barrier layer to generate the spin current. The writing current of the SOT-MTJ is in the lateral direction and does not require the "screening" effect of the barrier layer. The spin current is generated by spin-orbit coupling to control the flip of the magnetic moment of the FL. The magnetization switching by SOT was first demonstrated in a heavy metal/ferromagnet system. Currently, there are two main flipping mechanisms based on spin-orbit coupling due to the Rashba effect [53–55] and the spin Hall effect [56, 57]. Later, SOT is found to be an efficient way to drive domain wall motion.

The Rashba effect is an interface effect, which mainly exists in low-dimensional semiconductor structures. Through artificial control of the film structure to destroy the spatial inversion symmetry, the asymmetry of the macrostructure causes the spin split, called the Rashba split. For a two-dimensional system composed of heavy metal/FM films, if two adjacent materials have different spin orbit coupling strengths, the contact interface will produce a Rashba spin-orbit coupling due to the spatial inversion asymmetry.

Both the spin Hall effect and the Rashba effect are caused by spin-orbit coupling, but the spin Hall effect comes from the bulk effect of heavy metal materials. The spin Hall effect usually occurs in materials with strong spin-orbit coupling. When a current passes through heavy metals, spin-dependent scattering occurs due to strong spin-orbit coupling, and spin electrons with different spin directions will gather along both sides of the vertical direction to obtain a purely polarized current. The current has a spin transfer torque effect on the adjacent FM layer. When the polarization current is large enough, it can drive the magnetic moment of the FM layer to reverse.

The spin Hall effect in early experiments is small. In order to obtain the spin Hall effect enough to flip the FM layer, many different materials have been tried. In 2012, Liu et al. [57] discovered that β- Ta material has a huge spin Hall effect and can obtain a spin-polarized current that can reverse the spin magnetic moment. Based on this discovery, Liu first realized that the MTJ used SHE to reverse the magnetic moment. The spin Hall angle of β-Ta is 0.12–0.15. This result is better than the traditional high spin-orbit coupling material Pt, which is similar to Pt-doped Au material, but β-Ta does not significantly increase the magnetoresistance of the adjacent FM layer. As shown in Figure 7.5, Liu prepared a beta Ta/CoFeB/MgO/CoFeB/Ta typical MTJ structure. If materials with larger spin Hall angles can be further used, the required critical spin switching current can be further reduced. Mellnik [99] discovered that the topology insulator (TI) material of Bi_2Se_3 at room temperature, whose spin Hall angle is 2.5–3, is much larger than the common material of high spin-orbit coupling, which opens up the prospect of TI materials in magnetic random access memory. In addition to the common NM/FM structure, in 2016, Fukami et al. [36] also realized the spin-orbit coupling driving magnetic moment flip in the antiferromagnets/FM structure, which further broadened the SOT memory device in photon computing and other fields.

7.4.2 Different MTJ-Based Neuron Hardware

As mentioned in Section 1, the neuron, as an output device, has different types of transfer function according to its R-I characteristics and models they are applied in. Next, we will introduce MTJs designed for different transfer functions of neurons.

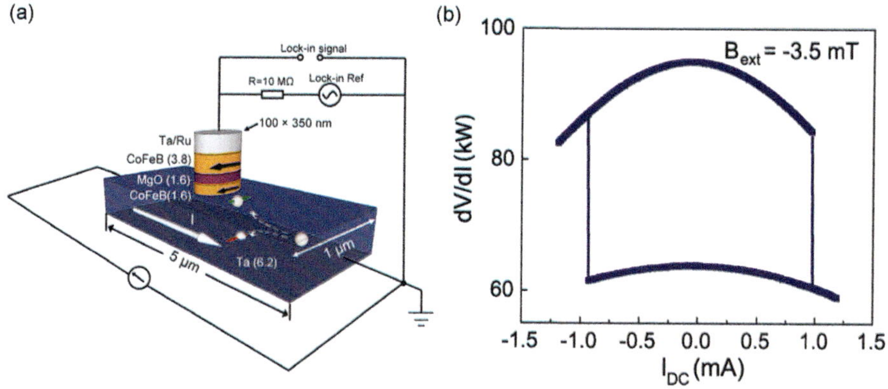

Figure 7.5 MTJ structure based on the SHE effect and its magnetization reversal characteristic curve under an external magnetic field [57].

7.4.2.1 Step Function

STT-MTJ was originally oriented to be developed for memory use. However, it was later discovered that the output of STT-MTJ during writing process is a natural approximate step function, so it can be used to design step function neurons. In the earlier stage in this area, Sengupta's group proposed an all STT-MTJ-based SNN device and circuit structure in 2015 [40]. As shown in Figure 7.6(a) and (b), the resistance of the MTJ is simultaneously modulated by the pulse width and amplitude of the write current. The shape of the voltage vs. current loop is a natural step function that can be used to realize the neuron function. Further, the neural computing architecture consists of the STT neuron and the resistive synapse in a crossbar array network. Using parameters by offline training in MNIST validate the data set, the results of this neural network show 5 to 6 bits of precision with a working frequency of 100 MHz. The recognition accuracy up to 87% can be achieved on the MNIST data set. Compared with the 45 nm node CMOS neural network architecture, the power consumption is saved by times of 1.63–1.79. In 2018, Jea-Gun Park [58] designed two terminal perpendicular MTJ devices. The device uses a dual-MgO interface MTJ with SAF as pinning layer, which is a typical MTJ film structure. However, they made some changes in MgO barrier. They chose poly-crystalline MgO with various grain sizes rather than a perfect single crystalline. This device has a different flipping mechanism from STT-MRAM. It takes advantage of the imperfections of MgO, so the required auxiliary magnetic field is much smaller than the coercive force required by the MTJ. The magnetic switching will first occur from boundaries, due to the lower exchange coupling and magnetic crystalline anisotropy. It can effectively realize the integration-fire operation as a neuron.

It should be noted that in order to simulate a step transfer function neuron, the synaptic resistance crossbar array must be operated at a higher voltage to ensure that the synthetic synaptic current input to the neuron falls outside the hysteresis curve. This constraint, coupled with the fact that the input synaptic current must flow through the oxide layer, causes the entire crossbar array to work at a higher voltage, resulting in greater power consumption. However, appropriate circuit-level technology can be used to reduce the power consumption of the array, thereby reducing the power consumption of the entire system. The synaptic cross array operated at a lower voltage was explored [40], where the MTJ is always reset to the AP state before operation and is exerted an input bias current close to the critical switching current. Therefore, it requires a very small voltage to switch the magnetization. However, since the write operation of STT- MTJ requires a very high critical current, the power and energy advantages are still limited.

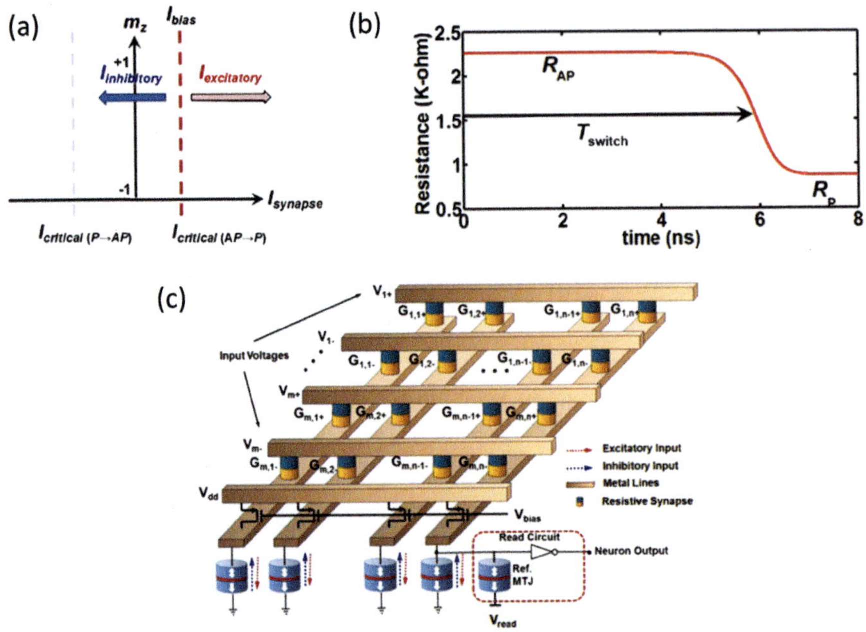

Figure 7.6 The relationship between the state of MTJ and (a) current intensity and (b) pulse time. (c) Resistive synapses and STT-neuron crossbar Array [40].

SOT-MTJ is a device that can resolve the high writing current issues. In 2015, Sengupta [41] proposed a neuron device based on SOT-MTJ. The device is characterized by a two-step switching scheme to achieve threshold operation, as shown in Figure 7.7. The first step, a spin current generated from the metal layers, was injected to the FL and directed the magnetization direction to the hard magnetization axis. In the second step, the current input from the synapse flows through the MTJ perpendicularly and reverses the magnetization direction of the FL. The direction of the synaptic current determines the final direction of the FL. This two-step operation scheme can effectively reduce the synaptic input current required to drive neurons. In addition, the two-step method of SOT+STT can avoid the problem of random magnetic orientation flip by SOT effect.

7.4.2.2 Nonlinear Activation Function

Nonlinear activation function mimics the function of neurons better than the step functions. Fan [38] in 2015 realized soft-limited nonlinear neurons that operate under ultra-low supply voltage and current, as is shown in Figure 7.8. They brought up a three-terminal STT device, where the STT effect is used to drive the lateral movement of the magnetic domain wall of the FL, so that continuous state changes can be realized. Contrary to previous spin-based neurons that can only implement hard-limited transfer functions, this STT-SNN shows continuous resistance changes with varying the input current, so that the soft limit neuron transfer function can be realized. The logical complexity of the soft limit function is high, so the use of neuron devices to implement the soft limit function can lead to higher network accuracy and lower network complexity. Furthermore, they proposed an ANN architecture that implements deep-triode current sources (DTCS) as interface circuits and a memristor crossbar array (MCA) [59] as a synapse and forms a convolutional layer of STT- SNN.

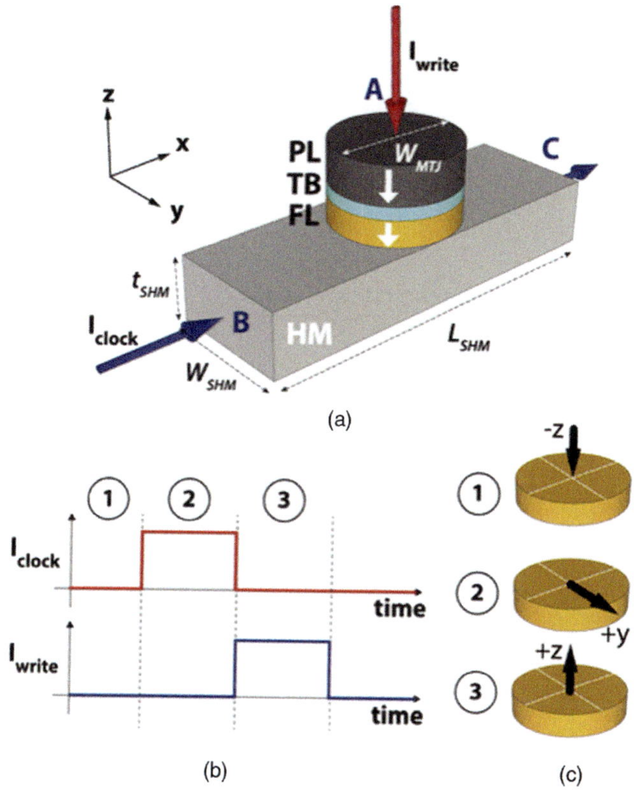

Figure 7.7 (a) Three-terminal neuron threshold device (b) Writing timing of two-step switching operation (c) Writing state of two-step operation [41].

Compared with a neural network using only CMOS circuits, this structure can achieve soft-limited nonlinear neurons with fewer areas, thereby reducing network complexity. In addition, the energy consumed by STT-based neurons is significantly lower. Sengupta et al. further proposed a neural network architecture based entirely on spintronics. The neurons and synapses are composed of DWM-MTJ driven by 3-terminal SOT.

SOT-MTJ can also be used to realize non-linear transfer functions such as soft limit transfer functions, i.e., sigmoid functions. Akhilesh Jaiswal [60] used the S-type probabilistic-characteristic magnetic flip of SOT–MTJ and designed neurons behaved as a sigmoid transfer characteristic. They proposed a pulse neural network and its accuracy was validated using the CIFAR-10 data set. From a biological point of view, biological neural signal transmission is not deterministic but mostly random, so the probabilistic flipping characteristics of SOT are consistent with the mechanism of neurons. Roy's results further show that neurons with sigmoid characteristics have the least impact on accuracy when the property uniformities were induced by the processing, which also highlights the robustness of neural networks based on SOT-MTJ.

7.4.2.3 Spike or Probability Based Neuron

Nonvolatile MTJ devices have been proved to be compatible with a spiking neural network [43, 61–63]. Considering a MTJ with uniaxial magnetic anisotropy, the magnetization of the FL tends to lie in the easy axis. The total energy of the free magnetic layer of MTJs can rise

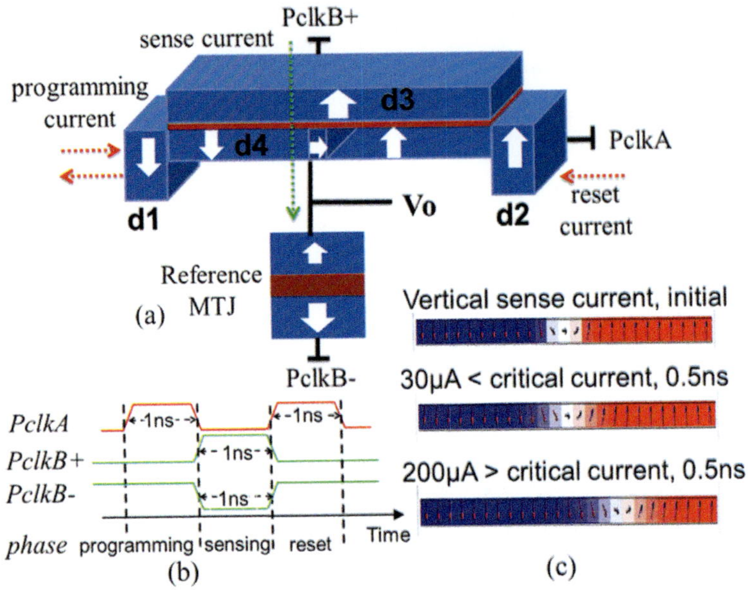

Figure 7.8 (a) Programming and sensing circuit based on a STT Neuron device. (b) Clock waveform diagram. (c) The free magnetic layer on a magnetic domain wall motion simulated at different drive currents [38].

during a specific period of writing pulse current, such as an integration operation, and at the same time it will continuously lose energy, like a leak operation. When reaching the threshold energy, the MTJ will "fire" a switching. That is why a MTJ is the ideal device to realize spiking neurons. Furthermore, MTJs can be influenced by thermal noise and happen to emit a stochastic switching. This feature is very interesting for the well imitation to the probabilistic firing nature of biological neurons for the neurons in brain can be inspired by bio-electricity signal from the synapse and stimulate with the certain probability when reach the threshold membrane potential [64]. Also, for a neuron based on CMOS circuits, it requires more than 20 MOSFETs to map a single probabilistic spiking neuron [65, 66]. Thus, using MTJ as a spiking neuron is very attractive in terms of area efficiency.

A SOT-MTJ device can be used as a spiking neuron device to achieve the leaky integrate and fire (LIF). The LIF neuron model is one of most popular models to mimic the dynamics of a neuron [67, 68]. This model uses resistance states to present the function of a membrane to simulate the integrate and leaky circuit functions. Neuron membrane potential will be charged until it reaches the threshold, and then generate a spike and reset to the threshold. LIF is a more effective model in spiking neuron implementation, for it is closer to the real neuron system than IF, which ignores the memory loss process. However, the analog LIF dynamic is complex and inefficient for CMOS to realize. So, the work has been carried out extensively to explore the potential of non-volatile devices such as MTJs. In 2016, Akhilesh Jaiswal designed a novel LIF spiking neuron using magneto-electric (ME) control of magnetic switching [69]. Since the ME-based MTJ device has a leaky-integrate behavior, it can be used to switch the underlying ferromagnet, and post-neuron will in turn emit a voltage spike according to the state of the ferromagnet. ME switching and a non-volatile ferromagnet are both energy-efficient; furthermore, it has been proved feasible for solving practical issues such as hand writing number recognition. In 2019, Ohno's group in Nagoya University [70] proposed a device using films of Ta (3 nm)/Pt (2.5)/PtMn (9.5)/Pt (0.6)/[Co (0.3)/ Ni $(0.6)]_2$/ Co(0.3)/MgO (1)/Ru (1) as spiking neural components. As shown in Figure 7.8, with the same pulse width and frequency, as the number of pulses increases, the switching probability continues to rise. When the threshold is reached, a deterministic flip can occur, indicating that the device can achieve IF operation. In addition, by setting different pulse frequencies, the higher the frequency, the higher the probability of flipping, indicating that the device has the "forgetting" characteristic, which can also be used in the design of neurosynaptic devices.

In the VLSI 2020 conference, Wu et al. proposed a Poisson neuron based on a novel Poisson probability coding method [71]. Using the back-hopping oscillation (BHO) features of STT–MTJ, such neurons encode the input information into a probability peak sequence with adjustable duty cycle. Poisson neurons make the full use of the STT effect and the random BHO caused by local Joule heating. Based on STT-MRAM compact design Poisson neuronal structures, no additional capacitor and a reset circuit is required, and therefore it takes only 1/6 of the area compared to the CMOS accumulation-firing neurons.

Inspired by the method of neuronal populations encoding, Mizrahi et al. [72] proposed using MTJs to achieve population coding. Neuron population coding uses groups of neurons to finish the transfer function instead of one neuron. It can solve diverse problems by different specific trainings using the same ANN. Each neuron has its own weight, and can occur in different excitation region ranges, as shown in Figure 7.9(a). The output result obtained by the neuron in the learning process is compared with the actual result to obtain the residual. The weight is continuously corrected to minimize the residual. The final output is a weighted summation of the feature functions of all neurons. This algorithm can be used in three-dimensional positioning, image recognition, motor control, and other fields. This work takes advantage of random oscillations caused by thermal disturbance to simulate the tuning curve of random spike neurons. A vertical direct current is applied to the MTJ to generate a spin transfer torque, which can modulate the frequency of the oscillations. The

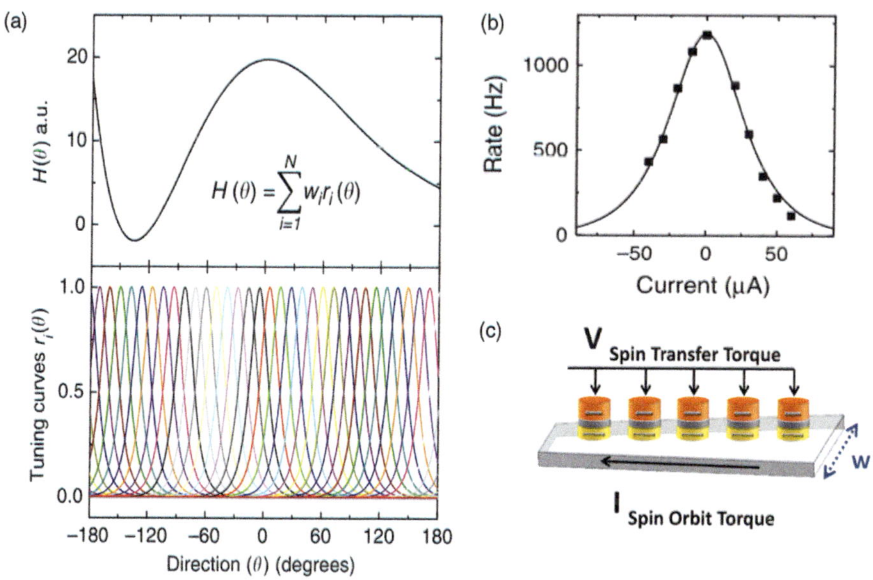

Figure 7.9 (a) Population coding neurons output curve and tuning curve population distribution. (b) The relationship between the random oscillation frequency of neurons and the direct current. (c) A schematic diagram of the hardware implementation of neuron group using spin-orbit torque.

relationship between the frequency and the current amplitude follows a Gaussian distribution, as shown in Figure 7.9(b). In addition, population coding also requires neurons to be distributed in different firing regions. The SOT effect is able to achieve this feature as shown in Figure 7.9(c), which is equivalent to introducing a bias current to modulate the tuning function of the neuron. The strength of the SOT effect is directly related to the writing current density. Injecting the same writing current, the write current density can be directly affected by adjusting the width of the write path. Therefore, this structure designs a bottom electrode with a gradual width, thereby realizing the distribution of the neuron's tuning curve in different regions. This structure introduces STT-MRAM to store weights, and cooperates with peripheral control and arithmetic logic to form a complete learning inference system.

7.5 Magnetic Tunnel Junctions for Synaptic Devices

For a multi-valued neural network, the neural synapse device requires plasticity to achieve a variable weighting input. Its essence is to realize the function of programmable resistance. The synapse device should have two key features. One should be resistor-programmable, i.e., a plurality of resistance states. And another feature is the wide range of resistance change. Simulations show that 13 to 19 resistive states are enough to deal with the task of pattern recognition and classification [73]. For binary state MTJs, it is difficult to achieve polymorphism. At present, researchers show many ways to realize this function in the form of SOT/STT combined with DWM [42, 74–76]. As shown in Figure 7.10(a), Sengupta [43] presents a SOT-MTJ synapse device using a multi-domain FM FL. The current flowing in heavy metal between T2 and T3 drives the domain wall motion, determining the resistance at the end of ports T2 and T3. The equivalent circuit can be described using three parallel resistors in Figure 7.10(b). Assuming the magnetic domain wall is located in x position in a device at a width of L, on the left side of the domain wall, the MTJ is at low resistance state with the M in two FM layers in parallel, and thus the conductance is G_P (x/L). On the right side, the MTJ is in the anti-parallel state and presents high resistance, so the conductance is G_{AP} (1− x/L). The conductance of the domain wall can be regarded as a constant of G_{DW}. The overall resistance changes linearly with the movement of the magnetic domain wall, and the relationship is shown below

$$G_{eq} = G_P\left(\frac{x}{L}\right) + G_{DW} + G_{AP}\left(1-\frac{x}{L}\right) \tag{7.2}$$

where G_{eq} is the overall equivalent conductance, G_P is the high conductance of the device, and G_{AP} is the low conductance of the MTJ. It can be seen that G_{eq} and x change linearly. Sengupta proved through simulation that the write current can drive the movement of the magnetic domain wall, so as to realize the control of the synaptic resistance, and finally realize the synaptic function. However, it should be noted that if the magnetic domain wall is too close to one side and the cross-sectional area of one of the equivalent junctions is too small, the read current may affect its state through the STT. This problem can be solved by appropriately increasing the area of the device. Cao reported recently [37] that they proposed a novel method to tune the binary ferromagnet to a multi-state field-free synapse with a Pt/FM1/Ta/FM2 structure, where the easy axis of FM1 and FM2 lies in perpendicular and in-plane directions, respectively. The multistate feature depends on the in-plane coupling field,

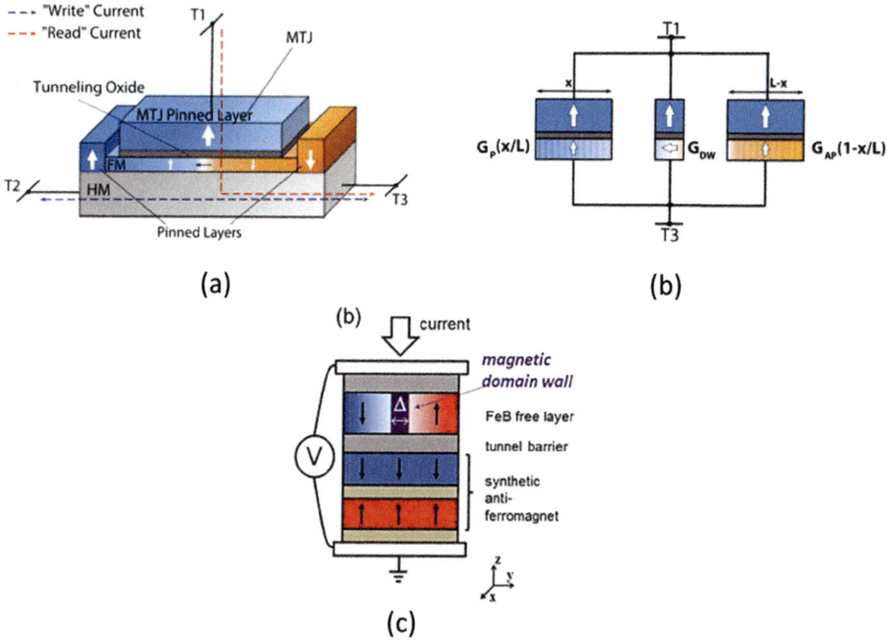

Figure 7.10 (a) SOT-MTJ Neurosynaptic device [43]. (b) Equivalent circuit diagram of MTJ during magnetic domain wall motion process. (c) STT-MTJ Neurosynaptic device [75].

the pulse number and pulse magnitude. This design demonstrated the possibility of implementing MTJ as synaptic electronics to realize the excitatory and inhibitory postsynaptic potentials and spike timing-dependent plasticity.

STT-MTJ can also adopt DWM to realize neurosynaptic devices [42, 74, 75, 77]. Comparison with SOT-MTJ, STT-MTJs experience small read and write windows since both the read and write currents flow through the junctions. It is necessary to consider the possible influence of a read current on the memory state. In 2016, Lequeux et al [75] proposed a STT-MTJ synapses device based on DWM, whose reading current is minimized to avoid read error rate, as shown in Figure 7.10(c). The device is a two-terminal device, under the control of the vertical current, the magnetic domain wall nucleates and can move left and right. If a negative current is applied, the output of MTJ tends to an anti-parallel state. A positive current is applied, it inclined to form a parallel state. During the expansion process, the domain walls will be randomly pinned by defects on the edge or surface of the sample, thereby achieving multiple resistance states. Compared with the synapse device by SOT-MTJ using magnetic domain wall motion, the domain wall motion by STT can be driven with only a small current, thereby reducing power consumption.

For deep neural networks, the role of the MTJ as a neurosynaptic device is mainly to realize the role of weight storage. In the past, the weights of digital-based neural networks were mostly stored in external memory, and it was necessary to frequently call the weights to read and write through the bus. This inevitably brought huge power consumption and the calculation rate was severely limited by the bus bandwidth. Therefore, a variety of solutions were proposed. One is to design the storage unit inside the neural network; that is, near-memory computing, which reduces power consumption by shortening the call path. The other is based on Analog in-Memory Computing (AiMC) [78]. It uses digital-to-analog conversion to convert a digital signal into an analog current signal, and uses Kirchhoff's law and Ohm's law to realize multiply-accumulate calculation (MAC), which greatly reduces the power consumption. This architecture requires multistate storage device state between a stable high resistance and low resistance. In addition, for analog circuits, IR-drop, current leakage and other issues also limit the integration scale [79], which also affects the application in the deep neural network (DNN) [80]. IMEC proposed solutions using the high-resistance SOT-MRAM as AiMC hardware acceleration weight storage [81]. High resistance MRAM devices are realized by adjusting the thickness of MgO. Due to the separation of read and write of SOT-MRAM, the adjustment of the thickness of MgO will not affect the writing process and therefore will not increase the writing power consumption. The researchers took into account the resistance distribution and IR-drop issues caused by the process, which affects the hardware performance. They conducted training tests on the MNIST and CIFAR-100 data sets, and the results also met expectations.

7.6 Learning Methods Suitable for MTJs

For spiking neural network, there are many challenges in coding and training methods. Coding is the one of key features of SNN due to time-dependent operation mechanism [63], because the information type received by spiking neuron is spike timing-series. The raw analog data should be coded into spikes so that they can be input into the SNN. The main stream coding methods is the rate method [18, 82] and the temporal method [83]. These years temporal coding is more favorable [84–88]. The temporal coding, also known as pulse coding, is widely used for MTJ-based SNN.

The main challenge of SNN implementation is the realization of learning methods on hardware. There are many learning methods for spiking neural networks: spiking-timing dependent plasticity (STDP) [19, 74, 89, 90], probability-based STDP method [76, 91] and unsteady learning method [92]. The STDP learning method evolves from the Hebbian weight learning rule [93]. According to STDP learning rules, whether to increase or decrease the synaptic strength is determined by comparing the time difference between the arrival of pre-synaptic pulses and post-synaptic pulses, whlie the common learning rules define that the conductance of synapses decreases exponentially with the increasing spiking time difference [19]. The probability-based STDP method is a further supplement to the STDP method. Vincent et al. [77] simulated the random switching of binary MTJs and showed that they can be used to achieve plasticity that depends on spike timing, a learning rule inspired by neural biology. Using an accurate physical model of MTJs, they demonstrated the super recognition ability of patterns in the video stream. Compared with traditional training, this method reduces the memory volume of the neural network, so that the weights require fewer bits. For some tasks, one bit per synapse is sufficient. Fundamentally, the MTJ can be programmed with short low current pulses, which greatly reduces the energy consumption during the learning process.

Different from the conventional constant probability update STDP method [77], probabilistic STDP method updates the probability of synapse's weight changing based on the pulse width, which is similar to the probabilistic switching mode of a MTJ. In the neural network cross-array structure, Srinivasan proposed a MTJ [76] trained based on probabilistic STDP method. In this work, the typical update probability of this training method does not exceed 0.1, so it can significantly reduce the power consumption required for learning.

In addition to the above methods, the unsteady learning method is also a feasible training program. Sengupta proposed a long term-short term (LT-ST) unsteady learning trained by a high-frequency short-term pulse [92]. The method uses a bistable magnetic tunnel structure as a nerve synapse. If the input current pulse width and intensity are not enough to turn over the barrier between the easy magnetization axes, the FL magnetic moment of the MTJ will flip to an intermediate stable state after the pulse, and gradually return to the original easy magnetization direction. If the pulse frequency of time is high enough, the transfer torque will gradually accumulate until it can be reversed. Therefore, the switching probability is controlled by the input pulse frequency. Thus, MTJ devices are also suitable for temporal information processing.

7.7 Summary and Outlook

With the great development of neuromorphological calculations based on neural networks, the idea and design of using MTJs to realize neural network acceleration chips with non-von Neumann architecture have been continuously proposed. For accelerating neural network chip, the most critical parts are neurons device and synapses device. The MTJ have the potential to replace CMOS neural network in these devices, and greatly reduce power consumption and increase computing speed. In addition, a major advantage of the MTJ lies in its randomness, polymorphism and other characteristics that can be applied to brain-like neural networks such as pulsed neural networks.

However, the application of MTJs to neural-like devices and neural network design structures is still in the preliminary stage. The neural network integration is still in the stage of proposal and design [19, 38, 94]. There are still many challenges in the application of MTJs in brain-like computing and neural networks. First, the MTJ as an analog device will face severe challenges in large-scale integration. Secondly, compared with other memristive devices, the MTJ has a disadvantage in the range of resistance change, which limits the number of resistance states it can achieve. In addition, adjusting the algorithm to make it suitable for all-spin circuits is also a difficult task. In response to these problems, there have been many effective solutions, such as a small size of the MTJ to improve the linearity, in order to meet the needs of the back-propagation gradient descent algorithm [95]. In addition, using low data precision [96] and binary weights [97, 98] to reduce hardware complexity can also effectively solve the problems of small resistance change range and high integration difficulty.

In general, MTJ devices hold broad prospects due to its low power consumption and non-volatility in neuromorphic calculation. Its ability of calculation in memory also further improves the integration of neuromorphic computing acceleration hardware. At present, there have been relatively mature simulation results and many meaningful experimental results on nerve synapses and neuron devices. Therefore, the research on devices and the design and manufacturing of large-scale neural network chips should continue to be promoted.

References

1 Mead, C. (1990). Neuromorphic electronic systems. *Proc. IEEE* 78 (10): 1629–1636.
2 Minsky, M. and Papert, S. (1969). An Introduction to Computational Geometry. MIT Press. Cambridge.

3 Rosenblatt, F. (1958). The perceptron: a probabilistic model for information storage and organization in the brain. *Psychol. Rev.* 65 (6): 386.

4 Rumelhart, D.E., Hinton, G.E., and Williams, R.J. (1986). Learning representations by back-propagating errors. *Nature* 323 (6088): 533–536.

5 Miller, G.F., Todd, P.M., and Hegde, S.U. (1989). Designing neural networks using genetic algorithms. ICGA.

6 LeCun, Y., Bottou, L., Bengio, Y., and Haffner, P. (1998). Gradient-based learning applied to document recognition. *Proc. IEEE* 86 (11): 2278–2324.

7 LeCun, Y., Boser, B., Denker, J.S., Henderson, D., Howard, R.E., Hubbard, W., and Jackel, L.D. (1989). Backpropagation applied to handwritten zip code recognition. *Neural Comput.* 1 (4): 541–551.

8 LeCun, Y., Bengio, Y., and Hinton, G. (2015). Deep learning. *Nature* 521 (7553): 436–444.

9 Hochreiter, S. and Schmidhuber, J. (1997). Long short-term memory. *Neural Comput.* 9 (8): 1735–1780.

10 Tetko, I.V., Livingstone, D.J., and Luik, A.I. (1995). Neural network studies. 1. Comparison of overfitting and overtraining. *J. Chem. Inf. Comput. Sci.* 35 (5): 826–833.

11 Piotrowski, A.P. and Napiorkowski, J.J. (2013). A comparison of methods to avoid overfitting in neural networks training in the case of catchment runoff modelling. *J. Hydrol.* 476: 97–111.

12 Hopfield, J.J. (1995). Pattern recognition computation using action potential timing for stimulus representation. *Nature* 376 (6535): 33–36.

13 Sejnowski, T.J. (1995). Time for a new neural code? *Nature* 376 (6535): 21–22.

14 Jo, S.H., Chang, T., Ebong, I., Bhadviya, B.B., Mazumder, P., and Lu, W. (2010). Nanoscale memristor device as synapse in neuromorphic systems. *Nano Lett.* 10 (4): 1297–1301.

15 Chicca, E., Stefanini, F., Bartolozzi, C., and Indiveri, G. (2014). Neuromorphic electronic circuits for building autonomous cognitive systems. *Proc. IEEE* 102 (9): 1367–1388.

16 Sengupta, A. and Roy, K. (2018). Neuromorphic computing enabled by physics of electron spins: prospects and perspectives. *Appl. Phys. Express* 11: 3.

17 Simonyan, K. and Zisserman, A. (2014). Very deep convolutional networks for large-scale image recognition. *arXiv preprint arXiv:1409.1556.*

18 Maass, W. (1997). Networks of spiking neurons: the third generation of neural network models. *Neural Netw.* 10 (9): 1659–1671.

19 Sengupta, A., Banerjee, A., and Roy, K. (2016). Hybrid spintronic-CMOS spiking neural network with on-chip learning: devices, circuits, and systems. *Phys. Rev. Appl.* 6: 6.

20 Zhao, W., Chappert, C., Javerliac, V., and Noziere, J.-P. (2009). High speed, high stability and low power sensing amplifier for MTJ/CMOS hybrid logic circuits. *IEEE Trans. Magn.* 45 (10): 3784–3787.

21 Lakys, Y., Zhao, W.S., Devolder, T., Zhang, Y., Klein, J.-O., Ravelosona, D., and Chappert, C. (2012). Self-enabled "error-free" switching circuit for spin transfer torque MRAM and logic. *IEEE Trans. Magn.* 48 (9): 2403–2406.

22 Horng, C.T., Tong, R.-Y., Torng, C.-J., and Liu, G. (2010). SyAF structure to fabricate Mbit MTJ MRAM.

23 Zhou, P., Zhao, B., Yang, J., and Zhang, Y. (2009). Energy reduction for STT-RAM using early write termination. *2009 IEEE/ACM International Conference on Computer-Aided Design-Digest of Technical Papers.*

24 Bottou, L. and Bousquet, O. (2008). The tradeoffs of large scale learning. *Advances in Neural Information Processing Systems.*

25 Locatelli, N., Vincent, A.F., and Querlioz, D. (2018). Use of magnetoresistive random-access memory as approximate memory for training neural networks. *2018 25th IEEE International Conference on Electronics, Circuits and Systems (ICECS).*

26 Senn, W. and Fusi, S. (2005). Convergence of stochastic learning in perceptrons with binary synapses. *Phys. Rev. E* 71 (6): 061907.

27 Bill, J. and Legenstein, R. (2014). A compound memristive synapse model for statistical learning through STDP in spiking neural networks. *Front. Neurosci.* 8: 412.

28 Lippmann, R. (1987). An introduction to computing with neural nets. *IEEE Assp Magazine* 4 (2): 4–22.

29 Zhu, J. G. J., and Park, C. (2006). Magnetic tunnel junctions. *Materials Today* 9 (11): 36–45.

30 Wang, W. G., Li, M., Hageman, S. et al. (2012). Electric-field-assisted switching in magnetic tunnel junctions[J]. *Nature Materials* 11 (1): 64–68.

31 Yuasa, S., Nagahama, T., Fukushima, A., Suzuki, Y., and Ando, K. (2004). Giant room-temperature magnetoresistance in single-crystal Fe/MgO/Fe magnetic tunnel junctions. *Nature Materials* 3 (12): 868–871.

32 Indiveri, G., Linares-Barranco, B., Hamilton, T.J., Van Schaik, A., Etienne-Cummings, R., Delbruck, T., and Renaud, S. (2011). Neuromorphic silicon neuron circuits. *Front. Neurosci.* 5: 73.

33 Chen, Y.-H., Krishna, T., Emer, J.S., and Sze, V. (2016). Eyeriss: an energy-efficient reconfigurable accelerator for deep convolutional neural networks. *IEEE J. Solid-State Circuits* 52 (1): 127–138.

34 Anwar, S., Hwang, K., and Sung, W. (2015). Fixed point optimization of deep convolutional neural networks for object recognition. *2015 IEEE International Conference on Acoustics, Speech and Signal Processing (ICASSP).*

35 Hyvarinen, A. (1999). Fast and robust fixed-point algorithms for independent component analysis. *IEEE Trans. Neural Netw.* 10 (3): 626–634.

36 Fukami, S., Zhang, C.L., DuttaGupta, S., Kurenkov, A., and Ohno, H. (2016). Magnetization switching by spin-orbit torque in an antiferromagnet-ferromagnet bilayer system. *Nat. Mater.* 15 (5): 535–541.

37 Cao, Y., Rushforth, A., Sheng, Y., Zheng, H., and Wang, K. (2019). Tuning a binary ferromagnet into a multistate synapse with spin–orbit-torque-induced plasticity. *Adv. Funct. Mater.* 29 (25): 1808104.

38 Fan, D., Shim, Y., Raghunathan, A., and Roy, K. (2015). STT-SNN: a spin-transfer-torque based soft-limiting non-linear neuron for low-power artificial neural networks. *IEEE Trans. Nanotechnol.* 14 (6): 1013–1023.

39 Sengupta, A. (2018). Efficient neuromorphic computing enabled by spin-transfer torque: devices, circuits and systems. Purdue University.

40 Sengupta, A. and Roy, K. (2015). Spin-transfer torque magnetic neuron for low power neuromorphic computing. *2015 International Joint Conference on Neural Networks (IJCNN).*

41 Sengupta, A., Choday, S.H., Kim, Y., and Roy, K. (2015). Spin orbit torque based electronic neuron. *Appl. Phys. Lett.* 106: 14.

42 Sharad, M., Augustine, C., Panagopoulos, G., and Roy, K. (2012). Spin-based neuron model with domain-wall magnets as synapse. *IEEE Trans. Nanotechnol.* 11 (4): 843–853.

43 Sengupta, A. and Roy, K. (2016). A vision for all-spin neural networks: a device to system perspective. *IEEE Trans. Circuits Syst. I: Regul. Pap.* 63 (12): 2267–2277.

44 Sengupta, A., Shim, Y., and Roy, K. (2016). Proposal for an all-spin artificial neural network: emulating neural and synaptic functionalities through domain wall motion in ferromagnets. *IEEE Trans. Biomed. Circuits Syst.* 10 (6): 1152–1160.

45 Sengupta, A., Han, B., and Roy, K. (2016). Toward a spintronic deep learning spiking neural processor. *Biomedical Circuits and Systems Conference (BioCAS), 2016 IEEE.*

46 Sengupta, A., Panda, P., Wijesinghe, P., Kim, Y., and Roy, K. (2016). Magnetic tunnel junction mimics stochastic cortical spiking neurons. *Sci. Rep.* 6: 30039.

47 Berger, L. (1978). Low-field magnetoresistance and domain drag in ferromagnets. *J. Appl. Phys.* 49 (3): 2156–2161.

48 Slonczewski, J.C. (1996). Current-driven excitation of magnetic multilayers. *J. Magn. Magn. Mater.* 159 (1–2): L1–L7.

49 Myers, E.B., Ralph, D.C., Katine, J.A., Louie, R.N., and Buhrman, R.A. (1999). Current-induced switching of domains in magnetic multilayer devices. *Science* 285 (5429): 867–870.

50 Wei, H.X., Qin, Q.H., Ma, M., Sharif, R., and Han, X.F. (2007). 80% tunneling magnetoresistance at room temperature for thin Al-O barrier magnetic tunnel junction with CoFeB as free and reference layers. *J. Appl. Phys.* 101 (9): 3. Article 09b501.

51 Ikeda, S., Miura, K., Yamamoto, H., Mizunuma, K., Gan, H.D., Endo, M., and Ohno, H. (2010). A perpendicular-anisotropy CoFeB-MgO magnetic tunnel junction. *Nat. Mater.* 9 (9): 721–724.

52 Yakushiji, K., Saruya, T., Kubota, H., Fukushima, A., Nagahama, T., Yuasa, S., and Ando, K. (2010). Ultrathin Co/Pt and Co/Pd superlattice films for MgO-based perpendicular magnetic tunnel junctions. *Appl. Phys. Lett.* 97 (23): 232508–232508–232503.

53 Miron, I.M., Garello, K., Gaudin, G., Zermatten, P.J., Costache, M.V., Auffret, S., and Gambardella, P. (2011). Perpendicular switching of a single ferromagnetic layer induced by in-plane current injection. *Nature* 476 (7359): 189–U188.

54 Miron, I.M., Gaudin, G., Auffret, S., Rodmacq, B., Schuhl, A., Pizzini, S., and Gambardella, P. (2010). Current-driven spin torque induced by the Rashba effect in a ferromagnetic metal layer. *Nat. Mater.* 9 (3): 230–234.

55 Miron, I.M., Moore, T., Szambolics, H., Budaprejbeanu, L.D., Auffret, S., Rodmacq, B., and Schuhl, A. (2011). Fast current-induced domain-wall motion controlled by the Rashba effect. *Nat. Mater.* 10 (6): 419.

56 Liu, L.Q., Lee, O.J., Gudmundsen, T.J., Ralph, D.C., and Buhrman, R.A. (2012). Current-induced switching of perpendicularly magnetized magnetic layers using spin torque from the spin Hall effect [Article]. *Phys. Rev. Lett.* 109 (9): 5. Article 096602.

57 Liu, L.Q., Pai, C.F., Li, Y., Tseng, H.W., Ralph, D.C., and Buhrman, R.A. (2012). Spin-torque switching with the giant spin Hall effect of tantalum. *Science* 336 (6081): 555–558.

58 Kondo, K., Choi, J.-Y., Baek, J.-U., Jun, H.-S., Jung, S., Shim, T.-H., and Park, J.-G. (2018). A two-terminal perpendicular spin-transfer torque based artificial neuron. *J. Phys. D: Appl. Phys.* 51 (50): 504002.

59 Jo, S.H., Kim, K.-H., and Lu, W. (2009). High-density crossbar arrays based on a Si memristive system. *Nano Lett.* 9 (2): 870–874.

60 Jaiswal, A., Agrawal, A., Chakraborty, I., Roy, D., and Roy, K. (2019). On robustness of spin-orbit-torque based stochastic sigmoid neurons for spiking neural networks. *2019 International Joint Conference on Neural Networks (IJCNN).*

61 Grollier, J., Querlioz, D., Camsari, K., Everschor-Sitte, K., Fukami, S., and Stiles, M. (2020). Neuromorphic spintronics. *Nat. Electron.*: 1–11.

62 Umesh, S. and Mittal, S. (2019). A survey of spintronic architectures for processing-in-memory and neural networks. *J. Syst. Archit.* 97: 349–372.

63 Ghosh-Dastidar, S. and Adeli, H. (2009). Spiking neural networks. *Int. J. Neural Syst.* 19 (04): 295–308.

64 Sobie, C., Babul, A., and de Sousa, R. (2011). Neuron dynamics in the presence of 1/f noise [Article]. *Phys. Rev. E* 83 (5): 11. Article 051912.

65 Rajendran, B. and Alibart, F. (2016). Neuromorphic computing based on emerging memory technologies. *IEEE Trans. Emerg. Sel. Topics Circuits Syst.* 6 (2): 198–211.

66 Indiveri, G. (2003). A low-power adaptive integrate-and-fire neuron circuit. *Proceedings of the 2003 International Symposium on Circuits and Systems (ISCAS'03)*. IEEE.

67 Liu, Y. H. and Wang, X. J. (2001). Spike-frequency adaptation of a generalized leaky integrate-and-fire model neuron. *J. Comput. Neurosci.* 10 (1): 25–45.

68 Burkitt, A.N. (2006). A review of the integrate-and-fire neuron model: i. Homogeneous synaptic input. *Biol. Cybern.* 95 (1): 1–19.

69 Jaiswal, A., Roy, S., Srinivasan, G., and Roy, K. (2017). Proposal for a leaky-integrate-fire spiking neuron based on magnetoelectric switching of ferromagnets. *IEEE Trans. Electron Devices* 64 (4): 1818–1824.

70 Kurenkov, A., DuttaGupta, S., Zhang, C., Fukami, S., Horio, Y., and Ohno, H. (2019). Artificial neuron and synapse realized in an antiferromagnet/ferromagnet heterostructure using dynamics of spin–orbit torque switching. *Adv. Mater.* 31 (23): 1900636.

71 Wu, M.-H., Huang, M.-S., Zhu, Z., Liang, F.-X., Hong, M.-C., Deng, J., Wei, J.-H., Sheu, S.-S., Wu, C.-I., Liang, G., and Hou, T.-H. (2020). Compact probabilistic poisson neuron based on back-hopping oscillation in stt-mram for all-spin deep spiking neural network. *2020 Symposium on VLSI Technology*.

72 Mizrahi, A., Hirtzlin, T., Fukushima, A., Kubota, H., Yuasa, S., Grollier, J., and Querlioz, D. (2018). Neural-like computing with populations of superparamagnetic basis functions. *Nat. Commun.* 9 (1): 1–11.

73 Querlioz, D., Bichler, O., Vincent, A.F., and Gamrat, C. (2015). Bioinspired programming of memory devices for implementing an inference engine. *Proc. IEEE* 103 (8): 1398–1416.

74 Krzysteczko, P., Münchenberger, J., Schäfers, M., Reiss, G., and Thomas, A. (2012). The memristive magnetic tunnel junction as a nanoscopic synapse-neuron system. *Adv. Mater.* 24 (6): 762–766.

75 Lequeux, S., Sampaio, J., Cros, V., Yakushiji, K., Fukushima, A., Matsumoto, R., and Grollier, J. (2016). A magnetic synapse: multilevel spin-torque memristor with perpendicular anisotropy. *Sci. Rep.* 6 (1): 1–7.

76 Srinivasan, G., Sengupta, A., and Roy, K. (2016). Magnetic tunnel junction based long-term short-term stochastic synapse for a spiking neural network with on-chip STDP learning. *Sci. Rep.* 6: 29545.

77 Vincent, A.F., Larroque, J., Locatelli, N., Romdhane, N.B., Bichler, O., Gamrat, C., and Querlioz, D. (2015). Spin-transfer torque magnetic memory as a stochastic memristive synapse for neuromorphic systems. *IEEE Trans. Biomed. Circuits Syst.* 9 (2): 166–174.

78 Cosemans, S., Verhoef, B., Doevenspeck, J., Papistas, I., Catthoor, F., Debacker, P., and Verkest, D. (2019). Towards 10000TOPS/W DNN inference with analog in-memory computing–a circuit blueprint, device options and requirements. *2019 IEEE International Electron Devices Meeting (IEDM)*.

79 Jung, M. and Lim, S.K. (2010). A study of IR-drop noise issues in 3D ICs with through-silicon-vias. *2010 IEEE International 3D Systems Integration Conference (3DIC)*.

80 Liu, B., Wen, W., Chen, Y., Li, X., Wu, C.-R., and Ho, T.-Y. (2015). Eda challenges for memristor-crossbar based neuromorphic computing. *Proceedings of the 25th edition on Great Lakes Symposium on VLSI.*

81 Doevenspeck, J., Garello, K., Verhoef, B., Degraeve, R., Van Beek, S., Crotti, D., Yasin, F., Couet, S., Jayakumar, G., Papistas, I.A., Debacker, P., Lauwereins, R., Dehaene, W., Kar, G.S., Cosemans, S., Mallik, A., and Verkest, D. (2020). SOT-MRAM based analog in-memory computing for DNN. *Inference 2020 Symposium on VLSI Technology.*

82 Gerstner, W. and Kistler, W.M. (2002). Spiking Neuron Models: Single Neurons, Populations, Plasticity. Cambridge University Press, Cambridge.

83 Charles, E.Y.A. (2006). Supervised and Unsupervised Weight and Delay Adaptation Learning in Temporal Coding Spiking Neural Networks. Cardiff University.

84 Thorpe, S.J. (1990). Spike arrival times: a highly efficient coding scheme for neural networks. In Eckmiller, R., Hartmann, G. and Hauske, G. (Editor): Parallel Processing in Neural Systems, 91–94. North-Holland. Elsevier.

85 Thorpe, S.J. and Imbert, M. (1989). Biological constraints on connectionist modelling. *Connectionism in Perspective*, 63–92.

86 Yu, Q., Tang, H., Hu, J., and Tan, K.C. (2017). Rapid feedforward computation by temporal encoding and learning with spiking neurons. In Eckmiller, R., Hartmann, G. and Hauske, G.(Editor): Neuromorphic Cognitive Systems, 19–41. Springer.

87 Brette, R., Rudolph, M., Carnevale, T., Hines, M., Beeman, D., Bower, J.M., and Harris, F.C. (2007). Simulation of networks of spiking neurons: a review of tools and strategies. *J. Comput. Neurosci.* 23 (3): 349–398.

88 Lobov, S.A., Chernyshov, A.V., Krilova, N.P., Shamshin, M.O., and Kazantsev, V.B. (2020). Competitive learning in a spiking neural network: towards an intelligent pattern classifier. *Sensors* 20 (2): 500.

89 Locatelli, N., Cros, V., and Grollier, J. (2014). Spin-torque building blocks. *Nat. Mater.* 13 (1): 11–20.

90 Borders, W.A., Akima, H., Fukami, S., Moriya, S., Kurihara, S., Horio, Y., and Ohno, H. (2017). Analogue spin–orbit torque device for artificial-neural-network-based associative memory operation. *Appl. Phys. Express* 10 (1).

91 Vincent, A.F., Larroque, J., Zhao, W., Romdhane, N.B., Bichler, O., Gamrat, C., and Querlioz, D. (2014). Spin-transfer torque magnetic memory as a stochastic memristive synapse. *2014 IEEE International Symposium on Circuits and Systems (ISCAS).*

92 Sengupta, A. and Roy, K. (2016). Short-term plasticity and long-term potentiation in magnetic tunnel junctions: towards volatile synapses. *Phys. Rev. Appl.* 5 (2).

93 Caporale, N. and Dan, Y. (2008). Spike timing–dependent plasticity: a Hebbian learning rule. *Annu. Rev. Neurosci.* 31: 25–46.

94 Srinivasan, G., Sengupta, A., and Roy, K. (2017). Magnetic tunnel junction enabled all-spin stochastic spiking neural network. *2017 Design, Automation & Test in Europe Conference & Exhibition (DATE).*

95 Nagaosa, N. and Tokura, Y. (2013). Topological properties and dynamics of magnetic skyrmions. *Nat. Nanotechnol.* 8 (12): 899–911.

96 Wu, S., Li, G., Chen, F., and Shi, L. (2018). Training and inference with integers in deep neural networks. *arXiv preprint arXiv:1802.04680.*

97 Hubara, I., Courbariaux, M., Soudry, D., El-Yaniv, R., and Bengio, Y. (2017). Quantized neural networks: training neural networks with low precision weights and activations. *J. Mach. Learn. Res.* 18 (1): 6869–6898.

98 Rastegari, M., Ordonez, V., Redmon, J., and Farhadi, A. (2016). Xnor-net: imagenet classification using binary convolutional neural networks. *European Conference on Computer Vision.*

99 Mellnik, A. R., Lee, J. S., Richardella, A., Grab, J.L., Mintun, P. J., Fischer, M. H., Vaezi, A., Manchon, A., Kim, E. -A., Samarth, N. and Ralph, D. C., (2014). Spin-transfer torque generated by a topological insulator. *Nature*. 551(7510) (1): 449-451.

8

Three-Dimensional Magnetic Structures of B20 Chiral Magnets

Kejing Ran[1,2], *Dongsheng Song*[3], *Weiwei Wang*[3], *Haifeng Du*[4], *and Shilei Zhang*[1,2]

[1]*School of Physical Science and Technology, ShanghaiTech University, Shanghai, China*
[2]*ShanghaiTech Laboratory for Topological Physics, ShanghaiTech University, Shanghai, China*
[3]*Institutes of Physical Science and Information Technology, Anhui University, Hefei, China*
[4]*The Anhui Key Laboratory of Condensed Matter Physics at Extreme Conditions, High Magnetic Field Laboratory and University of Science and Technology of China, Chinese Academy of Science, Hefei, Anhui Province, China.*

CHAPTER MENU

The study of magnetic skyrmions has been carried out for more than a decade since their discovery, and the magnetism community has witnessed tremendous advances in the idea of topological magnetism. The conventional way of describing a skyrmion is to simplify it as a two-dimensional object. However, recent advances in this field have emphasized that their three-dimensional structures are essential to be taken into account, in order to fully grasp the underlying physics of such a fascinating system. We review here the recent theoretical and experimental advances in the topic of three-dimensional topological structures, and expect the prospective of this research field.

8.1 Theoretical Development

The magnetic systems in nanoscale can be described either through the atomistic spin model [1] or through standard micromagnetics [2]. The atomistic spin model is particular appropriate for systems with high anisotropy or strong DMI that the spin may change discontinuously. The typical Hamiltonian for the atomistic spin model reads

$$H = -J\sum_{\langle i,j\rangle}\mathbf{m}_i\cdot\mathbf{m}_j + \sum_{\langle i,j\rangle}\mathbf{d}_{i,j}\cdot\left[\mathbf{m}_i\times\mathbf{m}_j\right] + \frac{K}{2}m_z^2 + \mu_s\sum_i\mathbf{H}\cdot\mathbf{m}_i \tag{8.1}$$

Spintronics: Materials, Devices and Applications, First Edition. Edited by Kaiyou Wang, Meiyin Yang, and Jun Luo.

where $\langle i,j\rangle$ represents a unique pair of lattice sites i and j, J represents the symmetric exchange strength, $\mathbf{m}_i$ is the unit vector of the magnetic moment and $\mu_s = \hbar\gamma S$ where S is the atomic spin length and γ is the gyromagnetic ratio. The antisymmetric exchange, also known as the Dzyaloshinskii-Moriya interaction (DMI), is represented by D_0. For the bulk DMI case, the DMI vector $\mathbf{d}_{ij}$ can be written as $d_{ij} = D_0\hat{\mathbf{r}}_{ij}$, where $\hat{\mathbf{r}}_{ij}$ is the unit vector between $\mathbf{S}_i$ and $\mathbf{S}_j$. K denotes the anisotropy and $\mathbf{H}$ is the external field. The Hamiltonian (1) is consistent with the non-centrosymmetric cubic B20 crystals such as MnSi, FeGe, Fe_{1-}xCoxSi and β-Mn-type Co-Zn-Mn alloys.

In the absence of the external field and anisotropy, the ground state is helical state with a spiral period $\lambda \sim 2\pi Ja / D_0$ where a is the lattice constant. The ground state can be obtained either from the direct energy minimization using gradient descent methods or by numerically solving the Landau–Lifshitz–Gilbert (LLG) equation

$$\frac{\partial \mathbf{m}_i}{\partial t} = -\gamma \mathbf{m}_i \times \mathbf{H}_{\text{eff}} + \alpha \mathbf{m}_i \times \frac{\partial \mathbf{m}_i}{\partial t}\phi \tag{8.2}$$

where α is the Gilbert damping and $\mathbf{H}_{\text{eff}} = -\left(\frac{1}{\mu_s}\right)\left(\frac{\partial_H}{\partial_{mi}}\right)$ is the effective field. In the presence of external field or anisotropy, magnetic skyrmions could be stabilized. It is worth noting that the thermal effect is not included in the energy minimization or the LLG equation. To capture the feature that the skyrmion phase is thermal equilibrium state closed to the Tc, Monte Carlo methods based on Hamiltonian equation (8.1) can be applied [3]. Especially, the next-neighbor interaction could play an essential role in recovering the correct phase diagram observed in experiments [3].

In micromagnetics the magnetic state is described by a continuum variable, i.e., the magnetization $\mathbf{M} = \mathbf{M}(\mathbf{r}, t)$. Similar to the atomistic spin model, the total energy density can be written as [2]

$$E_{\text{tot}} = E_{\text{ex}} + E_{\text{DMI}} + E_{\text{zeeman}} + E_{\text{an}} + E_{\text{demag}} + \ldots \tag{8.3}$$

where $E_{\text{ex}} = A(\nabla \mathbf{m})^2 = A\left[(\nabla m_x)^2 + (\nabla m_y)^2 + (\nabla m_z)^2\right]$, $E_{\text{DMI}} = D\mathbf{m}\cdot(\nabla \times \mathbf{m})$, E_{an} and E_{demag} represent the energy contributions from exchange, DMI, anisotropy, and demagnetization, respectively. Unlike the discrete spin model, numerical schemes such as finite difference method (FDM) [4] or finite element method (FEM) [5] are needed to solve the coupled micromagnetic equations including the spatial effective fields and LLG equation. While the FDM can be easily implemented with the speed advantages, the FEM managed to model complex 3D shapes with a high accuracy. With the improvement of computer hardware, as well as the enhanced parallel technologies, modern micromagnetic software is able to simulate micron-sized samples [6]. Moreover, graphical processing units (GPUs) are also used to speed up calculations, enabling the micromagntics to be a powerful tool to investigate the 3D structures.

The bulk DMI defined in equation (8.1) can be extended to three dimensional straightforwardly. The third dimension is essential for hosting new topological structures, such as skyrmion tubes [3, 7], bobbers [8, 9], hopfions [10, 11], and torons [12, 13]. Even the skyrmion tube is not a simple extrusion of its two-dimensional structure – an extra twist emerges due to the bulk DMI, resulting in a periodic modulation in z-direction, as shown in Figure 8.1

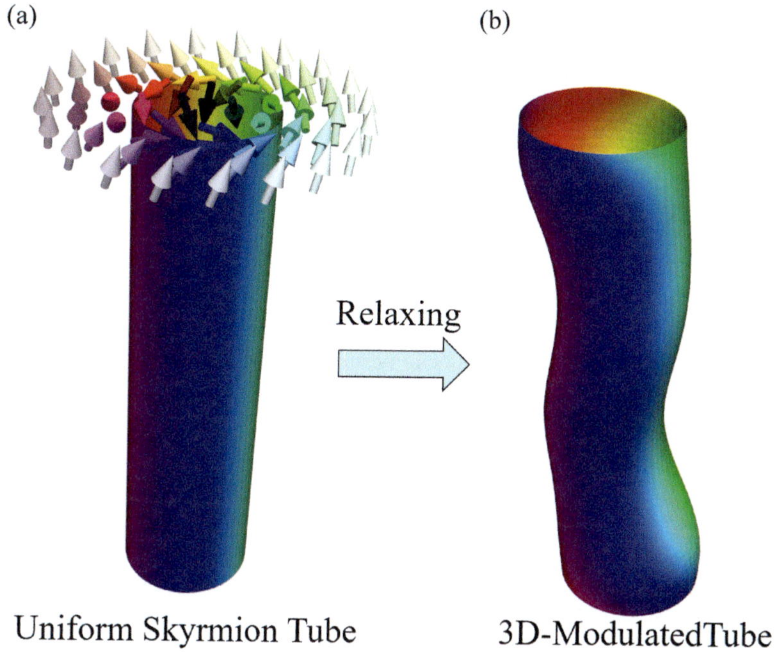

Figure 8.1 (a) A skyrmion tube which is obtained by simply extruding the corresponding 2D skyrmion profile. (b) Starting from profile (a), the skyrmion tube is relaxed to a twisted tube with a periodic modulation.

[14]. The periodic modulation can be understood by considering a typical 3D structure – cones, which is given by [14]

$$\cos\theta = H / H_D, \qquad \psi(z) = \psi_{cone}(z) = 2\pi z / L_D \tag{8.4}$$

where $H_D = D^2 / (2A \mid M)$ is the saturation field and L_D is the period of the helix and the unit vector of magnetization is expressed as $\mathbf{m} = (sin\theta cos\psi, sin\theta \sin\psi, \cos\psi)$. The skyrmion tube can thus be described using

$$\theta = \theta(\rho), \qquad \psi(\phi, z) = \phi + \pi / 2 + \psi_{cone}(z) \tag{8.5}$$

where ϕ and ρ represent the cylindrical coordinates, respectively.

Besides the skyrmion tube, magnetic bobbers could occur near the surface of chiral magnets [8, 9]. Unlike the skyrmion tube, the skyrmion core does not penetrate the entire sample along the thickness direction. Comparing with the normal skyrmion tube, the finite penetration depth of bobbers results in a weak contrast when observed using the quantitative off-axis electron holography. Moreover, the finite penetration depth means a magnetic singularity (Bloch Point) attached at the end of the bobber. The presence of the Bloch point introduces an extra pinning effect, thus making its dynamics different from that of a skyrmion tube when external driving forces such as electric currents are used.

Magnetic torons can also be stabilized in chiral magnets using external fields. A magnetic toron is a spatially localized 3D spin texture composed of skyrmionic layers with two Bloch points at its two ends [13]. In a film system with an in-plane current, the dynamic of the toron is similar to that for magnetic skyrmions. However, unidirectional motion emerges for torons in a cylindrical nanowire where the current is applied along the wire direction [13]. This unidirectional motion originates from the reflection symmetry broken due to the unique direction of the spin precession, which is confirmed by numerically using the modified LLG equation [13].

The magnetic hopfions are 3D topological solitons that can be classified using the so-called Hopf index Q_H [10, 11]. The hopfions were studied in many physical systems such as gauge theories, liquid crystals, Bose–Einstein condensates, fluids, and ferromagnets [15, 16]. Similar to the skyrmions, the magnetic hopfions can be manipulated using electric current through either spin-transfer torques (STT) or spin Hall torques [15, 16]. Similar to the skyrmion Hall effect, the transverse motion of hopfions also occurs under applied currents, accompanying a dilation depending on the ratio of β / α with β measures the nonadiabatic STT effect.

8.2 Observation Technique

8.2.1 Electron Holography

8.2.1.1 Historical Survey

Electron holography is a powerful technique for imaging the electric and magnetic fields of materials in the transmission electron microscopy (TEM), which has been developed for about 70 years. After the invention of TEM in 1930s, the concept of holography was raised by Dennis Gabor to increase the contrast of TEM images in 1948 [17]. The Nobel prize was awarded to Gabor in 1971 for his pioneering contribution to the technology and science of holography. In 1952, M.E. Haine and T. Mulvey succeeded in achieving the first TEM hologram at the spatial resolution of 1 nm [18]. The accelerating voltage of the TEM was 60 kV in their experiments and the morphology of the zinc oxide crystals were clearly observed after the reconstruction from diffraction images. Ten years later, E.N. Leith and J. Upatnieks present an off-axis light path to eliminate the disturbance of twin images [19]. The objective wave and the reference wave could interfere at an angle. Thus, the real and virtual images can be separated in the hologram. Based on this idea, the original off-axis electron holography was invented by G. Mollenst and H. Wahl in 1968 [20]. The current hardware of electron holography is basically the same as this.

From 1950 to 1980, electron holography was mainly used to enhance the quality of TEM images. Due to the development of the electron gun, especially the invention of the field emission electron gun, the coherence of the electron beam in TEM has been greatly improved. It means that the precise measurement of the electron-wave phase could be realized. Because the electric and magnetic fields can change the phase of the electron-wave during its propagation through the sample, the application of electron holography has gradually been extended to the measurement of the sample thickness, electric charge, and magnetic domains. In 1980, the fine structure of magnetic domain walls in a triangular cobalt particle was revealed by electron holography [21]. This experiment was carried by Nagoya University and Hitachi Ltd., which initiated a new period. In the following few decades, electron holography has become a hot technology in the research of heterojunction and magnetic materials. Especially in magnetism, the electron holography revealed plenty of novel domain structures, including the bubble domains, magnetic helical domains, and magnetic skyrmions. With the development of computer technology and digital image analysis techniques, the reconstruction process of hologram has also been upgraded. The optical reproduction technique through ordinary light source or laser source was obsoleted. The digital filtering technique (DFT) took the place and further increased the spatial resolution of the electron holography. The electron holography was always keeping the record of the highest spatial resolution (about 1 nm) of imaging the magnetic domain structure before the appearance of spin-polarized scanning tunneling microscope (STM).

8.2.1.2 Experimental Setup

Electron holography is based on the formation of interference hologram in the transmission electron microscope (TEM), from which the phase shift of exit electron wave could be recovered according to the Aharonov–Bohm effect. As the exit electron wave carries the information of electromagnetic potential in the specimen, the local variation of electrical or magnetic field could be measured through the retrieve of phase shift, which can be expressed as follows [22]

$$\phi(x) = C_E \int V(x,z)\,dz - \frac{e}{\hbar}\int\int B_{\perp}(x,z)\,dx\,dz \tag{8.6}$$

where $C_E = \dfrac{2\pi}{\lambda}\dfrac{E+E_0}{E(E+2E_0)}$ is the electron beam direction, x is the direction in the plane of the specimen, B is the component of the magnetic induction perpendicular to both x and z, V is the electrostatic potential, λ is the relativistic electron wavelength, and E and E_0 are the kinetic and rest mass energies of the incident electron, respectively.

The basic experimental setup of off-axis electron holography is schematically shown in Figure 8.2. The incident electron wave from the electron gun can be divided into two parts. One passes through the vacuum region as the reference wave while the other one passes through the specimen as the object wave. By applying the electrical voltage on the biprism (gold-coated quartz), the reference wave and object wave will overlap and interference with each other. Consequently, the representative hologram appears, which can be record with the charge coupling device (CCD). The phase shift is reconstructed by filtering out one sideband of the Fourier transform of the electron hologram. It should always be noted that the separation of phase shift from magnetic field and electrostatic potential are indispensable, in order to exclude the unwanted contributions from local variations of specimen thickness and compositions. At last, the real-space map of projected in-plane magnetic induction ($\mathbf{B}_{xy}$) is proportional to the gradient of magnetic phase shift. Additionally, the magnetization distributions are often pursued for resolving the local magnetic configurations. A model-based iterative reconstruction algorithm has been developed to reconstruct the in-plane projected magnetization ($\mathbf{M}_{xy}$) from the experimental magnetic phase images [23].

With the advance of TEM facilities, such as the coherence and brightness of electron gun and the sensitivity of CCD devices, the phase shift could be currently measured at the accuracy of $2\pi/1000$ [24], promising off-axis electron holography as a quantitative technique for the measurement of magnetic textures. The spatial resolution can approach 1 nm in the modern aberration-corrected TEM, in contrast the Lorentz Fresnel mode that the

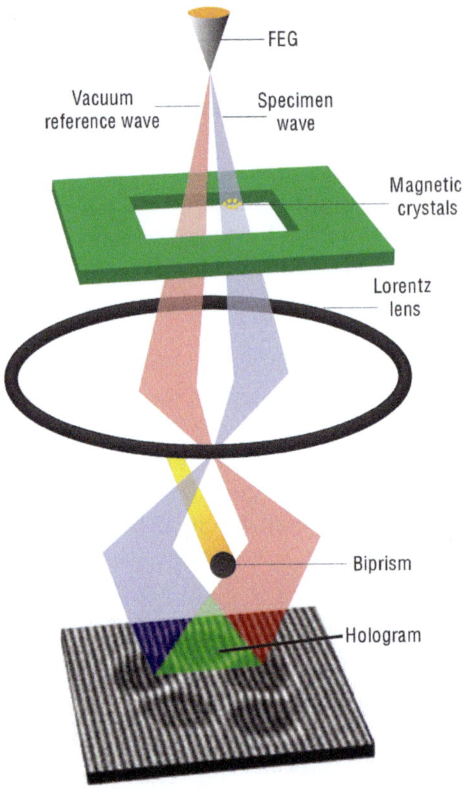

Figure 8.2 Schematics of off-axis electron holography in the transmission electron microscope. Figure reproduced with permission from [22].

delocalization is significantly increased owing to the out-of-defocus imaging condition. These competitive advantages make it very suitable to quantitatively investigate the magnetic nanostructures. One of most typical examples is its application in skyrmion-hosting nanostructures as shown in Figure 8.3. The magnetic edge twists in the nanostripe [25], target skyrmion in the nanodisk [26] as well as the elliptical skyrmion in the wedge-shaped nanostripe [27] are clearly resolved, which cannot be accessed through other magnetic characterization techniques. More importantly, the first experimental evidence of magnetic chiral bobbers [9] is demonstrated by off-axis EH, showing its incomparable advantage.

In addition, electron holography is also able to gain insight into the three-dimensional (3D) magnetic textures. Two different approaches are currently adopted with and without a series of specimen tilting, respectively. The former is based on the simple relationship that the magnetic phase shift should be linearly proportional to the specimen thickness if a uniform spin texture is formed through the thickness direction. Such an approach has already been used to infer the 3D properties of magnetic skyrmions in $Fe_{0.5}Co_{0.5}Si$ [28] and FeGe [25, 29]. However, it is an indirect method and the modulation of spin textures in the thickness direction are still missing. The latter one is electron tomographic holography technique, which is also called holographic vector field electron tomography [30–32], similar to the traditional tomographic technique while the specimen is rotated along two perpendicular tilt axes. The magnetic induction components of $\mathbf{B}_x$ and $\mathbf{B}_y$ in full 3D space can be separately reconstructed. The component of $\mathbf{B}_z$ is obtained from the Gauss law for magnetism. The

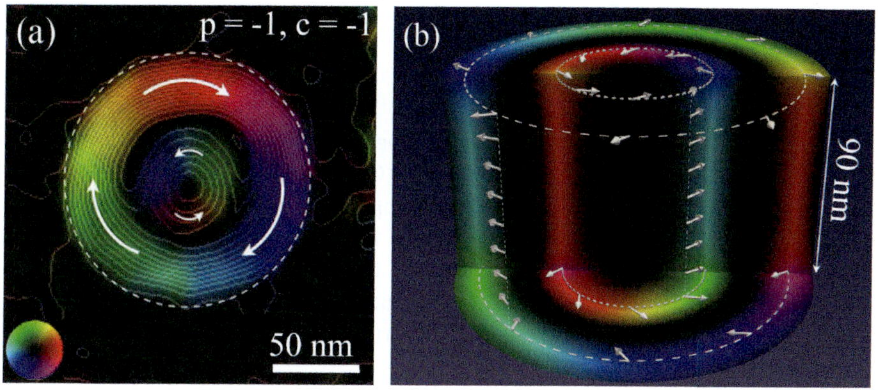

Figure 8.3 Applications of off-axis electron holography in skyrmion-hosted nanostructures. (a) Magnetic induction maps recorded using off-axis electron holography showing target skyrmions in an FeGe nanodisk. (b) Three-dimensional spin arrangement of a target skyrmion corresponding to (a). Figures reproduced from [26] licensed under CC BY 4.0.

further efforts of measuring the 3D magnetizations are still desired through advanced reconstructed algorithm.

8.2.2 Resonant Elastic X-ray Scattering

8.2.2.1 Historical Survey

The structural determination by x-ray diffraction is a well-established technique, utilizing the interaction between electromagnetic waves and the charge density of the electrons, called Thomson scattering [33]. The scattering process is clearly taking place between the photons and electrons, implying the sensitivity to the other two degrees of freedom of electrons, i.e., the orbitals and spins are accessible as well. The history of the studies on these interactions, reviewed in [34], can be traced back to the 1970s, when Platzman and Tzoar first showed the sensitivity of the x-ray scattering amplitude to magnetism in theory [35], followed by the experimental proof by De-Bergevin and Brunel [36]. This technique is termed as non-resonant magnetic scattering [37]. Nevertheless, the magnetic scattering cross-section is so weak that it was almost invisible in experiments at that time [38]. The milestone for resonant elastic process to be accessible to magnetism was set by Blume, Gibbs et al. [39, 40] and Hannon et al. [41], who did pioneering work on resonant x-ray scattering, and found the exact polarization dependence of the process. After almost three decades, this technique is mature and suitable to study many magnetic materials, thanks to the fast development of synchrotron radiation. Though the nomenclature may differ in the literature, here we call this process resonant elastic x-ray scattering, or REXS. For 3d transition metals at the $L_{2,3}$ edges, and rare earth metals at the $M_{4,5}$ edges, the resonant signal due to magnetism is comparable with the Thomson scattering, therefore providing excellent conditions for actually measuring this process [42, 43]. These resonance conditions require a photon energy range of 0.4–1 keV, which is the soft x-rays spectrum. In a REXS process using soft x-rays, the distribution, size, and shape of a 3D magnetic texture can be effectively probed in a designed experimental geometry. This offers great opportunities for the coverage of several blind areas from electron holography. For example, the topological winding number N and the helicity angle χ are the key parameters that govern the topological and chiral properties of the skyrmions.

Moreover, a surface-sensitive technique is required to study skyrmion structures on the surface level.

8.2.2.2 Theoretical Treatment

We start with the kinematic theory of surface diffraction in a three-dimensional lattice with a terminating surface, i.e., crystalline truncation rod (CTR) analysis. The magnetic CTR framework that treats a practical three-dimensional sample embraces other REXS scenarios that occur at two-dimensional systems. The scatterers in Cartesian real-space coordinates (not fractional coordinates) of $\mathbf{r}=(x,y,z)$ take the form factor of $f(\mathbf{r})$ and $\mathbf{q}=(q_x,q_y,q_z)$ is the reciprocal space point associated with the momentum transfer. For an ideal, defect-free surface, the CTR is expressed by the structure factor written as

$$F(\mathbf{q})=\sum_{z=0}^{\infty}e^{-z/L_z}\sum_{n}f_D(\mathbf{r})f(\mathbf{r})e^{iq\cdot r_n}, \tag{8.7}$$

where $z=0$ corresponds to the top surface, and the summation is carried out over all atomic positions n in the $x-y$ plane, L_z is the x-ray attenuation length, f_D is the Debye-Waller factor, which we did not take into account in our calculations. Note that here the Fourier transform is carried out for all scatterers, and the exact form of L_z depends on the photon energy $\hbar w$, the material, and the scattering geometry (described by wavevectors $\mathbf{k}$). Equation (8.7) leads to a scattering cross section $I=|F|^2$ for the crystalline truncation rod $\mathbf{q}$. Using the convolution theorem, equation (8.7) can equally be expressed in units of the Miller indices *(H, K, L)*.

In the REXS process, f carries real and imaginary parts

$$f=-Z r_0 \in_s^*\cdot\in_i+f'+if'', \tag{8.8}$$

where Z is the atomic number, and r_0 is the classical electron radius. $\in_i$ and $\in_s$ are the unit polarization vectors for incident and scattered x-rays, respectively. The first term describes the Thomson scattering process, which has been employed for standard CTR calculations. The second and third term describe 2^L-pole resonant scattering, which we will consider in surface diffraction

$$f'+if''=\frac{4\pi}{k}f_D\sum_{M=-L}^{L}\left[\in_s^*\cdot \mathrm{Y}_{LM}(\hat{\mathrm{k}}_s)\mathrm{Y}_{LM}^*(\hat{\mathrm{k}}_i)\cdot\in_i\right]F_{LM}(\omega), \tag{8.9}$$

where $\hat{\mathrm{k}}_i$ and $\hat{\mathrm{k}}_s$ are the unit wavevectors of incident and scattered x-rays. The radiation is expressed by multi-poles in vector spherical harmonics $\mathrm{Y}(\hat{\mathrm{k}})$ labeled by L. In the electric-dipole process that corresponds to the x-ray resonance at the Cu $L_{2,3}$ edge, $L=1, M=-L,\ldots,+L$. $F_{LM}(w)$ follows Fermi's golden rule, thus expressed by

$$F_{1M}(\omega)=\frac{e^2k^3}{18}\times\sum_{n'J'M'}\frac{\left|\langle n'J'M'\left|\sum_i r_i Y_{1M}(r_i)\right|nJM\rangle\right|^2}{E_{n'J'M'}-E_{nJM}-\hbar\omega+i\Gamma/2}, \tag{8.10}$$

where i runs over all electrons, Γ is the core hole lifetime, and E_{nJM} is the energy of the $|nJM\rangle$ level. Further reduction leads to

$$f' + if'' = F^{(0)} \in_s^* \cdot \in_i - iF^{(1)}(\in_s^* \times \in_i) \cdot \hat{\mathbf{n}} + F^{(2)} \left(\in_s^* \cdot \hat{\mathbf{n}} \right) \left(\in_i \cdot \hat{\mathbf{n}} \right), \tag{8.11}$$

where

$$\begin{aligned} F^{(0)} &= \frac{3}{4k}\left(F_{11} + F_{1-1}\right), \\ F^{(1)} &= \frac{3}{4k}\left(F_{11} - F_{1-1}\right), \\ F^{(2)} &= \frac{3}{4k}\left(2F_{10} - F_{11} - F_{1-1}\right). \end{aligned} \tag{8.12}$$

$\hat{\mathbf{n}}$ is the unit vector along the quantization axis of the local angular momentum $\mathbf{J}$ of the (magnetic) ion, described in the angular momentum basis $|\, nJM\rangle$. It links to the magnetic moment $\mathbf{m} = M_J \hat{\mathbf{n}}$ and $M_J = {}_{gJ\mu B}\sqrt{J(J+1)}$.

We now evaluate the x-ray attenuation process for a particular scattering geometry. For uniform crystalline materials, the factor e^{z/L_z} is replaced by

$$e^{z/L_z} \rightarrow \frac{1}{\Lambda} e^{-2z\sec\alpha/\Lambda}, \tag{8.13}$$

where α is the incidence angle with respect to the terminating surface and Λ is the x-ray attenuation length at normal incidence. For incommensurate non-collinear magnetic lattices, we drop the charge contributions [i.e., the first terms in equation (8.8) and equation (8.11)], as the crystalline Bragg peaks are well separated from the magnetic Bragg peaks. Also, higher-order terms [third term in equation (8.11)] will be neglected due to the reduced contribution to the measured signal. We write down the MTR structure factor as:

$$\mathbf{F}(\mathbf{q}) = -\hat{M}\left[iF^{(1)}\left(\in_s^* \times \in_i\right) \cdot \hat{\mathbf{n}} \right]. \tag{8.14}$$

The quantity $\hat{M}$ follows the truncation rod calculation:

$$\hat{M} = \frac{1}{\Lambda} \sum_{z=0}^{\infty} e^{-2z\sec\alpha/\Lambda} \sum_n e^{2\pi i \mathbf{q} \cdot \mathbf{r}_n}. \tag{8.15}$$

We calculate $I = |\mathbf{F}|^2$ using the density matrix formalism [34]

$$I = \mathrm{tr}(f_m\, u_i f_m^\dagger), \tag{8.16}$$

where u_i is the density matrix of the polarization state of the incident x-rays, and the Poincaré-Stokes representation of the polarization $\mathbf{P}$ is parameterized by the Pauli matrices σ as:

$$\begin{aligned} u &= \frac{1}{2}\left(1 + \mathbf{P} \cdot \sigma\right) \\ &= \frac{1}{2}\begin{pmatrix} 1 + P_1 & P_2 - iP_3 \\ P_2 + iP_3 & 1 - P_1 \end{pmatrix}. \end{aligned} \tag{8.17}$$

The corresponding form factor f_m can also be decomposed into a 2×2 matrix

$$f_m = -iF^{(1)} \begin{pmatrix} 0 & n_1 \cos\theta + n_3 \sin\theta \\ -n_1 \cos\theta + n_3 \sin\theta & -n_2 \cos 2\theta \end{pmatrix}, \quad (8.18)$$

where $\mathbf{n} = (n_1, n_2, n_3)$, θ is half of the scattering angle that relates to $\theta = \arcsin(q/2k)$, finally obtaining

$$I(\mathrm{q}) = \mathrm{tr}\left[\hat{M}\, f_m\, u_i\, \hat{M}\, f_m^{\dagger}\right]. \quad (8.19)$$

8.2.2.3 Experimental Setup

The goal of the experiments is to map out the intensity distribution spanning the reciprocal space. This activity is called reciprocal space mapping (RSM). Figure 8.4(a) illustrates the geometry of the experiment. In the Ewald representation, as shown in Figure 8.4(b), in order to bring a reciprocal space point **Q** into the diffraction condition, one has to rotate the sample such that **Q** "touches" the surface of the sphere. This rotation usually involves three axes, Ω, χ and Ψ, as shown in Figure 8.4(c). If a point detector is used, one not only has to move **Q** onto the sphere, but also bring it into the scattering plane. Therefore, at least two circles have to be implemented for an arbitrary **Q**. On the other hand, if an area detector is used, only Ω needs to be adjusted, because $q_h = \tau \approx 0.0158$ (r.l.u.) is very small. Thus, a very small volume of the reciprocal space has to be mapped. For example, for the skyrmion lattice phase, the six magnetic peaks are represented by the black arrows in Figure 8.4(d). For each of them, one only has to slightly move Ω (differently for each of them) to reach the diffraction condition. The CCD camera is mounted on the 2Θ-arm that can rotate in the scattering plane in order to capture the scattered x-rays. Each set of goniometer angles (Ω, χ, Ψ) selects a particular range of **Q**.

Further, we choose the coupled $\Omega - 2\Theta$ scan mode, i.e., 2Θ rotates twice as fast as Ω for the RSM scans, such that the specular position on the CCD camera is fixed. It is important to emphasise that for RSM only a small volume in reciprocal space needs to be covered, which is equivalent to employing the $\Omega - 2\Theta$ scan mode or the Ω scan mode (where only the Ω stage rotates, while the 2Θ position is fixed). However, there are several advantages of the $\Omega - 2\Theta$ method: (i) RSM along the l-direction will only appear in the center of the CCD camera. Therefore, it is more accurate and very efficient to carry out l-scans, which are important for mapping the helical and conical states. Moreover, as the structural (0,0,1) peak is extremely strong, leading to blooming especially of the vertical CCD pixels, it is difficult to define a peak in the CCD camera image, unless 2Θ is coupled to Ω. (ii) The adding-up of a full series of $\Omega - 2\Theta$ RSM scan images yields the symmetries of the magnetic satellites and their positions relative to the (0,0,1) Bragg peak. This integrated image is sufficiently accurate to discriminate between helical, conical, skyrmion, ferrimagnetic, and paramagnetic orders, thus allowing for an efficient mapping of the phase diagram. However, quantitative data processing that projects the camera information into reciprocal space has to be performed. (iii) The number of events for a magnetic satellite in an integrated image is equal to the integrated intensity under the "rocking curve" of that magnetic peak. This provides more accurate data for further quantitative analysis of the REXS intensity.

Here we take Cu_2OSeO_3 as an example. Cu_2OSeO_3 has a uniform magnetic phase diagram that is independent of the field-cooling history. Starting from a zero field-cooled system just below the ordering temperature of 57 K, the helical state should be stabilized, giving rise to

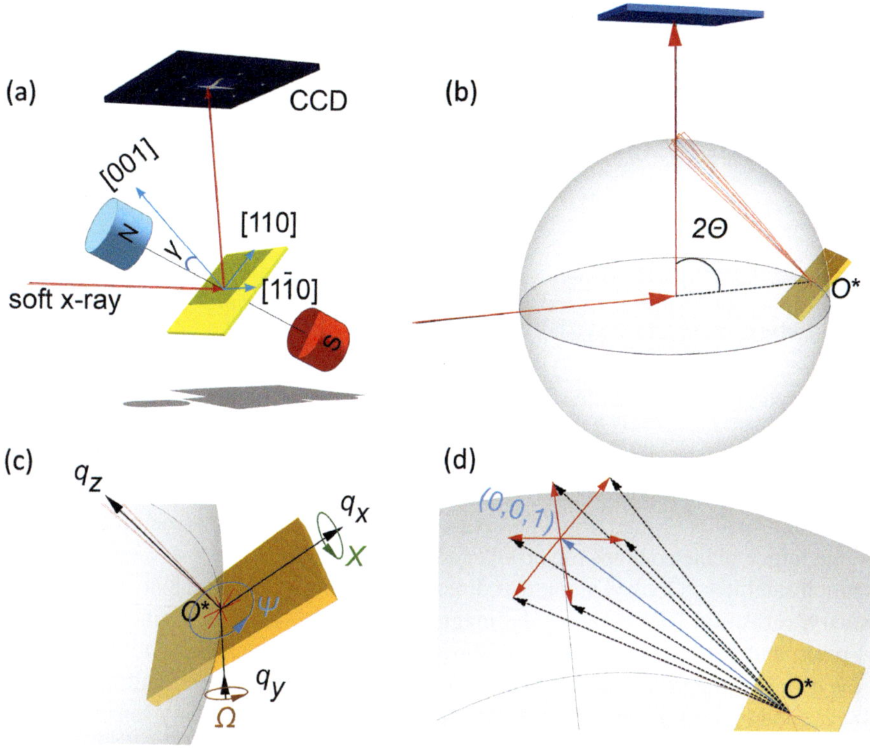

Figure 8.4 Experimental REXS setup. (a) Scattering geometry and field configuration in RASOR. The magnetic field is provided by permanent magnets in a rotatable variable field assembly. (b–d) Ewald sphere representation of REXS; (c) and (d) are the zoomed-in illustrations of (b). In (c), the standard four-circle diffractometer is presented in reciprocal space. The goniometer is driven by three axes, in which Ω is the major axis rotating within the scattering plane. The χ axis is coupled to Ω, while the Ψ axis is coupled to χ. This allows the three rotation matrices to be commutative. In (d), the (0,0,1) structure peak is marked as the blue arrow, while the magnetization propagation wavevectors for the skyrmion lattice phase is labeled by the six red arrows. This gives rise to the six magnetic peaks in reciprocal space, as labeled by black arrows. Figures reproduced with permission from [45].

three differently oriented helical domains, which should be locked along h, k, and l, respectively. Therefore, if a $\Omega - 2\Theta$-type RSM scan is carried out, the four wavevectors that lie on the hk-plane (l=1) should be visible in the camera image, while the two peaks along l should always be on the specular position of the CCD. The helical magnetic satellites, shown as the four spots in Figure 8.5(a), can be clearly identified. They correspond to two different helical domains, and each of them gives rise to a Friedel pair (i.e., $+\mathbf{q}_\mathrm{h}$ and $-\mathbf{q}_\mathrm{h}$ wavevectors). Detailed data processing projects the RSM scan images into reciprocal space hk1 plane.

By increasing the magnetic field slightly to the critical field B_{c1}, the helical domains undergo a spin-flip type first-order transition, forming a single-domain conical state, as shown in Figure 8.5(b). The conical propagation wavevector $\mathbf{q}_\mathrm{h}$ is along the field direction. Therefore, if the magnetic field is along [001], the magnetic satellites will lie along l, which is not visible in the camera. In order to show the Friedel pair of the conical peaks on the CCD image, a slight rotation of the field angle by γ has to be performed. By performing data

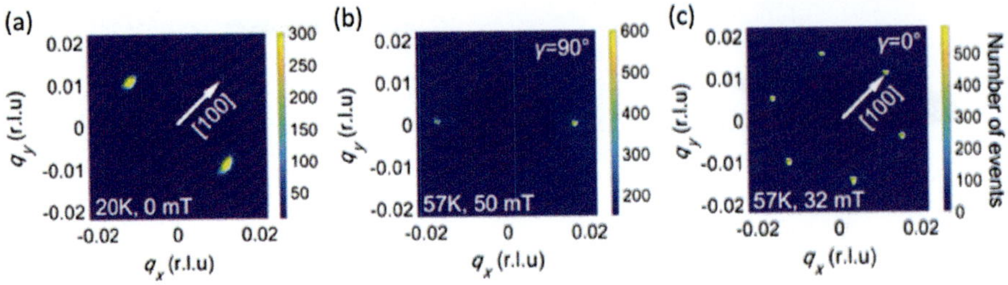

Figure 8.5 REXS data for a Cu_2OSeO_3 sample with opposite crystalline chirality. (a – c) Reciprocal space map in the hk1 plane for helical, conical ($\gamma = 90^0$), and skyrmion lattice phases, respectively. Figures reproduced with permission from [45].

processing, the real-space pitch of the conical modulation is essentially identical to that of the helical state.

By increasing the field at 57 K to 32 mT, i.e., above the critical field B_{A1}, a first-order transition from the conical state to the skyrmion lattice state will take place. This will manifest itself as a sudden change from two- to six-fold symmetric peaks in reciprocal space. The skyrmion lattice plane contains the [100] crystallographic orientation, so one pair of the three skyrmion satellites is locked along that direction, as shown in Figure 8.5(c).

In summary, reflection REXS shows the magnetic peaks are located around the (0,0,1) structural peak, and the diffraction patterns show a stark contrast, allowing to distinguish between the different magnetic phases. Second, this technique is sensitive to all three components q_x, q_y, and q_z, of a magnetic wavevector $\mathbf{q}_m$ (while neutrons are only sensitive to q_y and q_z components). This allows us to obtain the complete set of information about the magnetic structure. e.g., for the conical order, as well as the titling of the skyrmion plane, etc., which SANS and LTEM unable to provide. The probing area can vary from mm to few tens of μm, and the measurement time is very short in comparison. Third, reflection REXS is surface sensitive, therefore detailed information about the technologically relevant surface state can be obtained.

8.3 Experimental Results

8.3.1 Magnetic Bobbers

Concept of magnetic chiral bobber was inspired from the 3D structure of skyrmion tubes [14]. Putting the skyrmion tube as the initial state and artificially breaking off the tube, numerical results show the tube will evolve into a pair of new type of magnetic configurations that are characterized by stacking of skyrmions with shrinking sizes from the surface to the interior and a singularity Bloch point (BP) [44]. In analysis to the fish bobbers, they are named as magnetic chiral bobbers. Further theoretical analysis shows such structures are sometimes energetically more favorable than the chiral magnetic skyrmion tubes.

Experimental evidence for the existence of magnetic chiral bobber is first obtained using the EH technique. As magnetic bobbers are localized in the vicinity to the surface, they characterized by a weaker contrast compare to skyrmions in-plane component of the projected magnetization. Quantitative EH technique allows to obtain quantitative information about the projected magnetization. A wedged sample of FeGe nanostripe with varied thickness is used. The ground state is the spin helix, which will commonly transfer into magnetic skyrmions under the action of magnetic field normal to the stripe plane. If the magnetic field is slightly tilted from the plane, there is a chance to nucleate chiral bobber, as shown in Figure 8.6(a). The intensity of the phase shift signal is thickness dependent. It explains the reduction of the phase shift value with the reduction of the film thickness. However, three objects marked by arrows in Figure 8.6(b) for which the absolute value of the phase shift is approximately twice as small compare to the phase shift of other skyrmions at the same distance which in turn related to the thickness of the wedged sample. Thereby these objects can be identified as chiral bobbers. Quantitative calculations further confirm the conclusions [9].

The first experimental identification of chiral bobbers has brought about new concept for designing advanced racetrack memory based on topological magnetic orders. In the newly designed prototypical memory, skyrmion tubes, and chiral bobbers are encoded as binary information 1 and 0. The energy barrier between the two states maintains moderate, which is energetically friendly for bit-writing and bit-manipulation purposes. Therefore, the creation of chiral bobbers with ease has been a key challenge towards their applications. An effective way to nucleate chiral bobbers is through magnetic coupling effect in a heterostructure system, hybridizing a chiral magnet and a skyrmion-hosting multilayered thin film, form which skyrmions have comparable sizes. The mechanism can be quite intuitively understood in a framework of

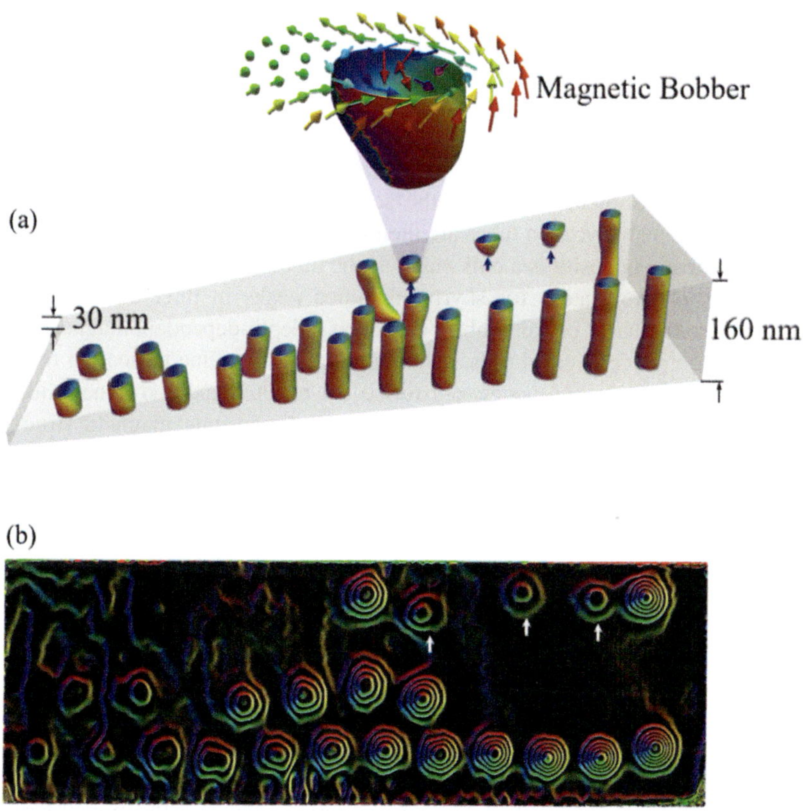

Figure 8.6 Experimental evidence for the magnetic chiral bobbers. (a) Schematic representation of magnetic configurations in a nanostripe with varied thickness. The bobbers are marked by blue arrows. (b) Magnetic-induction images that correspond to (a), the three bobbers are marked by white arrows.

demagnetization-indued closure-magnetization flux rule. At the near-interface region of the heterostructure, skyrmion tubes from two distinct species lock themselves in position. The dipolar field from the above skyrmion tubes resembles a bobber-like shape, through the flux lines are formed by the stray field, rather than magnetization density. Nevertheless, such bobber-shaped dipolar field cut the skyrmion tubes underneath, following by its signature profile. Consequently, chiral bobbers are stabilized at the near-surface region in the bulk chiral magnet.

From experimental point of view, the detailed structural properties of bobbers, such as their shape and penetration depth are important aspects to be considered for the in-depth investigations of such emerging topic. This demands advanced magnetic characterization techniques that are accessible for resolving the three-dimensional spin configurations, particularly along the depth of a finite-thickness sample. REXS seems to be the ideal technique to address the above issue, though significant upgrade of such method needs to be commissioned. The idea is in fact quite a historical one, which can be traced back in 1986, when the concept of surface diffraction was developed. The keyword of surface diffraction is called crystalline truncation rods (CTRs), which represents the abnormal scattering amplitude configuration in three-dimensional reciprocal space.

In short, if x-rays see finite-thickness of a sample, the Bragg peaks became Bragg rods that extend along the surface normal direction. Most interestingly, how does the rod look like strictly associates with the detailed real space atomic configuration. This forms the principle of CTR analysis – an important scattering technique that recovers the atomic configurations at near-surface region. For example, if certain atomic order is a bulk state, CTR appears to be rather a three-dimensional spot in reciprocal space. If the atomic order is buried underneath of another matrix lattice with finite penetration depth, CTR appears elongated along the surface normal. Next, one just has to replace the electron density in a CTR scenario with spin density, despite the fact that one needs to fundamentally modify the detailed theoretical treatment, and setup the experiments such that the scattering is in resonance. The key difference of chiral bobbers and skyrmion tubes are their depth-dependent profiles. For skyrmion tubes, the magnetization density is a bulk state, as the tubes go throughout the entire crystal. This leads to spot-like feature for skyrmion lattice peaks in three-dimensional reciprocal space. On the other hand, chiral bobbers exhibit as a depth-dependent magnetization density, directly leading to elongated rods, called magnetic crystalline truncation rods (MCTR). Further MCTR analysis can accurately refine the exact shape and penetration depth of a bobber, making an ideal tool to characterise the three-dimensional magnetic structures of the phase.

8.3.2 Surface Twists

It is demonstrated that the exact value of the helicity angle can be unambiguously determined making use of circular dichroism (CD) in a REXS experiment (CD-REXS). The hexagonally long-range ordered skyrmion lattice phase gives rise to the six-fold-symmetric diffraction pattern. Circular dichroism, the difference between the intensities obtained using left-and right-circularly polarized incident light, is known to be sensitive to the chirality of a material. They use the CD of the diffraction intensities from all six magnetic skyrmion lattice peaks, which effectively gives the dichroic information in reciprocal space. In this way, each helicity angle corresponds to a characteristic CD-REXS pattern, as shown in the movie. This novel method makes use of the dichroism extinction rule, a general measurement principle that was developed for determination of the helicity angle of a chiral structure.

Commonly, soft X-rays probe down to a finite depth below the surface, while averaging the signal. Intriguingly, by tuning the wavelength of the incident X-rays, it is possible to systematically vary the probing depth to perform a quasi-tomographic mapping of the magnetic state. At the $2p$ absorption edge of a $3d$ transition metal, the soft X-rays are particularly surface-sensitive as the absorption is large, whereas away from the absorption maximum, increasingly deeper layers are probed as well. In a CD-REXS experiment, the spectroscopic measurement therefore provides a strategy to obtain the helicity angle as a function of depth in a three-dimensional skyrmion crystal. Eventually, a clear 3D skyrmion structure can be reconstructed. Strikingly, very much like the tornado structure, an extra twist, i.e., the magnetization flux line that spirals around the skyrmion tube is unambiguously identified.

References

1 Evans, R.F.L., Fan, W.J., Chureemart, P., Ostler, T.A., Ellis, M.O.A., and Chantrell, R.W. (2014). Atomistic spin model simulations of magnetic nanomaterials. *J. Condens. Matter Phys.* 26: 103202.

2 Kronmller, H. (2007). In: *Handbook of Magnetism and Advanced Magnetic Materials*, edited by Kronmller, H., and Parkin, S., hmm201. Chichester, UK: John Wiley & Sons, Ltd.

3 Buhrandt, S. and Fritz, L. (2013). Skyrmion lattice phase in three-dimensional chiral magnets from Monte Carlo simulations. *Phys. Rev. B* 88: 195137.

4 Miltat, J.E. and Donahue, M.J. (2007). In: *Handbook of Magnetism and Advanced Magnetic Materials*, edited by H. Kronmller and S. Parkin, hmm202. Chichester, UK: John Wiley & Sons, Ltd.

5 Schrefl, T., Hrkac, G., Bance, S., Suess, D., Ertl, O., and Fidler, J. (2007). In: *Handbook of Magnetism and Advanced Magnetic Materials*, edited by H. Kronmller and S. Parkin, hmm203. Chichester, UK: John Wiley & Sons, Ltd.

6 Fernandez-Pacheco, A., Streubel, R., Fruchart, O., Hertel, R., Fischer, P., and Cowburn, R.P. (2017). Three-dimensional nanomagnetism. *Nat. Commun.* 8: 15756.

7 Birch, M.T., Corts-Ortuo, D., Turnbull, L.A., Wilson, M.N., Gro, F., Trger, N., Laurenson, A., Bukin, N., Moody, S.H., Weigand, M., Schtz, G., Popescu, H., Fan, R., Steadman, P., Verezhak, J.A.T., Balakrishnan, G., Loudon, J.C., Twitchett-Harrison, A.C., Hovorka, O., Fangohr, H., Ogrin, F.Y., Grfe, J., and Hatton, P.D. (2020). Real-space imaging of confined magnetic skyrmion tubes. *Nat. Commun.* 11: 1726.

8 Rybakov, F.N., Borisov, A.B., Blgel, S., and Kiselev, N.S. (2015). New Type of Stable Particlelike States in Chiral Magnets. *Phys. Rev. Lett.* 115: 117201.

9 Zheng, F., Rybakov, F.N., Borisov, A.B., Song, D., Wang, S., Li, Z.-A., Du, H., Kiselev, N.S., Caron, J., Kovcs, A., Tian, M., Zhang, Y., Blgel, S., and Dunin-Borkowski, R.E. (2018). Experimental observation of chiral magnetic bobbers in B20-type FeGe. *Nat. Nanotechnol.* 13: 451.

10 Tai, J.-S.B. and Smalyukh, I.I. (2018). Static Hopf Solitons and Knotted Emergent Fields in Solid-State Noncentrosymmetric Magnetic Nanostructures. *Phys. Rev. Lett.* 121: 187201, wOS: 000448756700007.

11 Liu, Y., Lake, R.K., and Zang, J. (2018). Binding a hopfion in a chiral magnet nanodisk. *Phys. Rev.* B 98: 174437, wOS: 000451602400003.

12 Leonov, A.O. and Inoue, K. (2018). Homogeneous and heterogeneous nucleation of skyrmions in thin layers of cubic helimagnets. *Phys. Rev. B* 98: 054404.

13 Hu, Q., Lyu, B., Tang, J., Kong, L., Du, H., and Wang, W. (2021). Unidirectional current-driven toron motion in a cylindrical nanowire. *Appl. Phys. Lett.* 118: 022404.

14 Rybakov, F.N., Borisov, A.B., and Bogdanov, A.N. (2013). Three-dimensional skyrmion states in thin films of cubic helimagnets. *Phys. Rev. B* 87: 094424.

15 Wang, X., Qaiumzadeh, A., and Brataas, A. (2019). Current-Driven Dynamics of Magnetic Hopfions. *Phys. Rev. Lett.* 123: 147203.

16 Liu, Y., Hou, W., Han, X., and Zang, J. (2020). Three-Dimensional Dynamics of a Magnetic Hopfion Driven by Spin Transfer Torque. *Phys. Rev. Lett.* 124: 127204.

17 Gabor, D. (1948). A New Microscopic Principle. *Nature* 161: 777.

18 Haine, M.E., Mulvey, T.,(1952). The Formation of the Diffraction Image with Electrons in the Gabor Diffraction Microscope. *J. Opt. Soc. Am.* 42: 763.

19 Leith, E.N., Upatnieks, J. (1962). Reconstructed Wavefronts and Communication Theory. *J. Opt. Soc. Am.* 52: 1123.

20 M¨ollenstedt, G. and Wahl, H. (1968). Elektronenholographie und Rekonstruktion mit Laserlicht. *Naturwissenschaften* 55: 340.

21 Tonomura, A., Matsuda, T., Endo, J., Arii, T., and Mihama, K. (1980). Direct Observation of Fine Structure of Magnetic Domain Walls by Electron Holography. *Phys. Rev. Lett.* 44: 1430.

22 Midgley, P. and Dunin-Borkowski, R. (2009). Direct Observation of Fine Structure of Magnetic Domain Walls by Electron Holography. *Nat. Mater.* 8: 271.

23 Caron, J., Diehle, P., Kovcs, A., Ungermann, J., and Dunin-Borkowski, R. (2017). Model-based Reconstruction of Magnetisation Distributions in Nanostructures from Electron Optical Phase Images. Ph.D. thesis, Universit¨aätsbibliothek der RWTH Aachen.

24 Voelkl, E. and Tang, D. (2010). Approaching routine 2pi/1000 phase resolution for off-axis type holography. *Ultramicroscopy* 110: 447.

25 Song, D., Li, Z.-A., Caron, J., Kovács, A., Tian, H., Jin, C., Du, H., Tian, M., Li, J., Zhu, J., et al. (2018). Quantification of Magnetic Surface and Edge States in an FeGe Nanostripe by Off-Axis Electron Holography. *Phys. Rev. Lett.* 120: 167204.

26 Zheng, F., Li, H., Wang, S., Song, D., Jin, C., Wei, W., Kovács, A., Zang, J., Tian, M., Zhang, Y., et al. (2017). Direct Imaging of a Zero-Field Target Skyrmion and Its Polarity Switch in a Chiral Magnetic Nanodisk. *Phys. Rev. Lett.* 119: 197205.

27 Jin, C., Li, Z.-A., Kovács, A., Caron, J., Zheng, F., Rybakov, F.N., Kiselev, N.S., Du, H., Blügel, S., Tian, M., et al. (2017). Control of morphology and formation of highly geometrically confined magnetic skyrmions. *Nat. Commun.* 8: 15569.

28 Park, H.S., Yu, X., Aizawa, S., Tanigaki, T., Akashi, T., Takahashi, Y., Matsuda, T., Kanazawa, N., Onose, Y., Shindo, D., et al. (2014). Observation of the magnetic flux and three-dimensional structure of skyrmion lattices by electron holography. *Nat. Nanotechnol.* 9: 337.

29 Schneider, S., Wolf, D., Stolt, M.J., Jin, S., Pohl, D., Rellinghaus, B., Schmidt, M., Büchner, B., G¨onnenwein, S.T., Nielsch, K., et al. (2018). Induction Mapping of the 3D-Modulated Spin Texture of Skyrmions in Thin Helimagnets. *Phys. Rev. Lett.* 120: 217201.

30 Simon, P., Wolf, D., Wang, C., Levin, A.A., Lubk, A., Sturm, S., Lichte, H., Fecher, G.H., and Felser, C. (2016). Synthesis and Three-Dimensional Magnetic Field Mapping of Co2FeGa Heusler Nanowires at 5 nm Resolution. *Nano Lett.* 16: 114.

31 Wolf, D., Rodriguez, L.A., Bèchè, A., Javon, E., Serrano, L., Magen, C., Gatel, C., Lubk, A., Lichte, H., Bals, S., et al. (2015). 3D Magnetic Induction Maps of Nanoscale Materials Revealed by Electron Holographic Tomography. *Chem. Mater.* 27: 6771.

32 Wolf, D., Biziere, N., Sturm, S., Reyes, D., Wade, T., Niermann, T., Krehl, J., Warot-Fonrose, B., B¨uchner, B., Snoeck, E., et al. (2019). Holographic vector field electron tomography of three-dimensional nanomagnets. *Commun. Phys.* 2: 87.

33 Als-Nielsen, J. and McMorrow, D. (2010). *Elements of Modern X-ray Physics*. Wiley.

34 van der Laan, G. (2008). Soft X-ray resonant magnetic scattering of magnetic nanostructures. *C. R. Physique* 9: 570.

35 Platzman, P.M. and Tzoar, N. (1970). Magnetic Scattering of X Rays from Electrons in Molecules and Solids. *Phys. Rev. B* 2: 3556.

36 de Bergevin, F. and Brunel, M. (1981). Diffraction of X-rays by magnetic materials. *I. General formulae and measurements on ferro- and ferrimagnetic compounds. Acta Cryst. A* 37: 314.

37 Blume, M. and Gibbs, D. (1988). Polarization dependence of magnetic x-ray scattering. *Phys. Rev. B* 37: 1779.

38 Lovesey, S.W. and Collins, S.P. (2002). *X-ray Scattering and Absorption by Magnetic Materials*. Oxford University Press.

39 Blume, M. (1985). Magnetic scattering of x rays (invited). *J. Appl. Phys.* 57: 3615.

40 Gibbs, D., Harshman, D.R., Isaacs, E.D., McWhan, D.B., Mills, D., and Vettier, C. (1988). Mobility Edge in a Model One-Dimensional Potential. *Phys. Rev. Lett.* 61.

41 Hannon, J.P., Trammell, G.T., Blume, M., and Gibbs, D. (1988). X-Ray Resonance Exchange Scattering. *Phys. Rev. Lett.* 61: 1245.

42 van der Laan, G. and Thole, B.T. (1991). Strong magnetic x-ray dichroism in 2p absorption spectra of 3d transition-metal ions. *Phys. Rev. B* 43: 13401.

43 Thole, B.T., van der Laan, G., Fuggle, J.C., Sawatzky, G.A., Karnatak, R.C., and Esteva, J.-M. (1985). 3d x-ray-absorption lines and the 3d94fn+1 multiplets of the lanthanidesPhys. *Rev. B* 32: 5107.

44 Rybakov, F.N., Borisov, A.B., Bl¨ugel, S., and Kiselev, N.S. (2015). New Type of Stable Particlelike States in Chiral Magnets. *Phys. Rev. Lett.* 115: 117201.

45 Zhang, S. (2018). Measurement of the Magnetic Long-Range Order. In: Chiral and Topological Nature of Magnetic Skyrmions. Springer Theses. Springer, Cham. https://doi.org/10.1007/978-3-319-98252-6_2

9

Multiferroelectric Materials

Xiaobin Guo[1] and Li Xi[2]

[1]*School of Physics & Optoelectric Engineering, Guangdong University of Technology, Guangzhou Higher Education Mega Centre, Guangzhou, China*
[2]*Key Laboratory for Magnetism and Magnetic Materials of Ministry of Education, Lanzhou University, Lanzhou, China*

CHAPTER MENU

In general physics, static magnetic fields and electric fields are described separately by Gauss's law and circuital theorem. They are linked each other via the four Maxwell equations [1]. From the theory of electromagnetism, magnetism and electricity are the two independent fundamental concepts. The polarization and magnetization were used to characterize the properties of dielectric and magnetic materials, respectively. Generally, a specific material only shows a dominated electric polarization under the electric field or magnetization response under the magnetic field. Ferroelectric materials refer to the crystal having a large ferroelectric response under an electric field with typical ferroelectric hysteresis loops, while the ferromagnetic materials refer to the crystal having a large magnetic response under a magnetic field with typical ferromagnetic hysteresis loops. Multiferroic materials refer to the materials simultaneously having more than one ferroic order parameter. It provides opportunities to tune and/or manipulate the order parameters with a non-traditional method; for example, control the electric polarization by magnetic field rather than an electric field, control the magnetic moments by electric fields rather than magnetic fields. The former was named the magnetoelectric (ME) coupling effect and the latter the inverse ME coupling effect [2, 3]. For single-phase multiferroic materials, they are rare and only present strong magnetoelectric coupling effect far below room temperature [4], except $BiFeO_3$ with large ferroelectric polarization in coexistence with notable weak anti-ferromagnetic (AFM) at room temperature [5]. Compared with single-phase multiferroic materials, artificial ferromagnetic/ferroelectric (FM/FE) heterostructures combining ferromagnetic and ferroelectric phases have attracted extensive attention due to the remarkable converse magnetoelectric response at room temperature. The widely accepted converse ME coupling mechanism mainly includes carrier modulation, the strain effect, exchange coupling, orbital reconstruction, and the electrochemical effect. A detailed discussion of each coupling mechanism can be found in Refs [6, 7].

Spintronics: Materials, Devices and Applications, First Edition. Edited by Kaiyou Wang, Meiyin Yang, and Jun Luo.

The driving force for multiferroic research comes not only from scientific curiosity but also from the potential applications for energy transformation, signal generation and processing, and information storage [8]. At present, for the developing of magnetic random access memory (MRAM), how to write the magnetic information "0" and "1" – i.e.,, manipulate the two distinguishable magnetization states in magnetic tunnel junctions (MTJs) – is still a challenging issue in in-depth physical mechanisms and applications. As a matter of fact, in traditional magnetic field-based MRAM, flaws such as broad magnetic field distribution and large current density hinder the further miniaturization and integration of memory cells [9, 10]. Based on the spin transfer torque (STT) effect [11–13] and spin-orbit torque (SOT) effect [14–16], STT-MRAM has entered the stage of industrialization [17] and SOT-MRAM has been realized in the laboratory [18]. However, they also face some technical difficulties: for instance, how to decrease the threshold current density (J_c) to about 10^5 A/cm^2 and realize magnetization switching without an in-plane magnetic field assistance in SOT-MRAM [19]. Actually, as a promising approach to modulate magnetism through converse magnetoelectric coupling effect, many researches have focused on electric field controlled magnetism in FM/FE heterostructures, aiming to realize high-density, non-volatile, and low power data storage [2, 20, 21]. Here we briefly present the progress for magnetization switching through converse ME effect–generated strain in FM/FE heterostructures, magnetization switching through changing the polarity of the exchange bias based on FE/FM heterostructures, and the Néel vector rotation of antiferromagnets for spintronics application at room temperature.

9.1 Electric Field-driven Magnetization Switching

In traditional magnetic recording media, a ferromagnetic layer was used to record "0" and "1" with the magnetization of the ferromagnetic layer setting into either of two stable potential wells determined by the uniaxial magnetic anisotropy (UMA). For FM films with UMA, the magnetic hysteresis (*M-H*) loop along the easy axis (EA) direction is square, whereas the hysteresis loop along the hard axis (HA) direction is slanted with a fairly small remanence ratio (M_r/M_s). To realize the magnetization switching, i.e., writing a bit, one should disturb the free energy landscape to drive the magnetization to the new minimum potential energy either by applying a magnetic field or spin polarization current. In FM/FE heterostructures, if one chooses both FE and FM materials respectively with a large piezoelectric coefficient and a large magnetostriction coefficient, electric field generated strain in FE will certainly induce another UMA in the FM layer through the magnetostriction effect. Assuming that the strain-induced UMA is along the original HA direction, the total free energy can be written as the sum of the UMA and the additional magnetoelastic energy, which is expressed as follows

$$F = \frac{V\left[-\mu_0 M_s H_K \sin^2(\phi) - 3\lambda_s \sigma \cos^2(\phi)\right]}{2} \tag{9.1}$$

where V is the volume of the FM film, M_s is the saturation magnetization, H_K is the effective UMA field, λ_s is the saturation magnetostriction, σ is the mechanical stress, and φ is the angle between the magnetization and HA. The extremum position can be calculated with $\varphi = 0°, 90°, 180°$ and $270°$ determined by the first-order derivative of equation (9.1) to zero.

In equilibrium, the magnetization always stays in a local minimum of free energy. The maximum and minimum positions of the free energy landscape respectively represent the magnetically hard and easy directions of the system. Figure 9.1 shows the angular dependence of the free energy. The angular dependence of the free energy under zero stress is typical for uniaxial ferromagnets with two minima at 90° and 270°, which can be the initial EA. One can see that, when the applied negative stress is increasing, there is no variation of the minimum positions of free energy valleys, indicating that the uniaxial anisotropy direction remains unchanged. However, at a particular positive stress, the free energy is constant. This indicates the isotropic properties of ferromagnetic thin films under this particular condition. When the positive stress is further increased, the free energy valleys rotate to 0° and 180° and become deeper with increasing positive stress. This free energy landscape clearly shows that the EA rotates 90° from the initial EA at large positive stress, indicating that the EA direction of a uniaxial anisotropy system can be rotated 90° by the electric field–induced strain. With this idea, a voltage-induced strain-controlled 90° magnetization switch was achieved in a FeNi/piezoelectric actuator hybrid system [22].

An electric field-induced reversible 90° magnetization rotation and manipulation were reported for a tunneling magneto-resistance junction at room temperature in a CoFeB/AlO_x/CoFeB tunnel junction fabricated on a (011) $Pb(Mg_{1/3}Nb_{2/3})_{0.7}Ti_{0.3}O_3$ (PMN-PT) ferroelectric single crystal, which shows a distinct, stable, and electrically reversible lattice strain due to its large piezoelectric coefficients (d_{31} =–3175 pC/N [100] and d_{32} = 1426 pC/N [01–1]) [23]. Figure 9.2(a) shows the schematic of the key magnetic tunneling junction cell with pinned layer magnetization along the [100] crystallographic direction of (011)PMN-PT. The *M-H* loop changes from a square-shaped loop to a slanted one when sweeping the magnetic field along [100] direction of PMN-PT under a +8 kV/cm electric field (see Figure 9.2(b)). It indicates that the electric fields drive the magnetization of the free layer through a rotation of 90°. The TMR ratio of this MTJ is around 30%, which is caused by the antiparallel

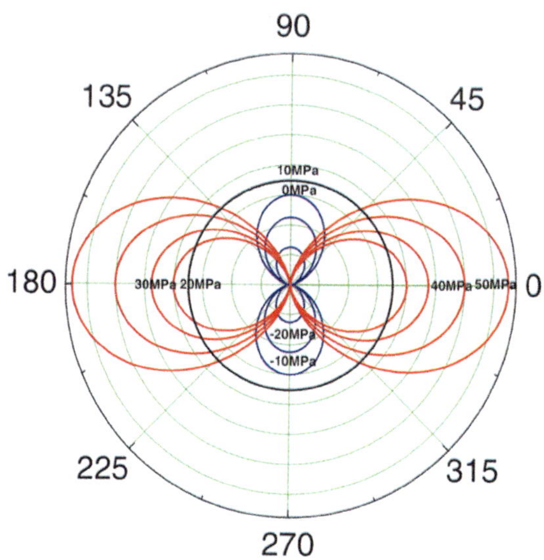

Figure 9.1 In the polar pattern, the energy decreases from the inner circle to the outer circle [22].

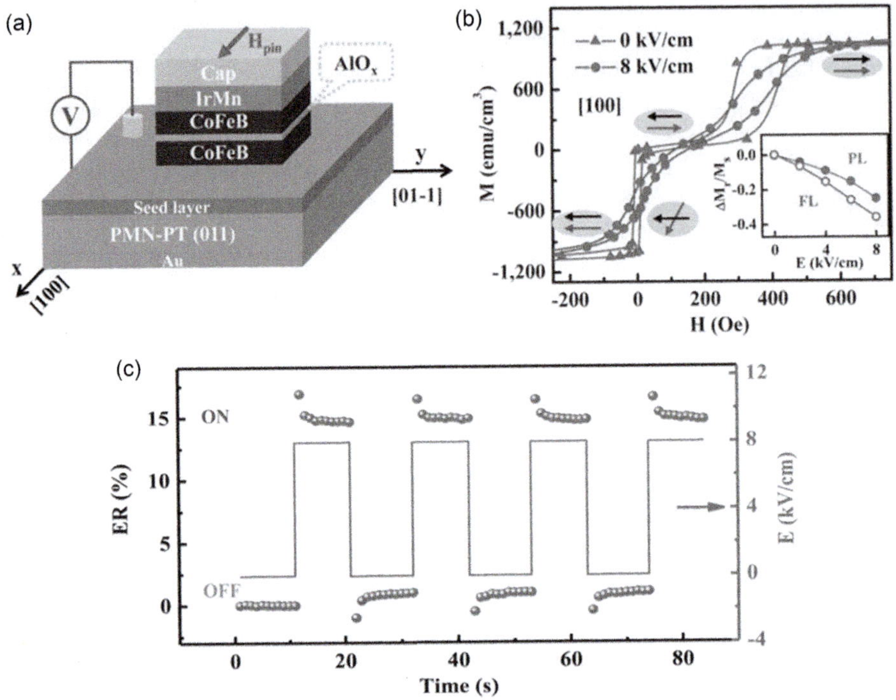

Figure 9.2 (a) Schematic of the key MTJ grown on a ferroelectric (011) PMN-PT single crystal substrate. (b) M-H curves measured under 0 and +8 kV/cm, respectively. (c) Repeatable high/low TMR states regulated by electric fields without the assistance of a magnetic field [24].

magnetization of a free layer and a pinned layer. In Figure 9.2(c), the electric field-induced electric resistance ratio is around 15%, which is around the half of the TMR ratio and is a reasonable value since electric field-induced magnetization switching of the free layer is only 90°[24]. These results provide a new way of exploring electric field-controlled spintronics devices.

The above achievements for 90° magnetization switching have some obstacles for application; for example, the magnetization switching is volatile and hard to integrate into spintronics devices. However, the 180° magnetization switching of the FM layer with UMA could be realized using a sequential electric field-induced strain. As shown in Figure 9.3(a), a MTJ deposited on FE layers shows a large piezoelectric performance and the free layer of MTJ shows a circular shape with uniaxial magnetic properties. Two pairs of electrodes AA' and BB' produce in-plane strains on FE [25]. For this system, the free energy can be expressed as

$$F = -K_U^0 \cos^2(\alpha) - K_U^{AA'} \cos^2(\alpha_{AA'} - \alpha) - K_U^{BB'} \cos^2(\alpha_{BB'} - \alpha) \tag{9.2}$$

where K_u^0 is the initial UMA constant of magnetic thin films, $Ku^{AA'}$ is the energy generated from stress-induced UMA when an AA' electrode is activated by voltage, $Ku^{BB'}$ is the energy

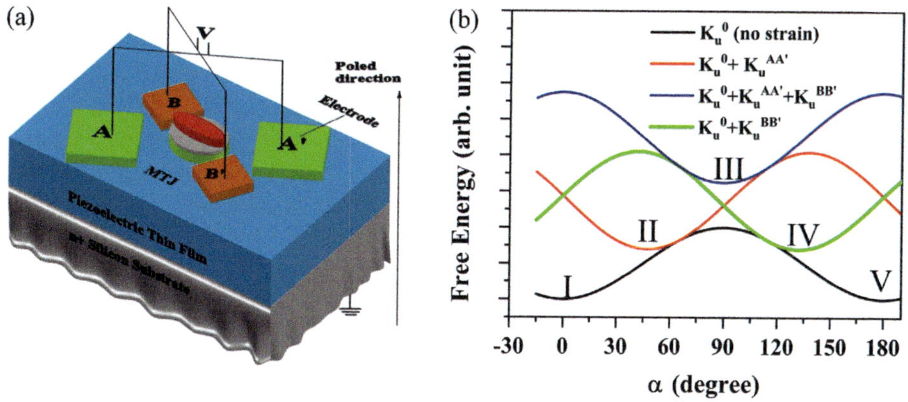

Figure 9.3 (a) Schematic of the memory element, it's poled with two pairs of electrodes AA′ and BB′ in the film plane [25] (b) Free energy profiles according to equation (9.2) with no strain (black line), with only an AA′ electrode activated by a voltage (red line), with both AA′ and BB′ electrodes activated (blue line), and with only BB′ electrodes activated (green line), respectively. The symbols I, II, III, IV, V represent the minimum position of free energy.

generated from stress-induced UMA when a BB' electrode is activated. α is the polar angle, and $\alpha_{AA'}$ ($\alpha_{BB'}$) is the angle between AA' (BB') and initial HA of the FM layer. The free energy based on equation (9.2) with the $\alpha_{AA'}$ and $\alpha_{BB'}$ around 65° and 115° was plotted in Figure 9.3(b). One can see that, if there is no stress, the minimum of free energy located at $\alpha = 0$, and π. Consider the case when the magnetization of the nanomagnet is initially in the stable State I with the initial stored bit "0." If the electrode pair AA' is activated, the uniaxial stress component is generated along the line joining that electrode pair, which will rotate the magnetization vector to State II, which is the only accessible global energy minimum. Next, keeping AA' activated, and also activating BB', causes a uniaxial stress component, which will rotate the magnetization clockwise to the new global energy minimum State III, which is the only accessible one. Next, deactivating AA' – i.e., the removal of stress – will drive the magnetization to State IV. Finally, deactivating BB' will drive the magnetization to State V since it is the only accessible global energy minimum at that point. It results in writing the new bit "1" through pure electric field-induced 180° magnetization switching. If we activate the electrode pairs in the opposite sequence, i.e., BB' first and then AA', the magnetization will rotate anti-clockwise from V to I, writing the new bit "0." Therefore, activating the electrodes in either sequence always flips the bit either by clockwise rotation or counter-clockwise rotation, depending on the sequence of applied voltage on the AA' and BB' electrode pairs. Thus, it is easy to realize the 180° magnetization switching by sequentially adjusting the free energy minima with a converse magnetoelectric effect.

We now refer to the exploration of non-volatile 90° magnetization switching in uniaxial anisotropy FM/PMN-PT heterostructures. The pursuit of non-volatile 180° magnetization switching at room temperature has become imperative owing to its direct relation with parallel and antiparallel magnetization in MTJs. In Co/(001) PMN-PT, Yang et al. reported a non-volatile in-plane 180° magnetization reversal with the aid of a small auxiliary magnetic field, as shown in Figure 9.4(a) [26]. Utilizing the exchange bias effect in FM/AFM structures, instead of small auxiliary magnetic field assistance, Chen et al. realized in-pane 180° magnetization switching in CoFeB/IrMn/(011)PMN-PT heterostructures. A remarkable angle range, with opposite magnetization orientation, was found upon the application of an 8 and 0 kV/cm electric field, which originates from an electrically controlled ratio between the strain-induced UMA and the unidirectional exchange bias field. For the typical angle of $\beta = 37°$, the separation of two magnetic hysteresis loops clearly reveals an in-plane 180° magnetization reversal under a zero magnetic field (see Figure 9.4(b, c)) [27]. Alternatively, with flower-shaped nanomagnets deposited on the FE layer, Wang et al. achieved a 180° magnetization switch by using phase field simulations and thermodynamic analysis; i.e., including two successive and deterministic 90° switching processes in a pure electric field [28].

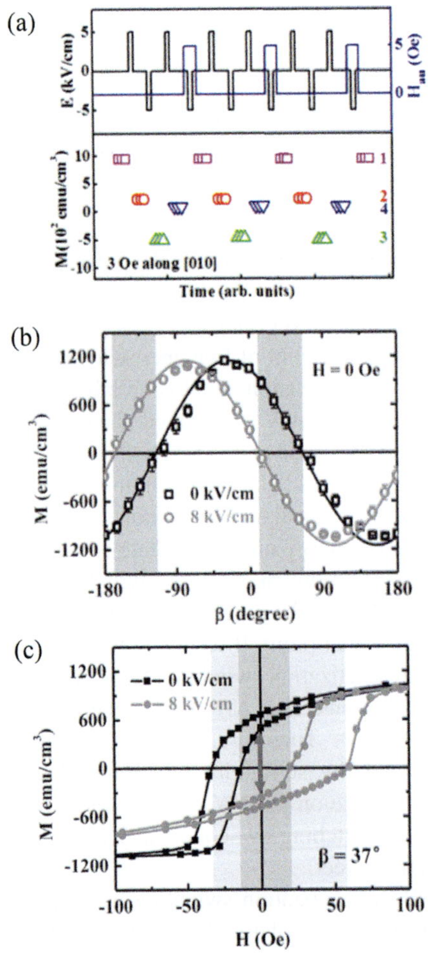

Figure 9.4 (a) Pulsed electrical operation with H_{au} (top) and the resulting magnetization (bottom) measured in a magnetic field of 3 Oe along[010][26]. (b) Angular dependence of M with and without an E-field at the zero magnetic field. (c) M-H loops at β = 37° with the E-field of 0 and 8 kV/cm, respectively [27].

For iron-based epitaxial films, the cubic magnetocrystalline anisotropy results in four magnetic easy axes and four magnetic hard axes in the film plane [29]. By applying an in-plane magnetic field, one can easily rotate the magnetization from one magnetic easy axis to the adjacent magnetic easy-axis direction; i.e., a deterministic 90° magnetization rotation. Conversely, the magnetoelectric coupling effect makes it possible to achieve multi-state magnetization switching through an electric field rather than a magnetic field. For the epitaxial FM/FE heterostructures, the total magnetic free energy is determined by an electric field-induced UMA and the original magnetocrystalline anisotropy. In the case of (001)-oriented epitaxial FM films, the simulated curves are shown in Figure 9.5(a, b). In equilibrium, the magnetic easy axis (MEA) always stays in a local minimum of free energy. In [010] ($\alpha = 45°$) and [100] ($\alpha = -45°$) directions, the free energy curves change from four-fold symmetry (where cubic magnetocrystalline anisotropy energy is dominant) to a two-fold symmetry (uniaxial anisotropy is dominant) when electric field-induced UMA increases from 0 to 1.5. But the minimum free energy position shifts by 90°, indicating that the MEA directions rotation in (001) crystallographic plane can be realized if there exists the above-mentioned strain-induced UMA in an electric field [30].

A piezoelectric transducer can produce strains upon the application of voltage, which is transferred to epitaxial films, changes the total magnetic anisotropy energy, and eventually leads to a 90° non-volatile magnetization switching. In (001)FeGa/(001)GaAs/piezoelectric transducer heterostructures, Parkes et al. reported a non-volatile 90° magnetization switching [31]. In Co_2FeAl/GaAs/PZT heterostructures, the planar Hall voltage of Co_2FeAl films can also be regulated by voltage-based strain-induced 90° magnetization switching. Moreover, the logic operation architectures of NOT and NOR logic gates were also proposed and demonstrated at room temperature without additional magnetic fields [32]. By means of FE substrate-induced strain, a non-volatile 90° MEA rotation and magnetic anisotropy transition was demonstrated.

In epitaxial (001)FeSi/(001)PMN-PT heterostructures, with epitaxial relationships of FeSi(001)[100]//PMN-PT(001)[110] (Figure 9.6), the magnetic anisotropy transition from four-fold CMA dominated to two-fold UMA dominated can occur after applying negative E-fields in $Fe_{86}Si_{14}$/(001)PMN-0.32PT heterostructures. Unfortunately, the magnetic anisotropy transition is non-volatile and irreversible [33]. After modifying the composition of the ferromagnetic layer and the ferroelectric substrate, in epitaxial $Fe_{80}Si_{20}$/(001)PMN-0.3PT,

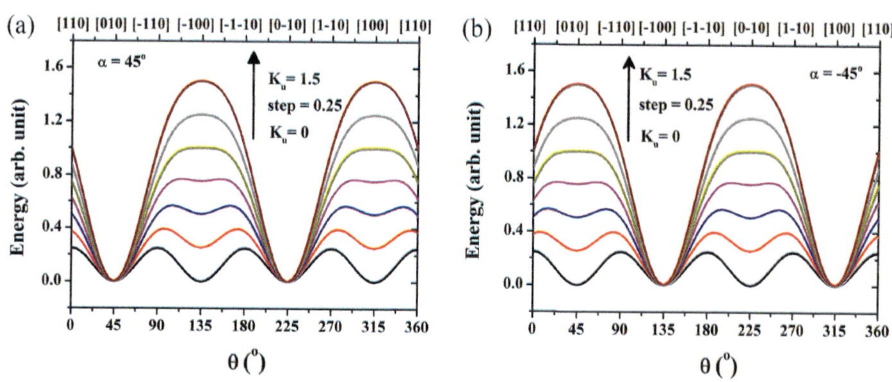

Figure 9.5 Theoretical calculations on the angular dependence of the free energy for strain-induced UMA energy along [010] (a) and [100] (b) directions [30].

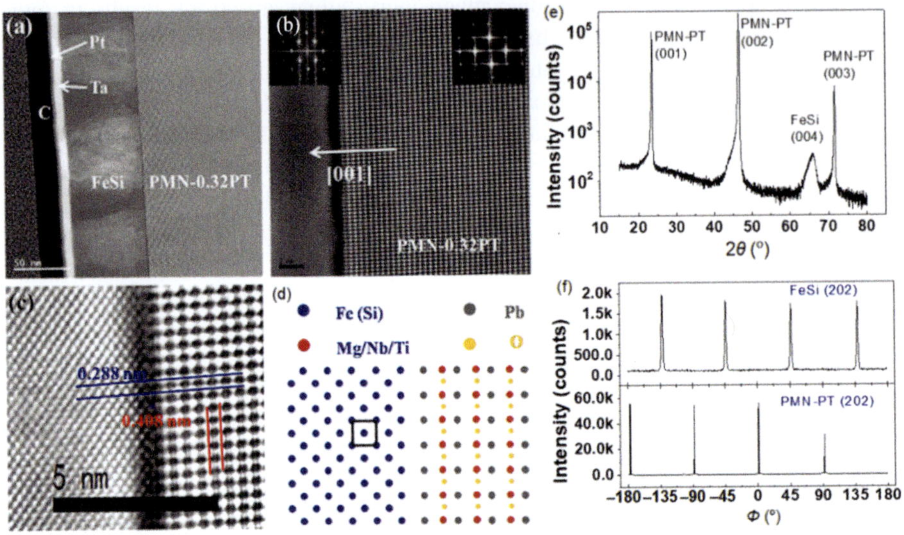

Figure 9.6 The cross-sectional bright-field TEM image of PMN-0.32PT/FeSi/Ta (a) high-resolution TEM image of FeSi/PMN-0.32PT sample along the [110] zone axis of PMN-0.32PT. (b) magnified TEM image at the interface area. (c) structure model of the atomic arrangement at the interface area (d) [33]. (e) θ-2θ XRD diffraction spectrum and (f) ϕ-scan of (202) crystal face in PMN-0.32PT/FeSi heterostructures. *Source:* Reproduced with permission of AIP Publishing LLC.

the angular dependence of the remanence (ARM) curves rapidly change from the original four-fold to two-fold symmetries in an E-field, and non-volatile and reversible 90° magnetization switching with the dominant UMA under ±6 kV/cm were observed in Figure 9.7(a, b). By measuring the micro-areas with a focused magneto optical Kerr magnetometer under E-fields, the internal competitive relationships between cubic magnetocrystalline anisotropy (CMA) energy and UMA energy are confirmed by changing the *M-H* loop's shape. It

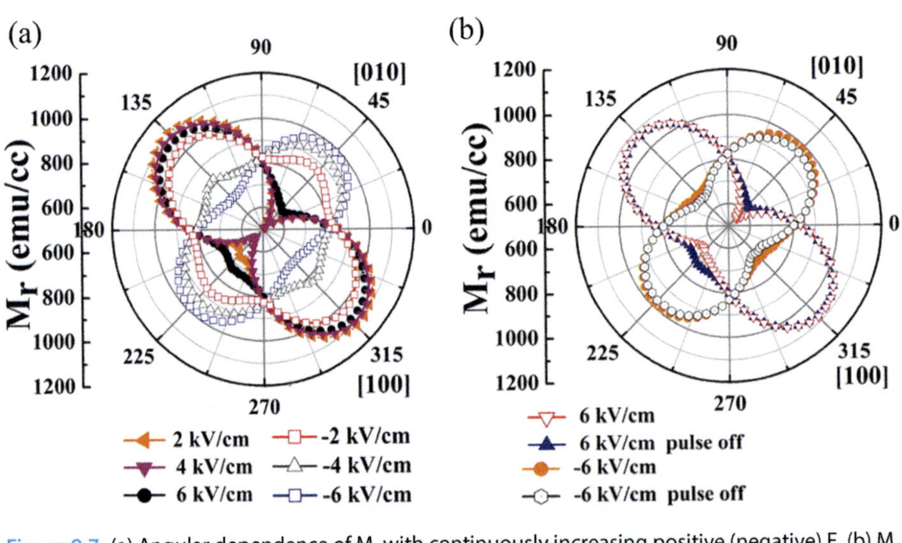

Figure 9.7 (a) Angular dependence of M_r with continuously increasing positive (negative) E. (b) M_r as a function of angle after ±6 kV/cm pulse [30].

illustrates that the strain-induced UMA overcomes the original barrier of CMA of epitaxial $Fe_{80}Si_{20}$ films and switches the magnetization direction with MEA to [010] ([100]) direction after a –6 kV/cm (+6 kV/cm) pulse [30].

9.2 Electric Field-driven Exchange Bias Reversal and Antiferromagnetic Domain Wall Motion

The ME effect of Cr_2O_3 was first predicted as a ME material by Dzyaloshinskii [34], and experimentally confirmed independently by magnetic field-induced polarization [35] and electric field-induced magnetization [36]. Cr_2O_3 having a hexagonal structure with lattice parameter of a = 0.49607 nm and c = 1.3599 nm, shows antiferromagnetic properties with two coupled antiparallel magnetic sub-lattices along c-axis, as shown in Figure 9.8. The specific spin alignment – i.e., Cr^{3+} spin lying along the c-axis – is suitable for inducing the perpendicular exchange bias in FM/Cr_2O_3 heterostructures [37]. Using the ME effect in Cr_2O_3, Borisov fabricated FM/Cr_2O_3 heterostructures to explore the antiferromagnetic order parameters of Cr_2O_3 and its linear ME effect. Usually, the polarity and the magnitude of the perpendicular exchange bias between FM and Cr_2O_3 due to the interfacial exchange coupling effect is determined by the interfacial AFM spin direction. Y. Shiratsuchi et al. [38, 39] have proven that the perpendicular exchange magnetic anisotropy constant up to 0.4 mJ/m^2 induced specific ion positions of Cr^{3+} and O^{2-} ions due to the different underlayers. The perpendicular exchange bias polarity could also be reversed by the so-called ME field cooling (MEFC) method, in which the temperature was raised to over the Néel temperature of Cr_2O_3 and decreased to room temperature under the different electric fields to change the direction of AFM spins.

Based on the above-mentioned spin alignment, two magnetic domains with the opposite Cr^{3+} spins are energetically equal with the free energy, as shown in Figure 9.8. When applying

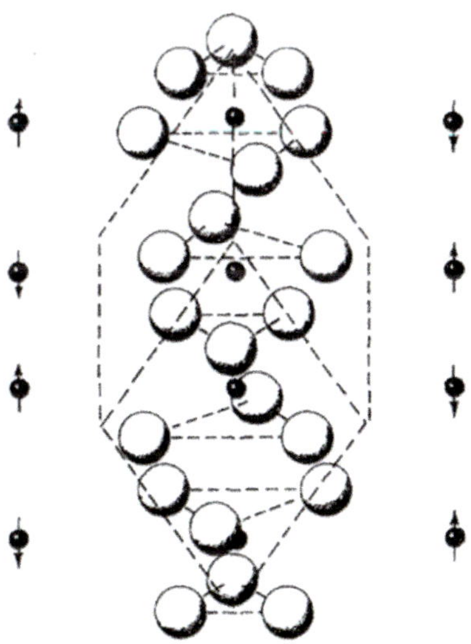

Figure 9.8 Crystal structure and spin alignment of Cr_2O_3. The open and solid spherical balls stand for O^{2-} and Cr^{3+}, respectively [37].

the electric field E and magnetic field H, the free energy for up and down antiferromagnetic domains can be expressed by

$$F^{up} = \chi_{ij} H_i H_j / 2 + \chi'_{ij} E_i E_j + \alpha_{ij} E_i H_j$$
$$F^{down} = \chi_{ij} H_i H_j / 2 + \chi'_{ij} E_i E_j - \alpha_{ij} E_i H_j \qquad (9.3)$$

where χ_{ij} and χ'_{ij} refer to the magnetic and dielectric susceptibility, respectively, and α_{ij} is the magnetoelectric coefficient. In the presence of magnetic and electric fields, the degeneration of free energy is broken and the energy difference between the two antiferromagnetic domains is

$$\Delta F = 2\alpha_{ij} E_i H_j. \qquad (9.4)$$

It indicates that the antiferromagnetic domain state will reverse in the simultaneous presence of suitable electric and magnetic fields. Since the polarity of perpendicular exchange bias in FM/Cr_2O_3 heterostructures is determined by AFM domain states, the polarity of the exchange bias could therefore be controlled by the ME effect.

In the MEFC process, magnetic and electric fields are simultaneously applied during the cooling process from a temperature higher than 307 K, which is the Néel temperature of bulk Cr_2O_3. The AFM domain state is determined by the energy competition between the interfacial exchange coupling at FM/Cr_2O_3 interface and the ME effect. In the MEFC process, there is a threshold value E_{th}. When $E < E_{th}$, the interfacial exchange coupling dominates the AFM domain state, correspondingly, and the exchange bias polarity is determined by the magnetic field; i.e., the magnetization direction of FM layers. When $E > E_{th}$, the ME dominates through the product of E and H. It results in the opposite exchange bias polarity compared with the previous case. Y. Shiratsuchi et al. demonstrated the MEFC process in a model structure with Pt/Co/Cr_2O_3/Pt fabricated in a (0001)-Al_2O_3 single crystal substrate [39, 40]. The optical image of the Hall structure for magnetization and exchange bias characterization is shown in Figure 9.9(a).

A negative exchange bias was observed for both the MEFC process with H = +9 kOe and E =–1000 kV/cm and the ME process with H = +9 kOe, as is shown in Figure 9.9(c). It indicates that the positive magnetic field favors the upward Co moment and the downward interfacial Cr moment since the Co and Cr spins are antiferromagnetically coupled at the interface. It results in a negative exchange bias. When $EH < 0$, the polarity of the exchange bias remains, indicating that negative EH favors the downward Cr spins. When $EH > 0$, the upward interfacial Cr spins are energetically favorable. When EH is high enough to overcome the interfacial exchange coupling, the interfacial Cr spins will be upward, consequently resulting in a positive exchange bias. Figure 9.10 shows the gradually changed polarity of exchange bias with the electric field gradually increasing from 0 to 600 kV/cm with the measurement technique of the polar magneto Kerr effect. One can see the gradually changed exchange bias and the corresponding remanence ratio at the threshold electric field around 420 kV/cm, at which the exchange bias is zero.

In the experiment, one finds that E_{th} is a function of temperature and applied magnetic field as shown in Figure 9.11. At a specific measuring temperature, E_{th} is proportional to $1/H$. The underlying physics can be understood as follows. Considering the Co and Cr

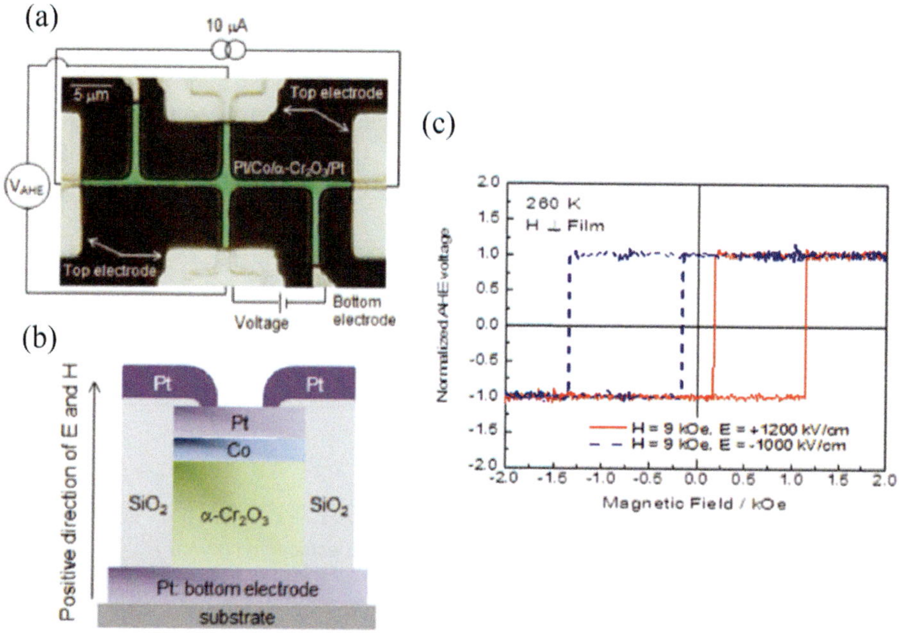

Figure 9.9 (a, b) The optical image of the Hall structure with the measurement setup and the schematic drawing of the cross-sectional view of the device [39]. (c) The normalized AHE loops after the MEFC of $H = +9$ kOe, $E = -1000$ kV/cm and $H = +9$ kOe, $E = +1200$ kV/cm, respectively (c) [40]. *Source:* Reproduced with permission of AIP Publishing LLC.

spins at the Co/Cr_2O_3 interface are antiferromagnetically coupled, the magnetic energy per unit area can be expressed as

$$F = K_{AFM} t_{AFM} \sin^2 \theta - \alpha_{33} E H t_{AFM} \cos\theta \text{-} J_K \cos\theta - M_{AFM} H \cos\theta \tag{9.5}$$

where K_{AFM} and t_{AFM} stand for the magnetic anisotropy energy and the thickness of the AFM layer, respectively. α_{33} is the ME coefficient along the c axis, and θ is the angle between the interfacial uncompensated AFM spin and the magnetic field direction. The first term in equation (9.5) represents the magnetic anisotropy energy, while the second term is the Zeeman energy from the electric field-induced magnetization of AFM due to the ME effect and the third term is the exchange anisotropy energy density with definition of $J_K = H_{EX} M_s t_{FM}$. The last term is the Zeeman energy of uncompensated antiferromagnetic moment. The switching condition of the electric field-induced magnetization can be derived from equation (9.5) by minimizing the total free energy as:

$$\frac{\partial F}{\partial \theta} = 0, \frac{\partial^2 F}{\partial^2 \theta} > 0. \tag{9.6}$$

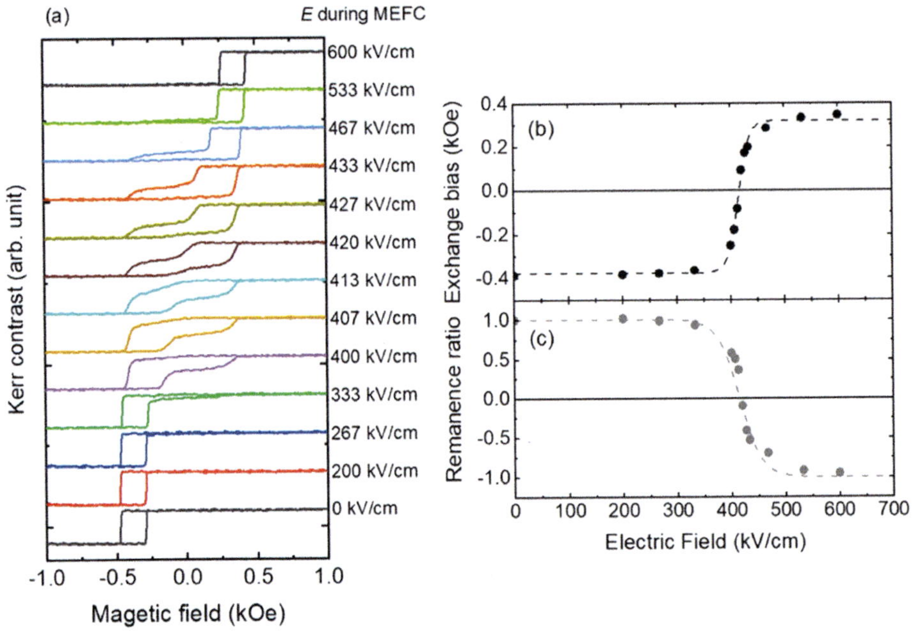

Figure 9.10 (a) Magnetic-field as a function of the Kerr contrast under different electric field. Variation of exchange bias field (b) and remanence ratio (c) under different electric field during MEFC [38].

For the perpendicular magnetized FM/Cr_2O_3 heterostructures, the equilibrium states are $\theta = 0$ or π. For each case, the energy condition can be simply expressed as:

$$2K_{AFM}\cos\theta = -\alpha_{33}EH - \frac{J_K}{t_{AFM}} - \frac{M_{AFM}}{t_{AFM}}H$$

$$(\alpha_{33}E + \frac{M_{AFM}}{t_{AFM}})H = -2K_{AFM} - \frac{J_K}{t_{AFM}}, \text{ (for } \theta\text{=0)}. \quad (9.7)$$

$$(\alpha_{33}E + \frac{M_{AFM}}{t_{AFM}})H = 2K_{AFM} - \frac{J_K}{t_{AFM}}, \text{(for } \theta = \pi)$$

According to this expression, E_{th} is proportional to $1/H$ when considering J_K and K_{AFM} are constant at a specific temperature.

The linear fitting of the experimental data according to equation (9.7) gives the slope and intercept values, from which the K_{AFM}/α_{33} and $J_K/(t_{AFM}\alpha_{33})$ can be calculated. Figure 9.12 shows the experimental data of a Pt(1.2)/Co(0.4)/Au(0.5)/Cr_2O_3(150)/Pt(20)/Al_2O_3 sample (thickness in nm). From the experimental data, the obtained K_{AFM} and J_K are around 4.5±0.6 kJ/m^3 and 1.5±0.2 × 10^{-2}mJ/m^2, respectively, assuming α_{33} = 3-4 ps/m. The estimated J_K is approximately twice the exchange anisotropy energy density calculated from the exchange bias at 280 K. The difference can be ascribed to the interfacial Cr spins which were not perfectly pinned but canted from the original direction with respect to the FM magnetization reversal.

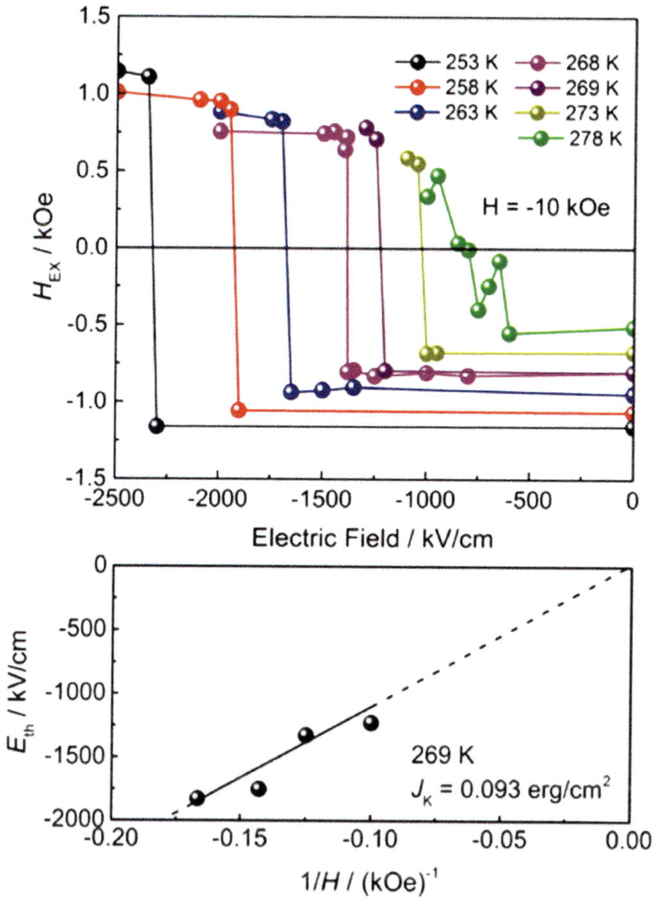

Figure 9.11 (a) Exchange bias field H_{EX} vs. the electric field under the magnetic field of–10 kOe, measured at various temperatures. (b)E_{th} as a function of 1/H, measured at 269 K [41].

Hence the remanence ratio between +1 and −1: i.e., magnetization switching between up and down states can be realized through changing the polarity of the exchange bias with the ME effect. To see the ME-induced magnetization process, Shiratsuchi et al. used X-ray magnetic circular dichroism (XMCD) to collect the spatial distribution of the XMCD intensity and to characterize the corresponding magnetic domains for Co and Cr spins independently [43]. Figure 9.13 shows the spatial distributions of the XMCD of Co and Cr for the Pt/Co/Cr_2O_3/Pt/Al_2O_3 multilayers measured at 205 K at the AC-demagnetizing state. The spatial distributions of the XMCD corresponding to the magnetic and antiferromagnetic domains were clearly observed by tuning the photon energy to either the ferromagnetic Co L_3 edge or the antiferromagnetic Cr L_3 edge. It can be seen that the ferromagnetic and antiferromagnetic domains are spatially coupled. The schematic drawing clearly demonstrates the orientation of Co and Cr spins at the two distinct areas, where the spins in the ferromagnetic and antiferromagnetic domains show antiparallel states. The ferromagnetic domain structures of Pt/Co/Cr_2O_3/Pt/Al_2O_3 multilayers at the AC demagnetizing state were also

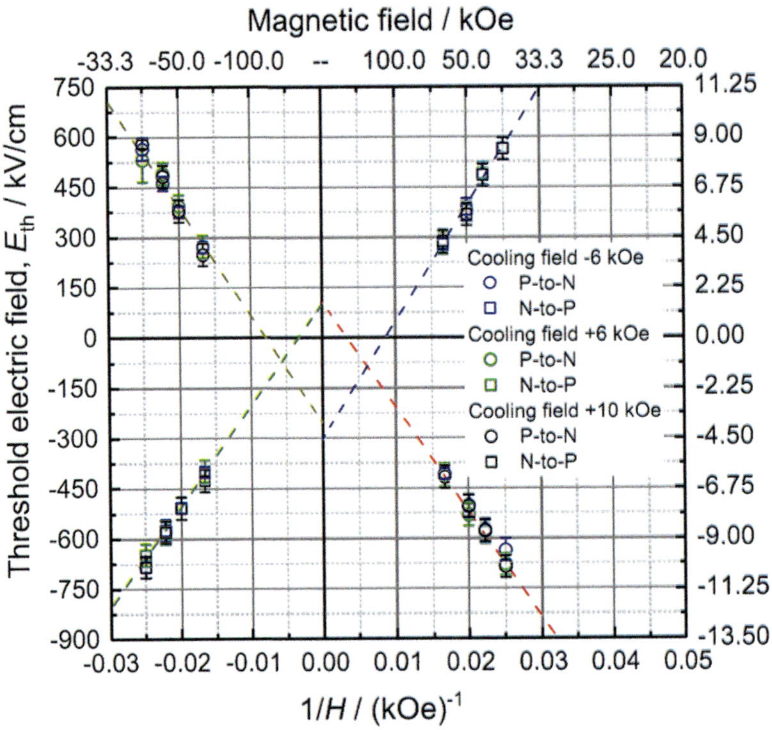

Figure 9.12 Magnetic field dependence of the threshold electric field for switching EB polarity for both positive to negative (P-to-N, i.e. $\theta = 0$) and negative to positive (N-to-P, i.e. $\theta = \pi$) processes, under different cooling fields of −6 kOe (blue marks), +6 kOe (red marks), and +10 kOe (black marks) for a Pt(1.2)/Co(0.4)/Au(0.5)/Cr_2O_3(150)/Pt(20)/Al_2O_3 sample (thickness in nm) [42].

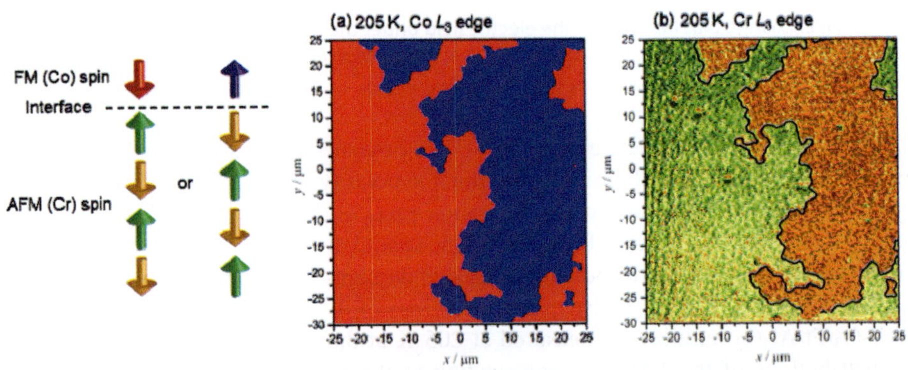

Figure 9.13 Spatial distribution of XMCD in Pt(1.5)/Co(0.5)/Cr_2O_3(100)/Pt(20)/Al_2O_3 film (thickness in nm) measured at 205 K. Photon energy used for the imaging was 778 eV (Co L3 edge) (a) and 576 eV (Cr L3 edge) (b). Orange and green contrast was induced by the down and up uncompensated Cr spins as shown in a left schematic drawing [43].

characterized by the polar magneto optical Kerr effect microscope (MOKE) [38]. In Figure 9.14, one can clearly see that the black and gray areas represent two magnetizations in those areas that are antiparallel with each other. The hysteresis loop of the whole area shows double hysteresis loops with both positive and negative exchange bias. The hysteresis loop in Area A, which is located in the black region, shows only one loop with a positive exchange bias, while the single loop in area B, which is located in the gray region, shows a negative exchange bias. It indicates that ferromagnetic domains are antiparallel coupled with the antiferromagnetic domain since the polarity of the exchange bias is linked with the orientation of antiferromagnetic spins. From the spatial distribution of XMCD and image of MOKE microscope, one may conclude that the AFM domain pattern can be observed by observing the FM domain, which is more easily detected by a regular method, at least in the $Pt/Co/Cr_2O_3/Pt/Al_2O_3$ system with perpendicular magnetic anisotropy and perpendicular exchange bias.

Although the ME switching of magnetization was realized in the $Pt/Co/Cr_2O_3/Pt/Al_2O_3$ system, the mechanisms of the switching process need to be elucidated. Figure 9.15 shows the evolution of the spatial distribution of XMCD intensity of a Hall device at the Co L_3 edge from the saturated state, which is obtained after applying fields of $H = +40$ kOe and $E = +13333$ kV/cm. The red color corresponding to the positive XMCD intensity represents that the Co spins are magnetized down with positive exchange bias. With the decrease of E from –600 to –753 kV/cm, one can see the color of the Hall device gradually changes to blue, which represents the up-magnetized Co spins with negative exchange bias. In the process, one can see the intermediate magnetic domain state from a positive exchange bias state to a negative exchange bias state – for example, when E =–667 kV/cm – while the intermediate domain states were also shown in the process from +1200 to +1300 kV/cm. It indicates that

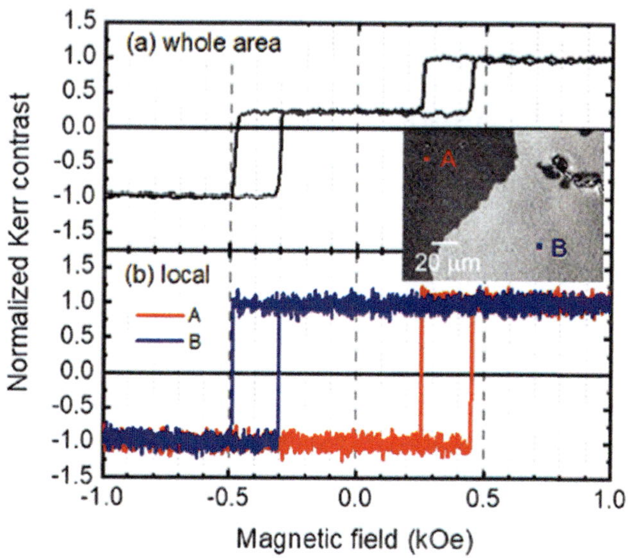

Figure 9.14 Magnetic-field dependence of the Kerr contrast collected from (a) whole observed area and (b) at local points after zero-field cooling in keeping the multi-domain state [38].

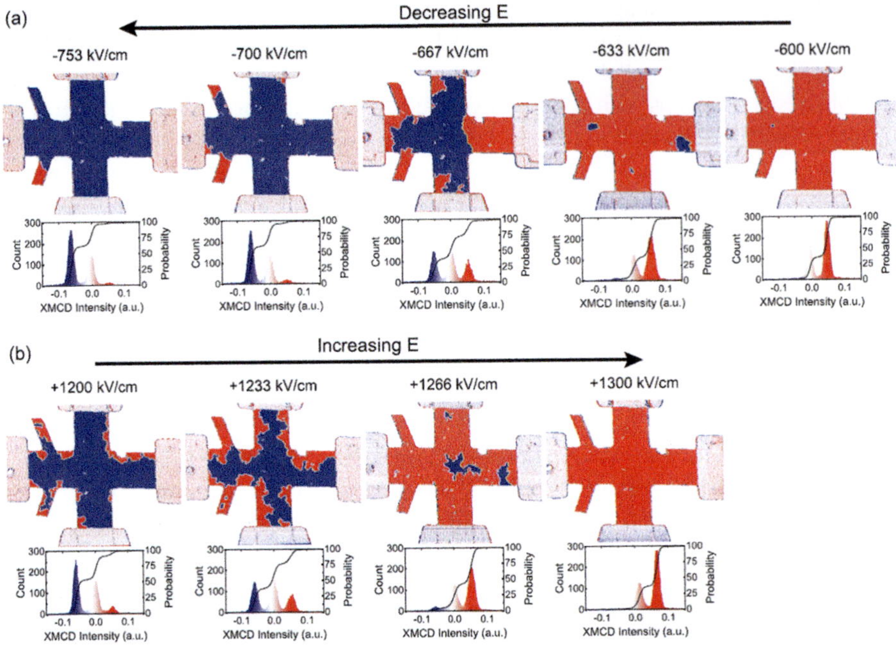

Figure 9.15 Spatial distributions of XMCD intensity at the Co L3 edge. (a) and (b) represent the change in the spatial distribution of the XMCD intensity in decreasing E from −600 to −753 kV/cm and increasing E from +1200 to +1300 kV/cm, respectively. Histograms of the XMCD intensity are also shown [44].

the magnetization switching process by ME effect is dominated by the nucleation and growth of the reversed magnetic domain. It is a pity that the XMCD intensity at the Cr L_3 edge was not reported. Thus, one may deduce that the reversal process of the perpendicular exchange bias polarity and the magnetization switching by the ME effect is the nucleation and growth of reversed interfacial uncompensated AFM spins and/or reversed AFM domains, since the ferromagnetic domain is coupled with the AFM domains. These results indicate that electric-field induced magnetization reversal in ME Cr_2O_3 is similar to exchange-biased FM magnetization.

The rotation of the antiferromagnetic domain through ferroelectric polarization was observed and used to make antiferromagnetic magnetoelectric random access memory in the multiferroic insulators Cr_2O_3 through the magnetoelectric coupling effect [45]. Figure 9.16(a) shows the writing and reading protocols of magnetic random access memory (MRAM), antiferromagnetic-MRAM, magnetoelectric random access memory (MERAM) and antiferromagnetic magnetoelectric random access memory (AFM-MERAM). The last one shows a new scheme that relies purely on antiferromagnetic components and does not require a ferromagnet for readout. Figure 9.16(b) shows an AFM-MERAM cell fabricated from an epitaxial layer stack of Al_2O_3(0001)/Pt(20 nm)/Cr_2O_3(200 nm)/Pt(2.5 nm). The thicker bottom Pt layer serves as the gate electrode and the thin Pt top layer is used to measure the

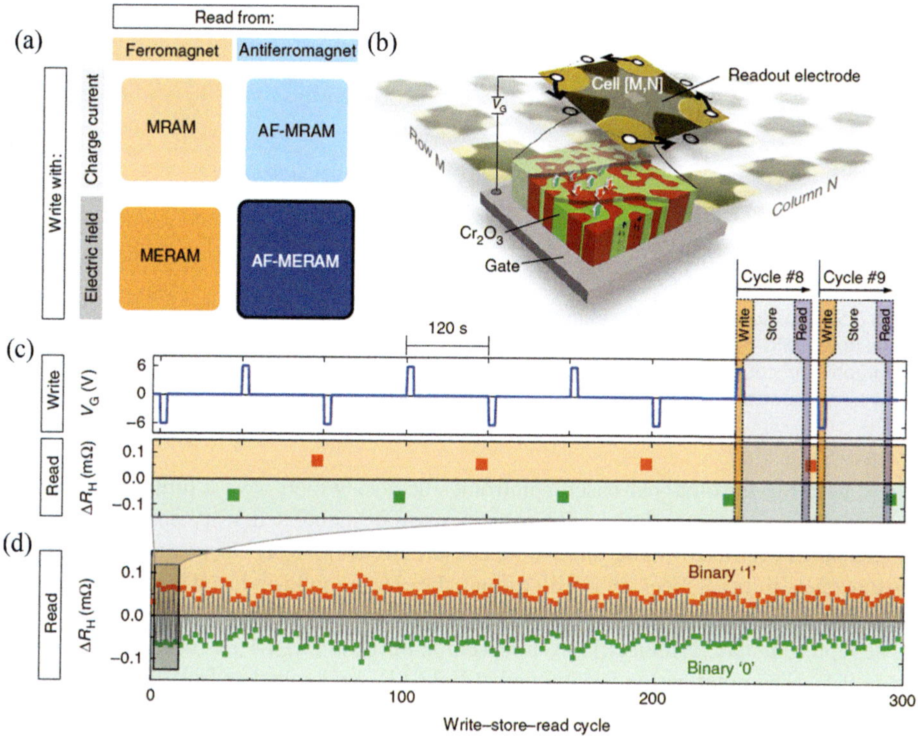

Figure 9.16 (a) Nonvolatile MRAM elements categorized according to their writing and readout interfaces. (b) Sketch of one memory cell within a matrix of devices. (c) Random access memory operation where binary information is written by a voltage pulse and stored in the antiferromagnetic order parameter. (d) Device behavior over 300 write–store–read cycles [45].

uncompensated boundary magnetization of Cr_2O_3 through proximity magnetization in the top Pt film. Figure 9.16(c) and (d) shows the essential write–store–read cycle up to 300 times of the AFM-MERAM cell. The reproducibility of this process reveals no performance degradation. This kind of AFM-MERAM provides substantially reduced writing thresholds over conventional MERAM prototypes, enabling further improvements in the energy efficient nonvolatile solid-state memory and logics. Beyond memory applications, the AFM-MERAM concept paves a way to find wide applicability in antiferromagnetic spintronics with an electric field as the driving force rather than the electric current of traditional spintronics devices.

9.3 Electric Field-driven Antiferromagnetic Vector Switching

Antiferromagnets show large potential application in high-density information storage because of their ultrafast spin dynamics and the absence of macroscopic magnetization. These unique properties make antiferromagnets suitable for faster and denser memory cells

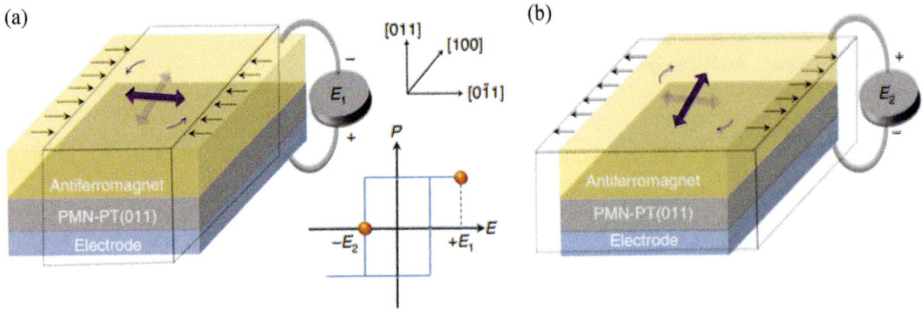

Figure 9.17 Schematic of ferroelastic strain switching of UMA driven by electric fields in antiferromagnet/PMN-PT(011) structure. Néel moments are switched towards the [0–11] and [100] crystallographic directions of (011)-oriented PMN-PT substrate in (a) and (b), respectively. The inset of (a) display a representation of the polarization vs. electric field (P–E) loop [46].

compared with the ferromagnet-based spintronics devices which have a limited switching speed of around a nanosecond and inhibit the increase of memory density due to their spatially extended stray fields. Except for the electric current-induced Néel vector rotation in antiferromagnets through Néel spin-orbit torque, the electric field-induced Néel vector rotation was also achieved in (011)PMN-PT/Mn_2Au heterostructures, as shown in Figure 9.17 [46]. The representative ferroelectric hysteresis loop of (011) PMN-PT is shown in the inset of Figure 9.17. The polarization is saturated at field $+E_1$, and switches from positive remanence to a negative saturated value at $-E_2$ upon depolarization process. E_2 was named the coercive electric field of PMN-PT. At E_1 the PMN-PT generates a compressive strain along the [0–11] direction and a tensile strain along the [100] direction. The Néel moment favors along the [0–11] direction according to the theoretical calculation. When the coercive field $-E_2$ was applied, the strain generated from (011)PMN-PT is reversed to tensile and compress one along the [0–11] and [100] directions, respectively, due to the change of the polarization of PMN-PT. In this state, The Néel moment favors along the [100] direction. Thus, the Néel vector can be switched by electric field through electric field generated strain. Thus, the writing and storage of information can be realized by an electric field and the Néel vector, respectively. The reading of the information could be realized by reading the planar Hall resistance due to the switching of the Néel vector of Mn_2Au. Figure 9.18 shows the measurement configurations and the Hall resistance. It can be seen that the variation of the Hall resistance changes between a high and a low resistive state due to the switching of the Néel vector between [0–11] and [100] after applying electric pulse fields between –2 and +4 kV/cm with a pulse width of 500 ms. Thus, using the magnetoelectric effect to write information in a pure antiferromagnet-based memory shows great potential for spintronics application with ultrahigh density and high energy efficiency.

In conclusion, multiferroelectric materials possess both ferroelectric and ferromagnetic ordering in one system, in which the electric and magnetic properties are highly coupled. It enables the control of magnetization (polarization) with electrical (magnetic) way. This review addressed the recent activities mainly focus on magnetic/ferroelectric heterostructures to show how to realize the magnetization switching of a ferromagnetic layer and the Néel vector rotation of an antiferromagnetic layer through the magnetoelectric effect by

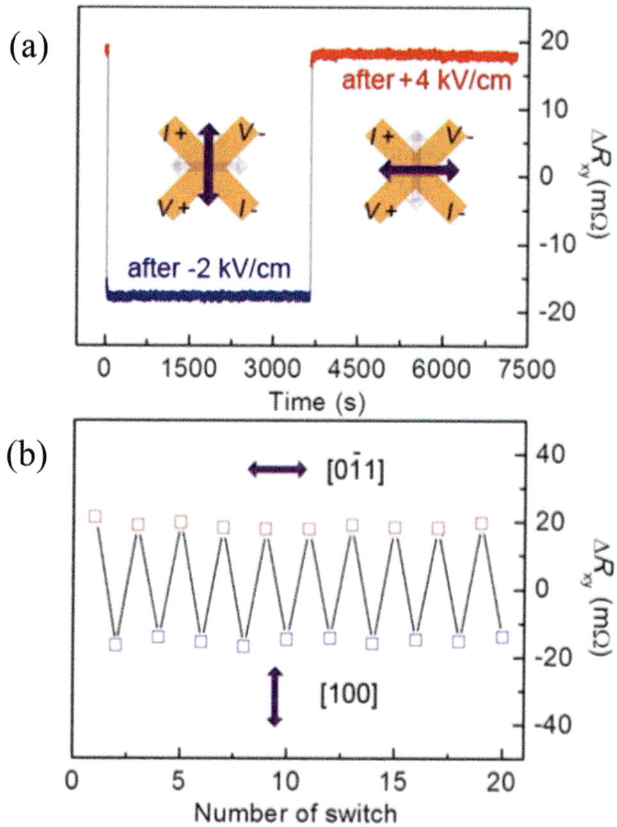

Figure 9.18 (a) The variations of nonvolatile Hall resistance variations after voltage pulses of −2 kV/cm and +4 kV/cm. (b) Reversible change of two Hall resistance states (after −2 kV/cm and +4 kV/cm) in ten circles [46].

applying quasi-static electric fields for application in spintronic devices. For the dynamic magnetoelectric effect in magnetoelectric composites, readers can refer to [47]. For the study of electric field control magnetism, we have realized the controlled magnetic anisotropy and magnetization switching in various FM/FE heterostructures. However, from the viewpoint of practical spintronic devices, in the following studies, we should further explore the precise controlling of the magnetization domain state and magnetization switching paths of FM layer by applying electric field timulation.

Acknowledgements

Financial support by the National Natural Science Foundation of China (No. 51671098, 11904056) Guangzhou Basic and Applied Basic Research Project (No. 202102020053) are gratefully acknowledged.

References

1 Maxwell, J.C. (1865). VIII. A dynamical theory of the electromagnetic field. *Philos. Trans. R. Soc. Lond.* (155): 459–512.
2 Ma, J., Hu, J., Li, Z., and Nan, C.W. (2011). Recent progress in multiferroic magnetoelectric composites: from bulk to thin films. *Adv. Mater.* 23: 1062–1087.
3 Hu, J.M., Chen, L.Q., and Nan, C.W. (2016). Multiferroic heterostructures integrating ferroelectric and magnetic materials. *Adv. Mater.* 28: 15–39.
4 Wang, K., Liu, J., and Ren, Z. (2009). Multiferroicity: the coupling between magnetic and polarization orders. *Adv. Phys.* 58: 321–448.
5 Sen, P., Dey, A., Mukhopadhyay, A.K., Bandyopadhyay, S.K., and Himanshu, A.K. (2012). Nanoindentation behaviour of nano BiFeO3. *Ceram. Int.* 38: 1347–1352.
6 Fiebig, M., Lottermoser, T., Meier, D., and Trassin, M. (2016). The evolution of multiferroics. *Nat. Rev. Mater.* 1: 16046.
7 Song, C., Cui, B., Li, F., Zhou, X., and Pan, F. (2017). Recent progress in voltage control of magnetism: materials, mechanisms, and performance. *Prog. Mater. Sci.* 87: 33–82.
8 Dong, S., Liu, J.-M., Cheong, S.-W., and Ren, Z. (2015). Multiferroic materials and magnetoelectric physics: symmetry, entanglement, excitation, and topology. *Adv. Phys.* 64: 519–626.
9 Chappert, C., Fert, A., and Van Dau, F.N. (2007). The emergence of spin electronics in data storage. *Nat. Mater.* 6: 813–823.
10 Jian-Gang, Z. (2008). Magnetoresistive random access memory: the path to competitiveness and scalability. *Proc. IEEE* 96: 1786–1798.
11 Slonczewski, J.C. (1996). Current-driven excitation of magnetic multilayers. *J. Magn. Magn. Mater.* 159: L1.
12 Berger, L. (1996). Emission of spin waves by a magnetic multilayer traversed by a current. *Phys. Rev. B* 54: 9353–9358.
13 Myers, E.B., Ralph, D.C., Katine, J.A., Louie, R.N., and Buhrman, R.A. (1999). Current-induced switching of domains in magnetic multilayer devices. *Science* 285: 867–870.
14 Miron, I.M., Gaudin, G., Auffret, S., Rodmacq, B., Schuhl, A., Pizzini, S., Vogel, J., and Gambardella, P. (2010). Current-driven spin torque induced by the Rashba effect in a ferromagnetic metal layer. *Nat. Mater.* 9: 230–234.
15 Liu, L., Lee, O.J., Gudmundsen, T., Ralph, D.C., and Buhrman, R.A. (2012). Current-induced switching of perpendicularly magnetized magnetic layers using spin torque from the spin Hall effect. *Phys. Rev. Lett.* 109: 096602.
16 Liu, L., Pai, C., Li, Y., Tseng, H., Ralph, D.C., and Buhrman, R.A. (2012). Spin-torque switching with the giant spin Hall effect of tantalum. *Science* 336: 555–558.
17 https://www.everspin.com/fifiles/everspin-releases-highest-density-mram-products-fifinal-041216.pdf.
18 Kim, Y., Fong, X., Kwon, K., Chen, M., and Roy, K. (2015). Multilevel spin-orbit torque MRAMs. *IEEE Trans. Electron Devices* 62: 561–568.
19 Brink, V.D.A.A., Vermijs, G.G., Solignac, A.A., Koo, J.J., Kohlhepp, J.T.J., Swagten, H.H., and Koopmans, B.B. (2016). Field-free magnetization reversal by spin-Hall effect and exchange bias. *Nat. Commun.* 7: 10854–10854.
20 Khomskii, D.I. (2009). Classifying multiferroics: mechanisms and effects. *Physics* 2: 20.

21 Caijiang, L., Ping, L., Yumei, W., Aichao, Y., Chao, Y., Decai, W., Wei, H., and Jitao, Z. (2014). Self-biased magnetoelectric responses in magnetostrictive/piezoelectric composites with different high-permeability alloys. *Chin. Phys. B* 23: 117503.

22 Xi, L., Guo, X., Wang, Z., Li, Y., Yao, Y., Zuo, Y., and Xue, D. (2013). Voltage-driven in-plane magnetization easy axis switching in FeNi/Piezoelectric actuator hybrid structure. *Appl. Phys. Express* 6: 015804.

23 Peng, J., Luo, H.-S., Lin, D., Xu, H.-Q., He, T.-H., and Jin, W.-Q. (2004). Orientation dependence of transverse piezoelectric properties of 0.70Pb(Mg1⁄3Nb2⁄3)O3-0.30PbTiO3 single crystals. *Appl. Phys. Lett.* 85: 6221–6223.

24 Li, P., Chen, A., Li, D., Zhao, Y., Zhang, S., Yang, L., Liu, Y., Zhu, M., Zhang, H., and Han, X. (2014). Electric field manipulation of magnetization rotation and tunneling magnetoresistance of magnetic tunnel junctions at room temperature. *Adv. Mater.* 26: 4320–4325.

25 Biswas, A.K., Bandyopadhyay, S., and Atulasimha, J. (2014). Complete magnetization reversal in a magnetostrictive nanomagnet with voltage-generated stress: a reliable energy-efficient non-volatile magneto-elastic memory. *Appl. Phys. Lett.* 105: 072408.

26 Yang, S.W., Peng, R.C., Jiang, T., Liu, Y.K., Feng, L., Wang, J.J., Chen, L.Q., Li, X.G., and Nan, C.W. (2014). Non-volatile 180 degrees magnetization reversal by an electric field in multiferroic heterostructures. *Adv. Mater.* 26: 7091–7095.

27 Chen, A., Zhao, Y., Li, P., Zhang, X., Peng, R., Huang, H., Zou, L., Zheng, X., Zhang, S., and Miao, P. (2016). Angular dependence of exchange bias and magnetization reversal controlled by electric-field-induced competing anisotropies. *Adv. Mater.* 28: 363–369.

28 Wang, J.J., Hu, J.M., Ma, J., Zhang, J.X., Chen, L.Q., and Nan, C.W. (2014). Full 180 degrees magnetization reversal with electric fields. *Sci. Rep.* 4: 7507.

29 Guo, X., Wu, K., Cui, B., Li, D., Yun, J., Zuo, Y., Zuo, H., Wang, T., and Xi, L. (2016). High temperature and full-in-plane-direction workable high-frequency soft magnetic epitaxial FeSi thin films on MgO (0 0 1). *J. Phys. D: Appl. Phys.* 49: 055005.

30 Guo, X., Zuo, Y., Li, D., Cui, B., Wu, K., Yun, J., Wang, T., and Xi, L. (2016). Electrical field control of non-volatile 90° magnetization switching in epitaxial FeSi films on (001) 0.7 [Pb (Mg1/3Nb2/3) O3]-0.3 [PbTi0. 3O3]. *Appl. Phys. Lett.* 108: 042403.

31 Parkes, D.E., Cavill, S.A., Hindmarch, A.T., Wadley, P., Mcgee, F., Staddon, C.R., Edmonds, K.W., Campion, R.P., Gallagher, B.L., and Rushforth, A.W. (2012). Non-volatile voltage control of magnetization and magnetic domain walls in magnetostrictive epitaxial thin films. *Appl. Phys. Lett.* 101: 072402.

32 Zhang, B., Meng, K.K., Yang, M.Y., Edmonds, K.W., Zhang, H., Cai, K.M., Sheng, Y., Zhang, N., Ji, Y., Zhao, J.H., Zheng, H.Z., and Wang, K.Y. (2016). Piezo voltage controlled planar Hall effect devices. *Sci. Rep.* 6: 28458.

33 Guo, X., Han, X., Zuo, Y., Zhang, J., Li, D., Cui, B., Wu, K., Yun, J., Wang, T., Peng, Y., and Xi, L. (2016). Electric field induced magnetic anisotropy transition from fourfold to twofold symmetry in (001) 0.68Pb(Mg1/3Nb2/3)O3-0.32PbTiO3/Fe0.86Si0.14 epitaxial heterostructures. *Appl. Phys. Lett.* 108: 152401.

34 Dzyaloshinskii, I.E. (1960). On the magneto-electrical effects in antiferromagnets. *Sov. Phys. JETP* 10: 628–629.

35 Astrov, D. (1960). The magnetoelectric effect in antiferromagnetics. *Sov. Phys. JETP* 11: 708–709.

36 Folen, V., Rado, G., and Stalder, E. (1961). Anisotropy of the magnetoelectric effect in Cr 2 O 3. *Phys. Rev. Lett.* 6: 607.

37 Martin, T. and Anderson, J. (1966). Antiferromagnetic domain switching in Cr 2 O 3. *IEEE Trans. Magn.* 2: 446–449.

38 Shiratsuchi, Y., Watanabe, S., Yonemura, S., Shibata, T., and Nakatani, R. (2018). Frustration and relaxation of antiferromagnetic domains reversed by magneto-electric field cooling in a Pt/Co/Au/Cr2O3/Pt-stacked film. *AIP Adv.* 8: 125313.

39 Shiratsuchi, Y., TVA, N., and Nakatani, R. (2018). Magnetoelectric control of antiferromagnetic domain of Cr2O3 thin film toward spintronic application. *J. Magn. Soc. Jpn.*, 42: 119–126.

40 Toyoki, K., Shiratsuchi, Y., Kobane, A., Harimoto, S., Onoue, S., Nomura, H., and Nakatani, R. (2015). Switching of perpendicular exchange bias in Pt/Co/Pt/α-Cr2O3/Pt layered structure using magneto-electric effect. *J. Appl. Phys.* 117: 17D902.

41 Toyoki, K., Shiratsuchi, Y., Kobane, A., Mitsumata, C., Kotani, Y., Nakamura, T., and Nakatani, R. (2015). Magnetoelectric switching of perpendicular exchange bias in Pt/Co/α-Cr2O3/Pt stacked films. *Appl. Phys. Lett.* 106: 162404.

42 Nguyen, T.V.A., Shiratsuchi, Y., Kobane, A., Yoshida, S., and Nakatani, R. (2017). Magnetic field dependence of threshold electric field for magnetoelectric switching of exchange-bias polarity. *J. Appl. Phys.* 122: 073905.

43 Shiratsuchi, Y., Kotani, Y., Yoshida, S., Yoshikawa, Y., Toyoki, K., Kobane, A., Nakatani, R., and Nakamura, T. (2015). Direct observations of ferromagnetic and antiferromagnetic domains in Pt/Co/Cr_2O_3/Pt perpendicular exchange biased film. *AIMS Mater. Sci.* 2: 484–496.

44 Shiratsuchi, Y., Watanabe, S., Yoshida, H., Kishida, N., Nakatani, R., Kotani, Y., Toyoki, K., and Nakamura, T. (2018). Observation of the magnetoelectric reversal process of the antiferromagnetic domain. *Appl. Phys. Lett.* 113: 242404.

45 Kosub, T., Kopte, M., Hühne, R., Appel, P., Shields, B., Maletinsky, P., Hübner, R., Liedke, M.O., Fassbender, J., and Schmidt, O.G. (2017). Purely antiferromagnetic magnetoelectric random access memory. *Nat. Commun.* 8: 1–7.

46 Chen, X., Zhou, X., Cheng, R., Song, C., Zhang, J., Wu, Y., Ba, Y., Li, H., Sun, Y., and You, Y. (2019). Electric field control of Néel spin–orbit torque in an antiferromagnet. *Nat. Mater.* 18: 931–935.

47 Bichurin, M.I. (2019). *Magnetoelectric Composites*. CRC Press.

10

Robust Manipulation of Magnetic Properties in (Ga,Mn)As

Hailong Wang[1,2] *and Jianhua Zhao*[1,2]

[1] *State Key Laboratory of Superlattices and Microstructures, Institute of Semiconductors, Chinese Academy of Sciences, Beijing, China*
[2] *Center of Materials Science and Optoelectronics Engineering, University of Chinese Academy of Sciences, Beijing, China*

CHAPTER MENU

10.1 Background and Introduction

Semiconductor spintronics aims at exploiting an electron's spin degree of freedom to realize the processing, storage, and communication of information, which requires the creation of net spins in certain materials as its first step [1–4]. Making nonmagnetic semiconductors magnetic is the way that magnetism and semiconducting properties can be combined into one material, which could achieve the goal of generating and manipulating spin-polarized carriers by following the wisdom of semiconductor technology [5–10]. In the past couple of decades, people have gradually turned their interest from concentrated magnetic semiconductors to dilute magnetic semiconductors (DMSs), and the growth and materials properties of Mn-doped III–V semiconductors have been extensively investigated [5–10]. It is now recognized that (III,Mn)V DMS acts as a fertile material platform for the discovery of many novel physical phenomena with rich spin-related physics, including the first experimental observation of electric field control of magnetism, tunneling anisotropic magnetoresistance, spin-orbit torque, and so on [9, 10].

One of the critical questions on (III,Mn)V DMS is whether the Curie temperature (T_C) of these materials can be raised above room temperature [5–12]. Accompanying the research on enhancing T_C, a lot of effort has also been devoted to exploring new spintronic concepts, among which robust manipulation of their magnetic properties is one of the most attractive subjects [9, 10]. Controlling magnetism efficiently is very important for the write operation in magnetic storage industry, in which the writing process usually consumes a large amount of energy, which also motivates the research on efficient manipulation of magnetism. Viewing from the stimulus-response theory, any external variables that interact with magnetic moments (spin or orbital) will induce some changes to them [13]. The stimuli might thus have the capability of manipulating a material's magnetic properties, which can be understood as collective behaviors manifested by a group of magnetic moments, including T_C,

Spintronics: Materials, Devices and Applications, First Edition. Edited by Kaiyou Wang, Meiyin Yang, and Jun Luo.

saturation magnetization (M_s), magnetic anisotropy, etc. In principle, the stimuli could be any physical quantity that has a direct or even indirect correlation with angular momentum, and various techniques have been employed to alter magnetism, such as using a magnetic field, an electric field, electric currents, heat (or temperature gradients), light, and mechanical strain. However, for quite a long time, only magnetic field and temperature were used to induce significant responses in magnetic systems, and it was commonly thought that magnetic properties of magnetic devices could not be changed once they are fabricated [14–16]. For real applications in magnetic memory, switching magnetization via Oersted magnetic field not only limits the enhancement of storage density, but also leads to unwanted energy dissipation, due to the non-local feature of magnetic field. This problem can be partly figured out by using the spin torque effects, in which magnetization reversal is accomplished by a torque generated by an electric current [13–16]. The spin torque technique is good for miniaturization of magnetic random access memory (MRAM) cell, and is now widely employed by the industry [17–20]. However, the required critical current density is on the order of 10^7 A/cm^2, still too large for high density integration [18, 19]. Therefore, searching new routes which are capable of manipulating magnetic properties much more efficiently is of great importance.

People naturally dreamed of using an electric field to control magnetism, in the consideration of its compatibility with modern semiconductor industry and low power dissipation [13–16]. This thought is reasonable and was conceived in the 1960s, but it is quite challenging since its realization should be built on the fast progress of film growth in the nanoscale [21, 22]. The successful synthesis of (III,Mn)V DMS by low-temperature molecular-beam epitaxy (LT-MBE) makes this dream feasible, due to its atomically flat surface, highly controllable thickness, and more importantly, its carrier-mediated ferromagnetism [5–10]. As a semiconductor featuring a mediate band gap, rich spin-related functionalities could then be imagined and implemented, leading to the emergence of many novel effects. In Figure 10.1, we present several key experimental achievements on the manipulation of magnetism,

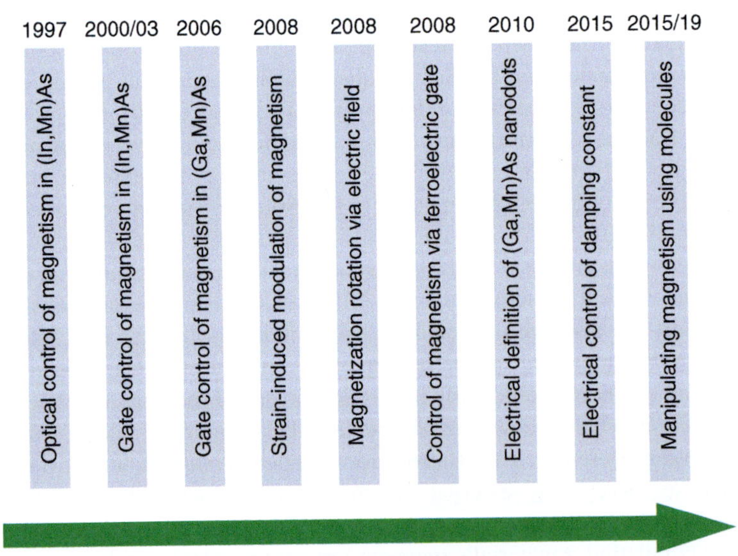

Figure 10.1 Timeline of the key experimental demonstrations on the manipulation of magnetism in (III,Mn)V DMS through diverse techniques, including electric field gating, photo-doping, strain, and forming hybrid structures with organic molecules.

arranged according to the time they were demonstrated [5–10]. Some of these methods seem to be unique for semiconductors, but more of these discoveries have already been borrowed and applied successfully to magnetic metals [23, 24].

This chapter reviews the basic ideas, discoveries, and progress in robust manipulation of the magnetic properties of (Ga,Mn)As. After presenting the exciting history of gate-controlled magnetism in (III,Mn)V DMS materials in the second section, we will then illustrate the methods that utilize light and mechanical strain in the third section. And the recent progress in utilizing self-assembled organic molecules to tune the magnetism will be given in the fourth section. We will finish this review by summarizing the challenges and discussing the opportunities in the future.

10.2 Electric Field Effects on the Magnetic Properties of (Ga,Mn)As

Electric field effects have been used a lot in modern semiconductor devices, having the huge advantages of low power consumption and high integration density. However, it was not until 2000 that electric field was proved to be capable of controlling magnetism, taking (In,Mn)As thin films as the conducting channel in a metal-insulator-semiconductor (MIS) device [14]. The success of this seminal experiment should be attributed to the carrier-mediated magnetism of magnetic semiconductors, while the carrier density of semiconductors is several orders of magnitude smaller than common metals [14–16, 23]. As is well known, the strong charge screening effect due to large electron density in metals immensely shortens the penetration depth of an applied electric field, and thus making it extremely difficult to change the bulk properties of a metallic ferromagnet by electric field. Although this problem partly facilitates the discovery of graphene in the hope of realizing "metallic electronics" by obtaining an atomic thin metal, whether real 2D materials are stable or not was doubted up until 2004 [25]; hence, the only feasible way that seems to be possible is to implement this thought in semiconductors.

In early works, transport measurements based on the anomalous Hall effect (AHE) were mainly used to detect the variation of magnetism by electrical gating, which is the result of several factors: firstly, the magnetic signal of ultrathin (III,Mn)V film is very weak due to their small saturation magnetization (usually less than 1/50 of Fe), which is quite challenging even for a magnetometer equipped with a superconducting quantum interference device (SQUID); secondly, an electric field which can induce an observable change of magnetism is approaching the breakdown strength of dielectrics (1 V/nm), which makes the fabrication of a MIS device with a large area, and suitable for magnetization measurements, rather difficult. Since the anomalous Hall resistance is usually proportional to the out-of-plane magnetization component, it is preferable to use samples with perpendicular magnetic anisotropy (PMA) [26, 27]. Fortunately, the magnetic anisotropy in (In,Mn)As and (Ga,Mn)As films can be tuned by strain, while in-plane tensile (compressive) strain usually leads to perpendicular (in-plane) magnetic easy axis [5–10].

The basic idea is that the accumulation (depletion) of holes in DMS would increase (reduce) the ferromagnetic exchange interaction among localized Mn ions, as intuitively shown in Figure 10.2(a). The anomalous Hall resistance will change correspondingly, and investigating its magnetic field dependence clearly verifies the gate tunable saturation magnetization and coercive field [14]. By carrying out the same measurements at different

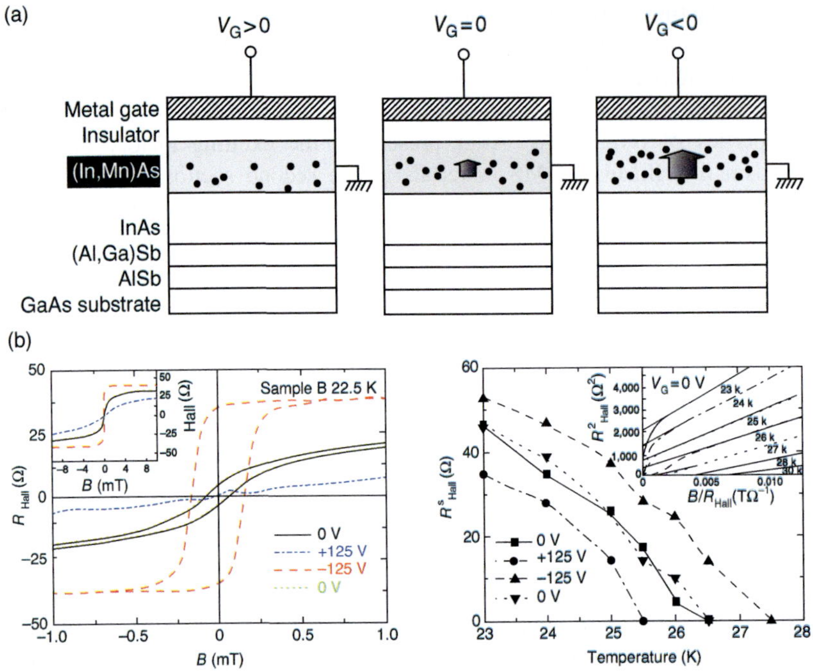

Figure 10.2 (a) Schematic diagram of the electric field effect on magnetism in a MIS device consisting of (In,Mn)As channel. (b) Evidence for the gate-controlled magnetism in (In,Mn)As probed by the anomalous Hall effect [14]/with permission of Springer Nature.

temperatures, the change of T_C of ~ 2 K was also confirmed by using the Arrott plots, which enables the electrical demagnetization of (In,Mn)As [14]. Subsequently, electrical switching of magnetization, assisted by a fixed magnetic field, was also demonstrated as a natural followup [15]. Importantly, this electric field effect was found to be reversible and reproducible to some extent, promising for potential application [15]. Nevertheless, the gate voltage was up to ~ 100 V while the sheet hole density change was only ~ 3×10^{12} cm^{-2}. Such high voltage is mainly due to the ~ 800 nm thick polyimide layer that was used as the gate insulator, and the magnitude of longitudinal resistance variation was mainly limited by the dielectric permittivity κ.

The success of gate-controlled magnetism in (In,Mn)As film was not quickly and easily copied to (Ga,Mn)As, and the reasons lie in that (Ga,Mn)As film tends to be insulating when decreasing its thickness to be less than 5 nm [28, 29]. In order to get a measurable AHE signal, thicker (Ga,Mn)As films should be used, which leads to higher sheet hole density and thus an unobservable change of T_C and minor modulation of magnetic hysteresis loop [29]. Although the electrical gating effect has already been reported in Mn δ-doped GaAs/(Al,Ga)As heterostructures, the attempts on (Ga,Mn)As films only showed minor modulation on coercivity without detectable change of T_C [30, 31]. This problem was then figured out either by using dielectrics with high κ such as Al_2O_3 or HfO_2, or by employing ionic liquid/gel that can produce a large interfacial electric field [32]. High quality Al_2O_3 or HfO_2 layer with thickness of ~ 50 nm is usually grown by atomic layer deposition (ALD), which also reduces the applied gate voltages when the same electric field is generated. The experimental results now clearly show similar electric field effects as that observed in (In,Mn)As films, with the change of T_C increased to ~ 5 K in a 7 nm thick (Ga,Mn)As film [29]. Systematic study of the (Ga,Mn)As channel thickness dependence showed enhanced gating effect, with the largest change of T_C increasing to ~ 15 K for an ultrathin film with thickness of 3.5 nm [33]. In terms of the modulation of coercivity, magneto-optical Kerr effect (MOKE) images implied that the magnetization reversal process was not coherent rotation but domain wall nucleation and propagation, so understanding how electric fields affect the nucleation field was crucial but still remains elusive [29].

In addition to the modulation of coercivity, whether an electric field can efficiently rotate the magnetization is also of highly interest, which has been studied in (Ga,Mn)As-based MIS structure as well [34]. Transport measurements are still very powerful in detecting the magnetization rotation, exploiting the planar Hall resistance (also known as the transverse anisotropic magnetoresistance R_{PHE}). R_{PHE} is closely related to the angle between the electric current and the magnetization directions, which is determined by the external magnetic field and the internal magnetic anisotropic fields [34]. For this reason, by deducing the magnetization angle (φ) as a function of the magnetic field angle (θ) under various gate voltages from the planar Hall measurements, the magnetic anisotropic fields of (Ga,Mn)As could then also be obtained by fitting the $\varphi(\theta)$ curves, as shown in Figure 10.3(b) [34]. Notice that the magnetization vector would change very slowly when the magnetic field sweeps across the magnetic easy axis, leading to a gradual change of R_{PHE}. Conversely, R_{PHE} will change rapidly when the magnetic field direction is near the magnetic hard axis, due to the quick rotation of the magnetization vector [35]. Therefore, the electric field influence on the magnetic anisotropy can be judged more intuitively from the derivative of $|dR_{PHE}/d\theta|$ in Figure 10.3(c), in which the peak corresponds to the direction of the magnetic hard axes, and their heights indicate how hard they are. Clear evolution of the relative height of the two peaks with the change of gate voltages were observed, implying the 90° switching of the uniaxial magnetic easy axis by the electric field. Quantitatively, the direction of magnetization can be manipulated by 10° through changing the gate electric field from ~ 0.3 V/nm to –0.4 V/nm (Figure 10.3d), which stimulates more interest in dreaming up a new scheme for switching magnetization. However, magnetization switching cannot be induced by an electric field solely since it does not break the time reversal symmetry, which calls for additional perturbation that could play this role, such as a small magnetic field [34, 36, 37].

Nevertheless, stochastic magnetization in a (Ga,Mn)As film by an electric field in the absence of an external magnetic field, was demonstrated experimentally [38]. The mechanism is thought to be the gate-controlled energy profile through the modulation of magnetic anisotropy, which is a strong support for conceiving a new scheme of switching magnetization with low power input by electric field.

Another interesting functionality that is inaccessible before is to define ferromagnetic nanodots electrically, simply by using a nanoscale meshed gate [39]. Once again, anomalous Hall resistance as a function of magnetic field was employed to detect the modulated-magnetism in (Ga,Mn)As. By designing a MIS device with both unpatterned and meshed gates (arrays of square windows in nanoscale), the gating results were compared and presented in Figure 10.4. For (Ga,Mn)As under the unpatterned gate, hole accumulation resulted in the enhancement of ferromagnetism as manifested by the increasing coercivity due to the modulated domain-nucleation field, while hole depletion suppressed ferromagnetism and finally led to a zero coercivity. However, for the portion of (Ga,Mn)As under the meshed

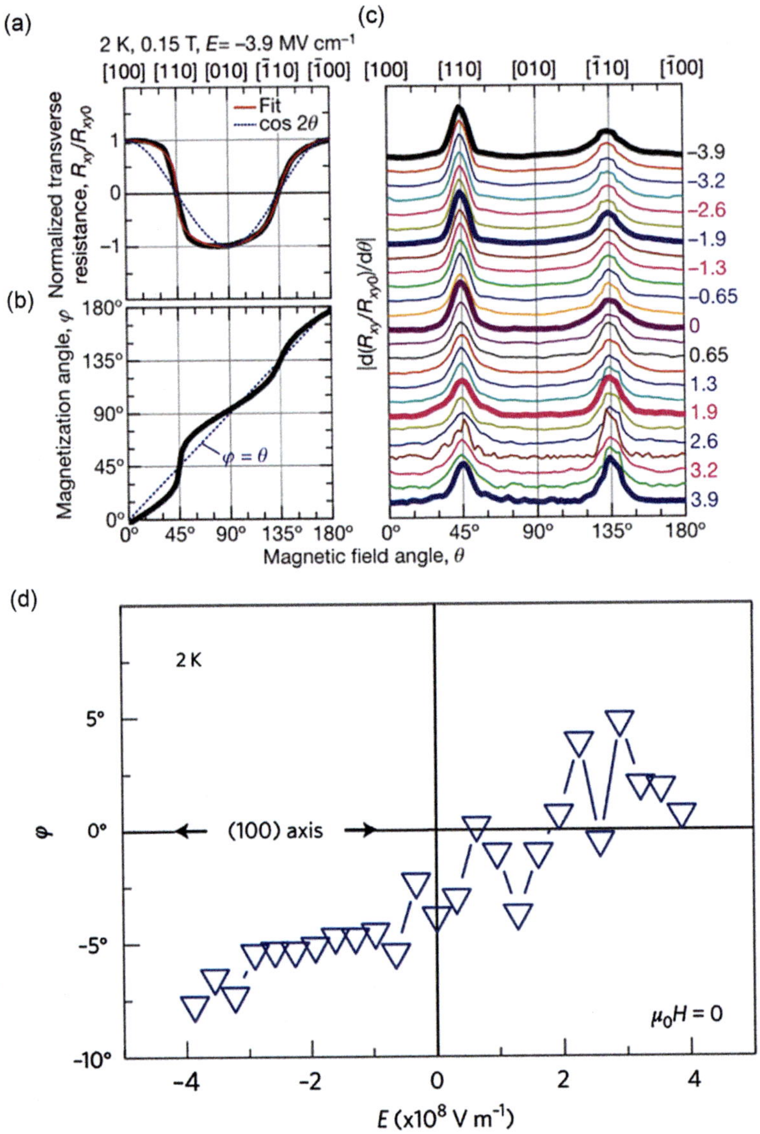

Figure 10.3 (a) Schematic diagram of the device for demonstrating the magnetization rotation in (Ga,Mn)As through electric field. (b) Dependence of planar Hall resistance and magnetization angle on magnetic field angle. (c) The absolute value of the derivative of the normalized transverse resistance as a function of the magnetic field angle, in which the two peaks are corresponding to the magnetic hard axis, and the height indicates how hard the axis is. (d) The calculated magnetization angle as a function of a gate electric field in the absence of a magnetic field [23, 34].

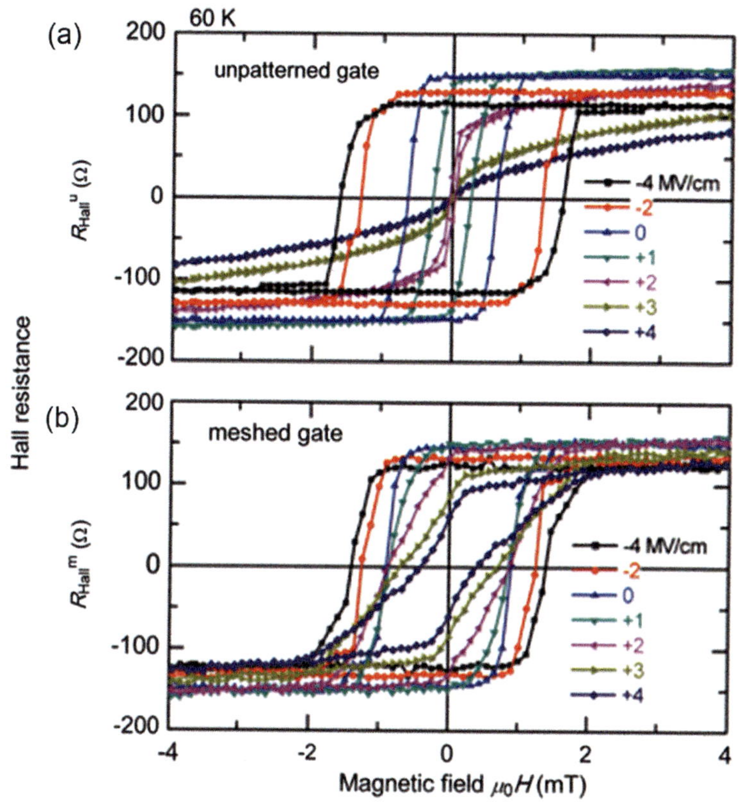

Figure 10.4 (a) Magnetic hysteresis curves probed by Hall resistance of unpatterned gate part and (b) the meshed gate part at 60 K [39]/with permission of American Chemical Society.

gate, clear hysteresis loops with non-zero coercivity were always observed, indicating the contribution of (Ga,Mn)As underneath the windows, or equivalently, the electrically defined nanodots. Besides, the skewed hysteresis loops for the latter case could be attributed to the distribution of coercive field of these nanodots, and the relatively large H_c might be caused by the suppression of domain wall propagation between nearby "nanodots."

Furthermore, gate-controlled magnetism could also be a powerful tool for the understanding of novel physical effects and collective phenomena, such as superconductivity, magnetoelectric coupling, and AHE. It is now established that AHE has three mechanisms: namely, the skew scattering, side jump, and Berry phase contributions. The third reflects the information of electronic band topology, and is often deemed to be an intrinsic term. A lot of effort has been devoted to separating these three contributions, through transport experiments dealing with the anomalous Hall and longitudinal resistivities. However, the results seem to be very confusing. This situation should be partly ascribed to the same scaling form for both the intrinsic and side jump contribution, while various scattering sources behaving differently would be another reason [26, 27, 40]. By controlling various scattering sources separately, a proper scaling law has been proposed that seems to explain most of the data quite well [40]. However, further investigation of materials with higher electrical conductivity has shown deviation and therefore invites a new model [41]. Obviously, a device with controllable carrier density and resistance would add an extra degree of freedom for studying AHE, in which (Ga,Mn)As-based MIS structure could make its contribution by providing more information on how the AHE behaves when the carrier density varies continuously. The experimental results are quite striking: In the low conductivity regime, the results could be described by a scaling law in the form of a power function with a constant exponent; however, when the conductivity was increased, the scaling relation broke down [42]. In the latter regime (i.e., with high conductivity), the anomalous Hall conductivity σ_{xy} rapidly decreased with decreasing temperature, and even changed sign at low temperature in some cases [42]. This is rather abnormal, considering that both the longitudinal resistance and the magnetization did not change too much with temperature. These unexpected results indicate the importance of disorder and dimensionality on AHE.

In addition to AHE, gate-tunable magnetism also extends the capability of probing the interplay between ferromagnetism and localization, in which the question of how Anderson-Mott localization affects magnetic correlations is critical [43]. In terms of (Ga,Mn)As, this issue is more intriguing due to the debate on its band structure evolution with the Mn concentration, the so-called battle of the band [44, 45]. Specifically, whether an Mn-relevant impurity band resides inside the band gap and detaches from the valence band of (Ga,Mn)As is the focus. Unfortunately, previous experiments based on (Ga,Mn)As samples gave contradictory conclusions, which might be attributed to the diverse magnetic properties sensitive to growth conditions [43–45]. This artifact could be avoided by using a single sample with electric field control of magnetism, during which the charge occupation situation of the electronic energy levels can be tuned by the electric field. Experimentally, direct magnetization measurements were used to detect the variation of magnetism as a function of the hole density, and the gate voltage modulation of T_C, saturation magnetization, and the electrical demagnetization were confirmed firstly by a SQUID magnetometer [43]. By depleting the (Ga,Mn)As channel, a monotonic decrease of T_C was observed, a negative evidence for the impurity band model. Further study indicated that a superparamagnetic-like (SPM-like) spin arrangement emerged as the ferromagnetism of (Ga,Mn)As was weakened, and the coexistence of these two magnetic regions in the sample was verified by the field-cooled measurements of the magnetic moments. The anisotropy of this SPM-like portion and its gate tunability were also revealed and quantitatively investigated as shown in Figure 10.5, and the phase separation was assigned to the quantum critical fluctuations in the local density of hole states. Hence, the loss of spontaneous magnetization was induced by the increase of the electronic disorder, in the vicinity of the Anderson-Mott metal to insulator transition (MIT) point.

The above research is mainly concerned with the static magnetic properties of (Ga,Mn)As: electric fields have been found to have significant influence on them. From the viewpoint of spin dynamics, whether gating method can still play an important role needs to be clarified. In spin dynamics, Gilbert damping constant α as a fundamental parameter is crucial for the performance of spintronic devices, which determines the coherence and relaxation of collective spins. Despite intensive research, the mechanism that determines α remains elusive, and only the materials-dependent feature of α is established. If an electric field can be used to control α, it would be helpful to elucidate the mechanism since it could provide a set of data from the same piece of sample. Based on similar reasons, (Ga,Mn)As ultrathin film is also an ideal material for this kind of study, during which the gate voltage effect on α in (Ga,Mn)As MIS structure could be determined by the technique of ferromagnetic resonance [46]. It was found that the modulation magnitude of α was much larger than other magnetic parameters in the same device, including the saturation magnetization, magnetic anisotropic

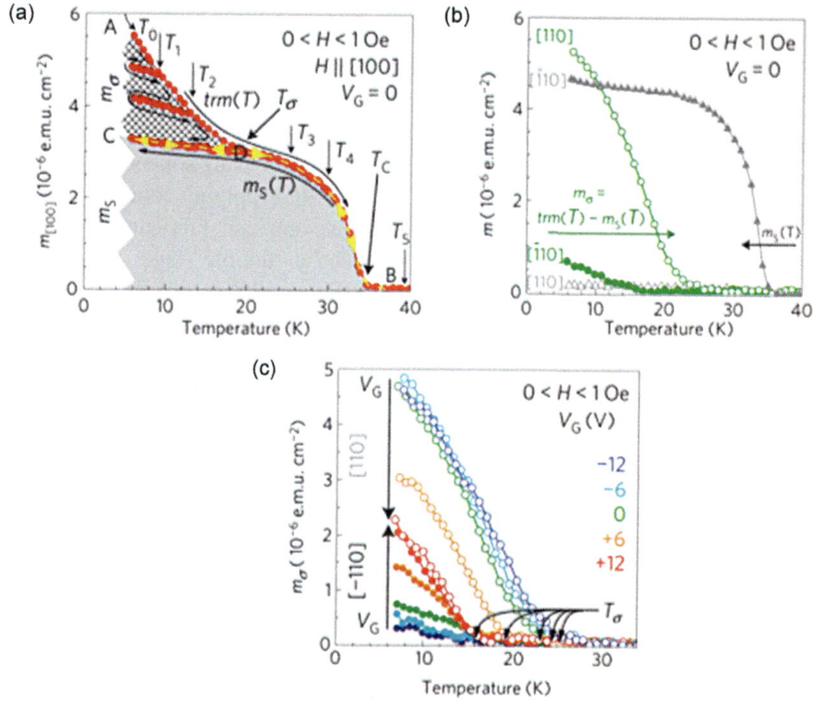

Figure 10.5 (a) Temperature-dependent magnetization of a field cooled (Ga,Mn)As film, in which the arrows indicate the trace that temperature was swept. (b) Temperature dependence of spontaneous and SPM-like magnetization. (c) Demonstration of gate-controlled transition temperature and the anisotropy of SPM-like magnetism [43]/with permission of Springer Nature.

fields, as well as the Lande g factor (see Figure 10.6). This result can be explained by the change of the SPM-like region as well, which is again determined by the disorder related to the Anderson-Mott MIT process.

As can be seen, most of the important magnetic parameters have been proved to be tunable by electric field. However, the applied voltage is still too high, in the range of several tens of volts to hundreds of volts, depending on both the employed dielectrics and hole density of the channel materials. To decrease the gate voltage to several volts without weakening the electric field effect on magnetism, designing heterostructures consisting of the channel material is then essential. For instance, by incorporating (Ga,Mn)As into a pn junction, a large depletion of holes was demonstrated at only a few volts [47, 48]. Figure 10.7(a) shows the schematic of the device, a Corbino disk is used to for the accurate evaluation of the anisotropic magnetoresistance (AMR), and the main idea here is to using the built-in electric field inside the pn junction [47]. Numerical simulation of the hole density profile in (Ga,Mn)As is presented in Figure 10.7(b–c), which shows significant change of the carrier distribution in (Ga,Mn)As while the gate voltage changes from −1 V to 3 V. Electrical modulation of T_C and AMR was observed, and magnetization switching induced by short voltage pulses of a few volts were realized (not shown) [47].

Another method resorts to the ionic liquid that allows the generation of a large interfacial electric field without applying a large voltage. The mechanism lies in that relatively low

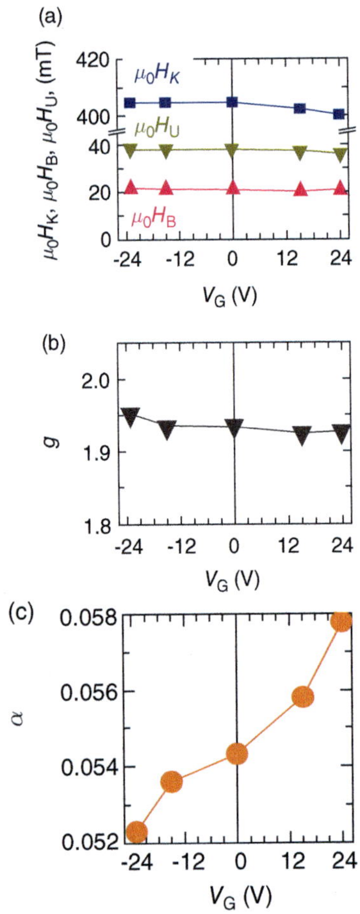

Figure 10.6 Gate-voltage dependence of (a) magnetic anisotropy fields, (b) Lande g factor, and (c) damping constants α [46]/with permission of American Physical Society.

voltage can drive the movable ions to the interface and then reside there, resulting in a short distance between the channel and the external charges. Previous reports prove that huge carrier variation up to 8×10^{14} cm^{-2} can be reached, about one order larger than the value obtained using solid state dielectrics [23, 29, 30, 32, 49]. Applying ionic liquid to (Ga,Mn)As thin films, one can fabricate a device common called electric double layer transistor (EDLT), as shown in Figure 10.8(a) [32]. Temperature dependence of the sheet resistance curves under various gate voltages shown in Figure 10.8(b) indicated an up to 14 K change of T_C, while the voltage changed from −1 V to 3 V. The modulation magnitude by using ionic liquid enlarges the gating range, and is meaningful for more systematic and thorough investigation of the mechanism of gate-tunable magnetism.

It is clear that the electric field effects on magnetism in (Ga,Mn)As are mainly limited by three factors, i.e., the lower-bound thickness of (Ga,Mn)As films, the breakdown field strength, and the electrical permittivity of dielectrics. Using compact and high-κ dielectrics enable people to observe the gate-controlled magnetism in (Ga,Mn)As, while the introduction of ionic liquid makes first demonstration of electric field control of coercivity possible in ferromagnetic metals FePd and FePt [50]. Nevertheless, the modulation range of T_C and magnetization via an electric field in (Ga,Mn)As is still moderate, in the range of ~ 10 K. In order to see whether it is possible to extend the modulation magnitude and even completely demagnetize the sample in the whole

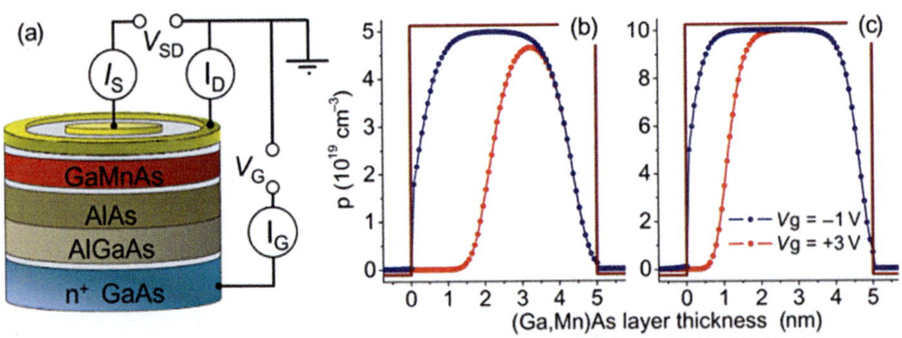

Figure 10.7 (a) Schematic of the ferromagnetic pn junction MIS structure and the Corbino disk geometry of the source and drain contacts. (b) and (c) Numerical simulations of the hole density profile at −1 V (accumulation) and + 3 V (depletion), considering a 2×10^{19} cm^{-3} electron doping in the n-GaAs and 5×10^{19} and 10^{20} cm^{-3} hole doping in the p-(Ga,Mn)As [47]/with permission of IOP Publishing.

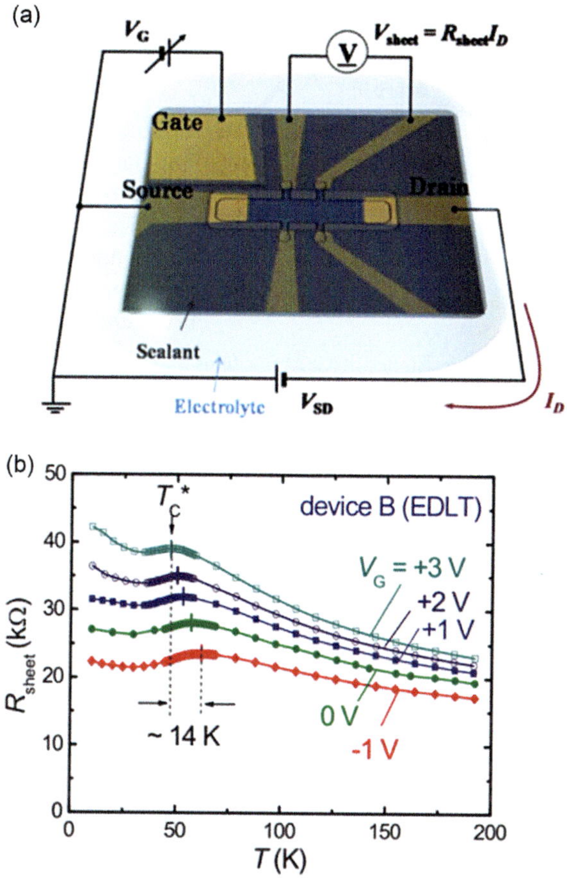

Figure 10.8 (a) Schematic EDLT device structure and measurement configuration. (b) Temperature dependence of sheet resistance for an EDLT device under various gate voltages [32]/with permission of AIP Publishing.

temperature range, one method is decreasing the thickness of (Ga,Mn)As without destroying its magnetism. Notice that (Ga,Mn)As is partially depleted at the interface between (Ga,Mn)As and the GaAs underlayer, grown at low temperature, so the effective thickness of the (Ga,Mn)As thin film that contributes ferromagnetism is actually smaller than its nominal thickness. This might also be one of the reasons that make (Ga,Mn)As lose its magnetism and tend to be insulating while its thickness is less than ~ 3 nm [29, 42].

To verify the above analyses, the basic magnetic and transport properties of (Ga,Mn)As ultrathin films (~2.5 nm), grown on a heavily Be-doped GaAs buffer layer (~3 nm), were investigated [51]. The substitution of a *p*-type GaAs buffer layer was found to be helpful for maintaining ferromagnetism in (Ga,Mn)As ultrathin films, which was then covered with ALD-grown HfO_2 and fabricated to EDLT devices. Temperature-dependent longitudinal resistance was used to inspect the variation of T_C with different gate biases, as shown in Figure 10.9. For the sample with high Mn doping concentration (~11%), gate voltages of ± 2 V induce obvious variation of T_C, with its magnitude up to ~ 56 K. This large modulation range of T_C was not realized in thicker (Ga,Mn)As films before. Moreover, for the sample with nominally 8% Mn content, a 27 K enhancement of T_C was achieved by applying a

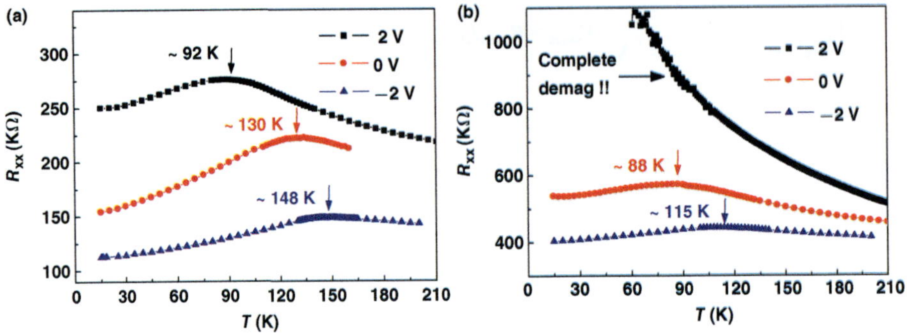

Figure 10.9 Temperature dependence of longitudinal resistance in (Ga,Mn)As films under various gate voltages, with Mn contents of (a) 8% and (b) 11% [51]/with permission of IOP Publishing.

voltage of −2 V (increasing hole density), while the channel became insulating with + 2 V gate voltage at temperatures lower than ~ 60 K. This insulating state might be the signature of complete demagnetization of (Ga,Mn)As layer in the whole temperature range, which means an unprecedented modulation magnitude of T_C (~115 K) [51].

Furthermore, based on (Ga,Mn)As quantum dots (QDs), superparamagnetism (SPM) has also been demonstrated [52]. It is well known that SPM emerges with the miniaturization of a magnetic unit cell, and it is therefore of particular interest to investigate the electric field effects in this regime [53]. The ensemble of (Ga,Mn)As QDs naturally shows superparamagnetic behaviors at temperatures higher than the so-called blocking temperature (T_B), and therefore acts as a good material system for studying electric field effects on SPM. (Ga,Mn)As QDs were self-assembled grown by MBE employing the droplet method, during which the heavily doped p-type Si substrate was used [54, 55]. A thin layer of HfO_2 was subsequently deposited on the sample, and an EDLT device was fabricated. By applying gate voltages, a 15 K change of T_B measured by zero field cooling (ZFC) was observed, indicating the modulation of SPM, as shown in Figure 10.10. This result could be explained by the electric field-enhanced magnetic anisotropic field, which was a dominant factor that determined T_B.

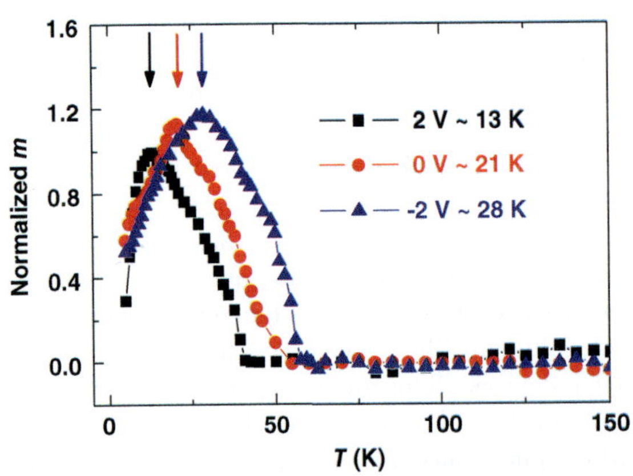

Figure 10.10 Zero field cooling curve of (Ga,Mn)As QDs under three gate voltages, indicating the electric field control of superparamagnetism [52].

Finally, integrating ferroelectric and ferromagnetic materials together to form a multiferroic structure was used to demonstrate nonvolatile modulation of magnetism by an electric field [56–60]. Figure 10.11(a) shows the cross section of the device consisting of a ferroelectric layer P(VDF-TrFE) in direct contact with (Ga,Mn)As channel and a top gate Au electrode. The 200 nm thick ferroelectric layer was simply spin-coated onto the surface of (Ga,Mn)As, and the applied gate voltage is ± 30 V, relatively low compared with the one using solid state dielectrics. In this case, the applied voltage induces a change of the electric polarization of P(VDF-TrFE), leading to an accumulation of effective holes or electrons at the interface. These accumulated charges at the interface would then result in an increase or decrease of holes in (Ga,Mn)As layer, resembling the mechanism of using ionic liquid. Similarly, depletion (accumulation) of holes in (Ga,Mn)As suppressed (enhanced) its ferromagnetism, as indicated by the temperature-dependent sheet resistance curves shown in Figure 10.11(b). The interesting aspect is the nonvolatile feature of this technique, which means that turning off the gate voltage back to zero would not change the magnetic state of (Ga,Mn)As layer, a very promising way for magnetic storage and spintronic devices.

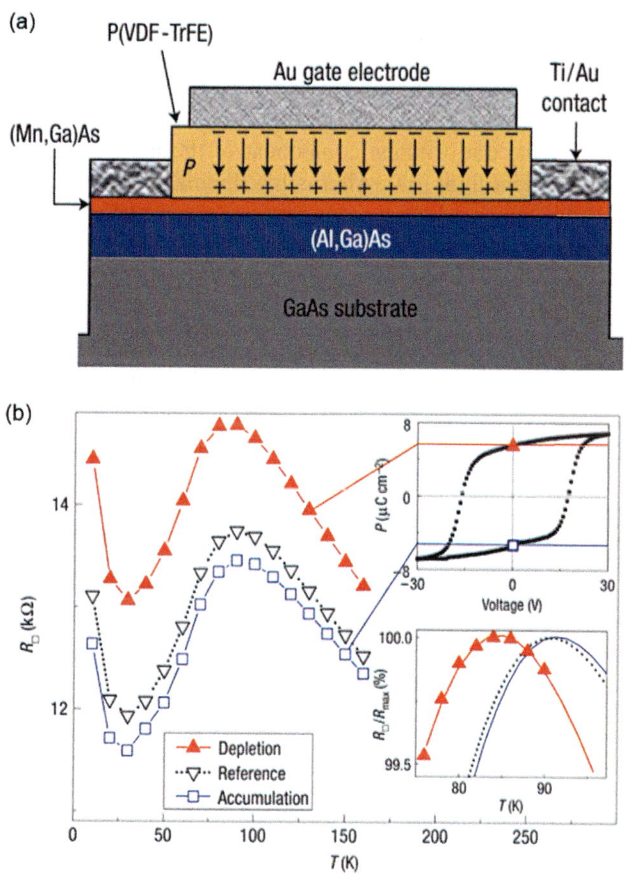

Figure 10.11 (a) Schematic of a MIS device consisting of a ferroelectric layer acting as the gate insulator, providing a nonvolatile feature. (b) Temperature-dependent sheet resistance of (Ga,Mn)As film after setting three different initial magnetic states, demonstrating the nonvolatile characteristic of the device [56]/with permission of Springer Nature.

10.3 Manipulation of the Magnetism in (Ga,Mn)As by Light and Strain

Light can excite carriers, one typical result would be the persistent photoconductivity (PPC) in some semiconductors [61]; light carries energy and is capable of heating materials locally, a viable method for crystallization of materials in nanoscale [62]; light can also transfer angular momentum to electrons, which could be used as an efficient way to create net spins (and the related net magnetic moments) in nonmagnetic materials [63]. As for magnetic semiconductors, light was readily used to manipulate their magnetic properties at the early stage [63–66]. For instance, reversible light tuning of magnetism was reported in II–VI DMS in modulation-doped p-type (Cd,Mn)Te-based heterostructures, whose mechanism was ascribed to the light-induced hole transfer inside the device [63, 64]. In the case of (III,Mn)V DMS, (In,Mn)As/GaSb heterostructure was one of the first reports on the tunable magnetism by light, during which the light increased the hole density in (In,Mn)As and enhanced its ferromagnetism, as shown in Figure 10.12 [66]. The PPC in (In,Mn)As/GaSb caused by light involved the trapping of photoelectrons at deep levels and was usually destroyed through increasing the temperature, which was therefore not a reversible process.

Circularly polarized light were shown to be capable of manipulating the magnetization in (Ga,Mn)As and Mn δ-doped GaAs reversibly, which was evidenced by Hall measurement [30, 67]. A ~ 15% change of the magnetization was found due to the optically generated spin-polarized holes, which means that the out-of-plane component of the magnetization was dependent on the helicity of the incident light [67]. Alternating illumination of σ^+ and σ^- polarized light induced an alternating change of the longitudinal resistance and magnetization as well, as indicated by the response of the Hall resistance (see Figure 10.13).

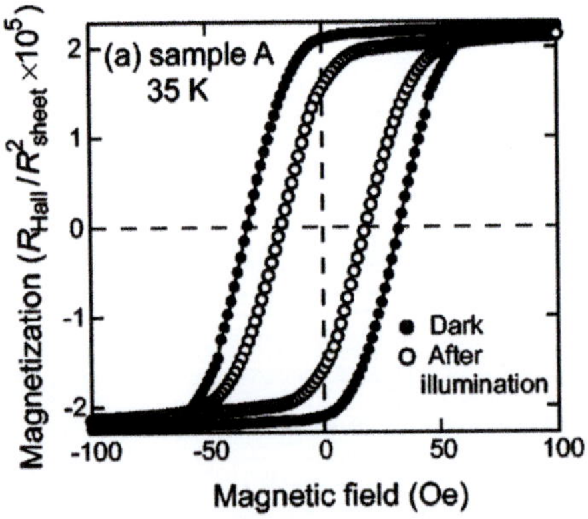

Figure 10.12 Magnetization hysteresis curves before and after light illumination in the sample of (In,Mn)As/GaSb [66]/with permission of AIP Publishing.

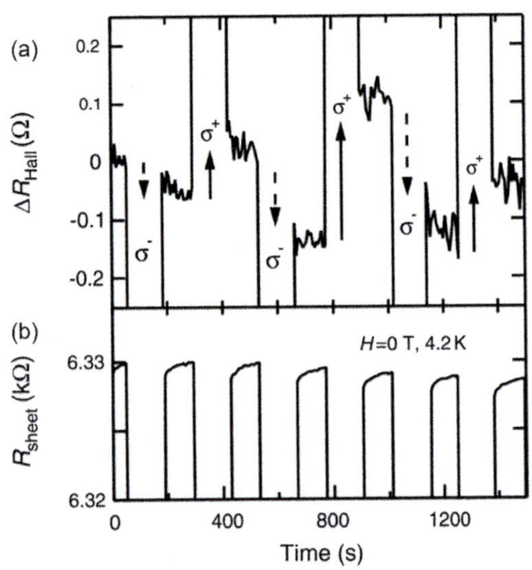

Figure 10.13 Responses of (a) Hall resistance and (b) sheet resistance in (Ga,Mn)As film to an alternating polarized light illumination [67]/with permission of American Physical Society.

Moreover, all-optical switching of magnetization was also realized in (Ga,Mn)As with cubic anisotropy, in which the technique of scanning laser magnetooptical microscope was employed [68]. In this experiment, the lithography-induced magnetic anisotropy played a critical role, and the main result of shedding light was heating the sample. By setting a proper initial state of the magnetization and precisely controlling the heating effect, a 90° switching of magnetization was observed in the absence of magnetic field. All-optical switching of magnetization would be useful for generating artificial magnetic structures that were difficult for other methods, considering its local manipulation feature [68].

Ultrafast manipulation of magnetization in (Ga,Mn)As is particularly interesting, and has been demonstrated in the form of transient Kerr effect [69]. Nonequilibrium spin polarization in (Ga,Mn)As could be excited optically, whose relaxation time was about 30 ps. The spectral characteristics of the Kerr effect induced by photoexcitation and via the magnetic field were observed to be very similar, and a pulsed photoexcitation with the influence of 10 $\mu J/cm^2$ was estimated to be ~ 10 mT. Furthermore, all-optical ultrafast switching of magnetization between four metastable magnetic states in (Ga,Mn)As was theoretically predicted [70]. In this proposal, the magnetization switching could be initiated within 100 fs and then completed ~ 100 ps later. Another scheme of ultrafast all-optical switching of magnetization using the spin selective optical Stark effect has also been proposed, whose details can be seen in reference [71]. There are also many other theoretical and experimental works that focus on light-induced magnetization manipulation, based on the transfer of energy and angular momentum from light to atoms or electrons [72–80].

Another appealing characteristic of (III,Mn)V DMS is their strain-sensitive magnetic anisotropy, which has been widely proved by lattice constant engineering during epitaxial growth and by post-growth lithography [5–10, 81–86]. Especially interesting is the electrical controllable magnetic anisotropy in the hybrids consisting of a (Ga,Mn)As film and a piezoelectric actuator layer [87–92]. Similar with the method that integrating (Ga,Mn)As with ferroelectrics, these hybrid devices are also nonvolatile.

Experimentally, significant change of AMR was used to manifest the magnetic anisotropy variation by *in situ* voltage-controlled strain [87], and then the magnetization rotation was calculated. It was found the strain resulted in a rotation angle of up to ~ 70° of the magnetic easy axis, while the applied voltage was in the range of ± 200 V [87]. In addition, the magnetization was then switched via sequential application of magnetic fields and voltages, although in an irreversible manner [93]. More intriguingly, ultrafast picosecond strain pulses were generated by using a femtosecond laser, which was theoretically analyzed and experimentally used to excite spin waves and cause magnetization precession in (Ga,Mn)As and (Ga,Mn)(As,P) films [94–97]. The frequency of the induced resonant precession of the magnetization in (Ga,Mn)As (311) films was in the order of 10 GHz, and the modulation of out-of-plane magnetization component was ~ 10%, which provided a promising way for the ultrafast magnetization control [97].

10.4 Giant Modulation of Magnetism via Organic Molecules

As mentioned above, the modulation capability of carrier density via an electric field is immensely restricted by the breakdown strength and permittivity of dielectrics. Although ionic liquid as well as its quasi-solid state (ionic gel) extend the upper limit of this method, the need for broadening the modulation range further is still urgent. Interestingly, a large family of organic molecules have already come into the field of metal spintronics, acting as an efficient spin filter, charge donors or acceptors. For instances, spin polarization was significantly enhanced at the interface of a hybrid junction consisting of ferromagnetic metal and nonmagnetic molecules [98–100]. And it is also found that some molecules can even modify the surface magnetic anisotropy or induce interfacial magnetic moments [101, 102]. In addition, a variety of molecules exhibits the capability of tuning carrier density to a large extent, exceeding the range reached by traditional dielectrics and even ionic liquid. So far, most of the previous works focus on the influences of organic molecules on the magnetism in magnetic or nonmagnetic metals, among which the relevant effects are usually small. Considering that the carrier density of semiconductors is several orders of magnitude lower than that of metals, the tuning effects would be much larger by combining molecules and (Ga,Mn)As together. In detail, two kinds of organic molecules, i.e., 16-mercaptohexadecanoic acid (MHA) and 2,3,5,6-tetrafluoro-7,7,8,8-tetracyanoquinodimethane (F4-TCNQ), were employed as electron donors and acceptors, respectively. Figure 10.14 shows the molecular structures of MHA and F4-TCNQ, in which the thiol and CN radicals play important roles in transferring charges to the adjacent layers [103].

The main obstacle for creating an effective molecule/(Ga,Mn)As heterostructure is to prepare a clean and properly treated (Ga,Mn)As surface, which is one of the prerequisites for forming an ideal interface [103]. For the decoration of self-assembled MHA molecules, the native oxide on (Ga,Mn)As samples is usually removed by ammonium sulfide solution, which not only has good etching selectivity but also leaves a sulfur passivation layer bonded to the sample surface. This passivation layer is crucial for the successful assembly of MHA molecules, which is characterized by the uniformity of the molecule coverage. Large area deposition of MHA molecules could simply be accomplished by immersing (Ga,Mn)As film into the MHA-ethanol solution with precisely controlled duration, while nanoscale patterning was achieved by a technique called dip-pen nanolithography (DPN). DPN allows one to write down various patterns with specially designed tips, which could also facilitate the study of nano-magnets, as demonstrated in the work of defining magnetic nanodots via making (Ga,Mn)As films into devices with a mesh gate [39]. As for the fabrication of F4-TCNQ/(Ga,Mn)As hybrid structures, diluted HCl solution was used to remove the surface oxide, for

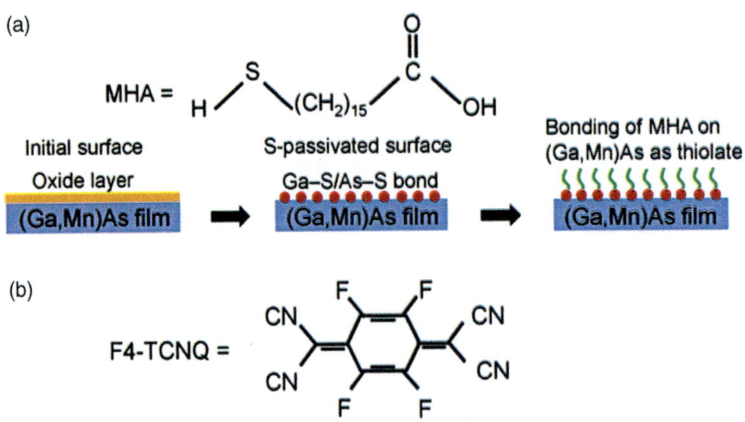

Figure 10.14 (a) Molecular structure of MHA and schematics of its self-assembly process. (b) Molecular structure of F4-TCNQ [103]/with permission of John Wiley & Sons.

the aim of avoiding any unwanted chemical residuals. Thermal evaporation with relatively small deposition rate can cover a compact layer of F4-TCNQ on (Ga,Mn)As with good uniformity. The interface bonding situations can be detected by x-ray photoelectron spectroscopy (XPS), and the double peaks appearing at specific energy can be used to recognize the existence of the Ga-S and As-S bonds, as well as the replacement of sulfur by the thiol radicals of MHA. To verify the electron transfer from F4-TCNQ to (Ga,Mn)As, the valence state of N was measured by XPS, and the appearance of N^{-1} anion state implied the existence of Ga-N bonds (N species were neutral in F4-TCNQ). In addition, the neutral N^0 state weight more and more when the thickness of F4-TCNQ was increased, indicating that the charge transfer effect is restricted to the interface.

Similar with the electrical gating results, molecules could also alter T_C and saturation magnetization of (Ga,Mn)As, as shown in Figure 10.15(a–b). The film thickness is ~ 9 nm with nominal Mn concentration of ~ 4%. As can be seen directly from the temperature-dependent magnetization curves, T_C and Ms of (Ga,Mn)As sample covered by MHA were decreased, while both of them were enhanced by F4-TCNQ. These results are interesting, since electric field gating experiments, using such thick (Ga,Mn)As films with relatively high carrier density, usually come out with an unobservable change of magnetism. Such results have not been reported before. As for the transport measurements, MHA (F4-TCNQ) increased (decreased) the resistivity of the film as shown in Figure 10.17 (c–d), which was consistent with the above magnetization data. The hole density in (Ga,Mn)As could be determined by fitting the high field Hall data, which gave direct

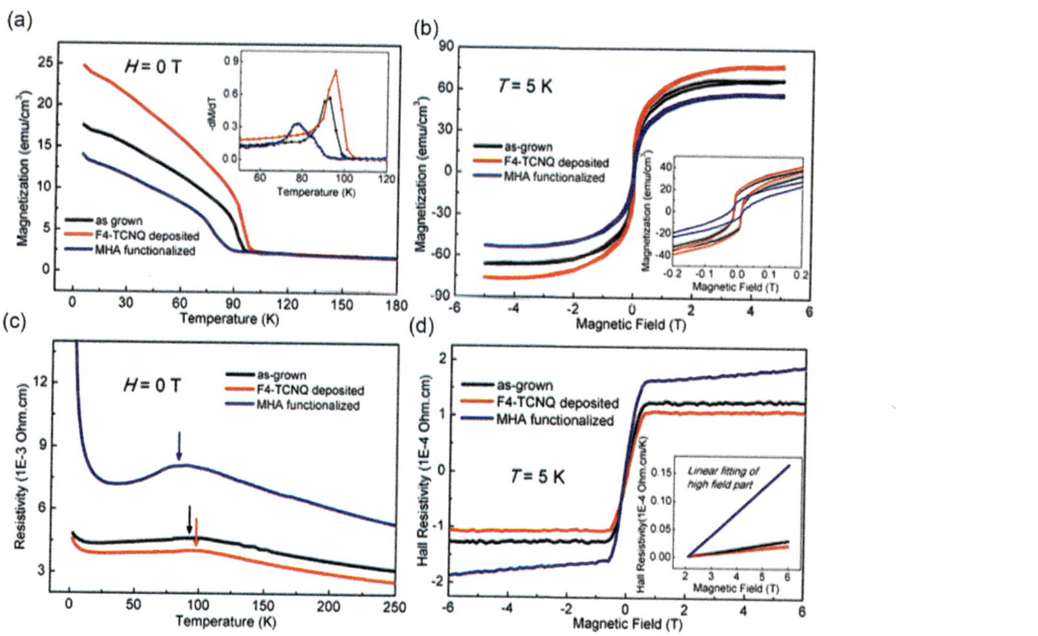

Figure 10.15 (a) Temperature dependence of the remnant magnetization in a 9 nm thick (Ga,Mn)As film with or without molecule decorations. The inset is the temperature derivative of the remnant magnetization as a function of temperature. (b) Magnetization hysteresis loops measured at 5 K for the same three pieces of samples in (a). (c) Corresponding sheet resistivity and (d) Hall resistivity measured at 5 K. The inset in (d) shows the linear fittings of the high field data, indicating a clear change of the hole concentrations [103]/ with permission of John Wiley & Sons.

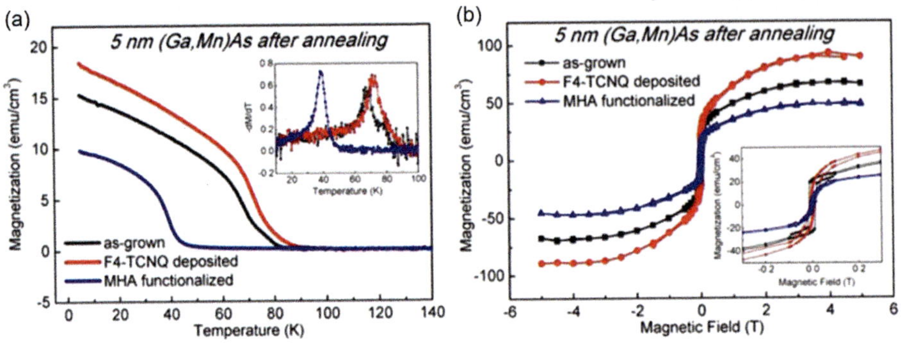

Figure 10.16 (a) Temperature dependence of remnant magnetization and (b) magnetic hysteresis loop of a 5 nm thick (Ga,Mn)As film with and without molecule decorations. The inset in (a) is the temperature derivative of the magnetization, in which the variation of T_C is determined to be ~ 36 K [103]/with permission of John Wiley & Sons.

evidence for the depletion (accumulation) of holes by these molecules. The variation of hole density was as large as ~ 10^{15} cm^{-2}, which was more than one order larger than the largest value realized by solid dielectrics [23]. It is worth mentioning that the larger anomalous Hall signal of MHA-decorated (Ga,Mn)As is mainly due to the increase of anomalous Hall coefficient (closely related to the longitudinal resistivity), although the magnetization is weakened by MHA.

It was found that suppressing the magnetism in (Ga,Mn)As films was easier than its enhancement, which might mainly be ascribed to the large hole density that was sufficient for maintaining the long range ferromagnetic order. And when decreasing the film thickness from 9 nm to 5 nm, a largest variation of T_C up to ~ 36 K was achieved, which was several times of the value obtained in gating experiments (see Figure 10.16) [32, 39, 104]. The much larger modulation effects might originate from the huge amount of charge transfer through the direct bonding between the interfacial atoms of (Ga,Mn)As and the radicals of molecules, in which the strong internal electric field seems to be bearable for these bonds. Deeper understanding of the detailed mechanism calls for further investigations.

According to the theory of carrier-mediated ferromagnetism in (III,Mn)V DMS, the magnetic anisotropy of (Ga,Mn)As will also be altered by these two molecules, as verified by a recent experimental work [5–10, 105]. Again, the methods based on the planar Hall data are used to analyze the change of magnetic anisotropy by decorating (Ga,Mn)As with different molecules [106]. As expected, the molecules indeed alter the magnetic anisotropy significantly in a large temperature range, which is summarized in Figure 10.17. The magnetization angles of (Ga,Mn)As film with different conditions are shown in Figure 10.17(d), and the largest rotation angle of the magnetization direction is up to 27°, which is almost three times larger than the gating results [106].

10.5 Conclusion and Outlook

In this review chapter, we have looked back on the motivation, original idea, experimental discovery, and subsequent developments in the field of manipulating magnetism in (Ga,Mn)As. Based on the coupling between various physical quantities and magnetic moments, diverse

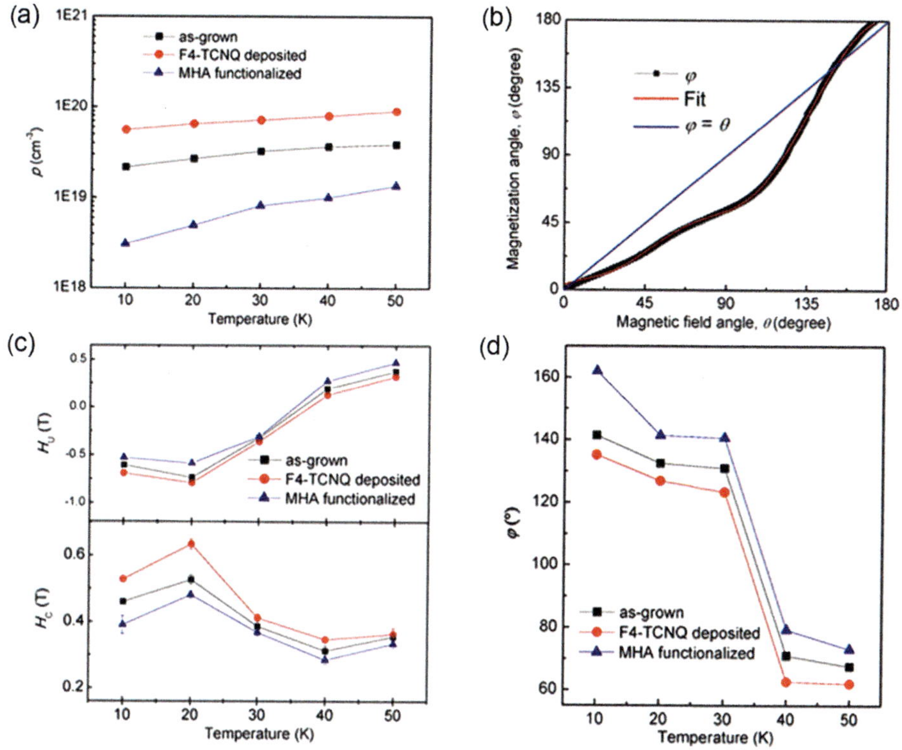

Figure 10.17 (a) Manipulation of carrier concentration by F4-TCNQ deposition and MHA functionalization at different temperatures. (b) Magnetization angle φ as a function of magnetic field angle θ and its fitting curve (red), the $\varphi = \theta$ (blue) line is shown for comparison. (c) Temperature dependence of the uniaxial and cubic magnetic anisotropic field, showing the modulation of magnetic anisotropy by the two types of molecules. (d) Temperature dependence of the magnetization angle through molecular decoration [106]/with permission of American Chemical Society.

methods – such as an electric field, light, strain, and even forming hybrids with organic molecules – were explored to manipulate the magnetic properties of (Ga,Mn)As. Basic magnetic parameters, including T_C, magnetization magnitude, and magnetic anisotropy (and thus the magnetization direction), have been shown to be controllable. The effective tuning of magnetism provides us with opportunities to investigate some phenomena which were inaccessible before, facilitating the understanding of many debated topics in spintronics.

Despite all this achievement and progress in (III,Mn)V DMS, and especially in (Ga,Mn)As, robust and effective manipulation of magnetism in (Ga,Mn)As – or, more generally, in magnetic semiconductors and metals – is still challenging.

Firstly, new mechanisms should be established to achieve an enhanced modulation magnitude of magnetism. For instance, imagine that T_C of (III,Mn)V DMS is raised to above room temperature at the price of very high hole density. The electric field will then be strongly screened, resulting in a minor change of magnetic properties in the scheme of tuning magnetism simply by changing carrier density. The developments of modulating magnetism in ferromagnetic metals might provide some inspiration, in which new ideas of varying the occupation situation of electronic orbitals by electric field could be utilized to effectively manipulate magnetism, in spite of their larger electron density [50, 107–115].

Secondly, combining the diverse manipulation methods more effectively would be useful, as one technique cannot fully meet the requirements of application. This might be difficult at first glance, but could possibly be adopted in the final product. Heat-assisted magnetic recording (HAMR) is a good example, which was initially considered infeasible, due to its requirement of local heating and cooling in nanoseconds at nanoscale, but then proved to be very promising for improving data storage density [116, 117]. One such effort in spintronics is switching magnetization by magnetic field, assisted by a pulse of electric field, as already demonstrated in magnetic tunnel junctions [118–122].

Thirdly, how to efficiently transfer the already-established manipulation techniques in ferromagnets to other materials with different spin-related long range ordering is still unknown. This is essential and urgent as numerous new storage schemes, based on other collective phenomena, are quickly emerging. For instance, antiferromagnetic materials, such as CuMnAs, Mn_2Au, and NiO, were found to be very interesting due to the discovery of electrical magnetization switching, although the critical current density is still too high [123–126]. In this sense, methods of effectively manipulating Neel vectors (or antiferromagnetism) in a variety of ways, are crucial for the application of these new materials [127–135].

Great opportunity always accompanies big challenges. We hope and believe that there will be rapid progress in this field.

Acknowledgements

This work was supported by NSFC (Grants No. 12174383 and 11834013) and CAS Project for Young Scientists in Basic Research (YSBR-030).

References

1 Prinz, G.A. (1998). Magnetoelectronics. *Science* 282: 1660–1663.
2 Wolf, S.A., Awschalom, D.D., Buhrman, R.A. et al. (2001). Spintronics: a spin-based electronics vision for the future. *Science* 294: 1488–1495.
3 Zutic, I., Fabian, J., and Sarma, S.D. (2004). Spintronics: fundamentals and applications. *Rev. Mod. Phys.* 76: 323–410.
4 Awschalom, D.D. and Flatte, M.E. (2007). Challenges for semiconductor spintronics. *Nat. Phys.* 3: 153–159.
5 Ohno, H. (1998). Making nonmagnetic semiconductors ferromagnetic. *Science* 281: 951–956.
6 Dietl, T., Ohno, H., Matsukura, F., Cibert, J., and Ferrand, D. (2000). Zener model description of ferromagnetism in zinc-blende magnetic semiconductors. *Science* 287: 1019–1022.
7 Jungwirth, T., Sinova, J., Masek, J., Kucera, J., and MacDonald, A.H. (2006). Theory of ferromagnetic (III,Mn)V semiconductors. *Rev. Mod. Phys.* 78: 809–864.
8 Sato, K., Bergqvist, L., Kudrnovsky, J. et al. (2010). First-principles theory of dilute magnetic semiconductors. *Rev. Mod. Phys.* 82: 1633–1690.
9 Dietl, T. and Ohno, H. (2014). Dilute ferromagnetic semiconductors: physics and spintronic structures. *Rev. Mod. Phys.* 86: 187–251.

10 Jungwirth, T., Wunderlich, J., and Novak, V. (2014). Spin-dependent phenomena and device concepts explored in (Ga,Mn) As. *Rev. Mod. Phys.* 86: 855–896.

11 Jungwirth, T., Wang, K.Y., Masek, J. et al. (2005). Prospects for high temperature ferromagnetism in (Ga,Mn)As semiconductors. *Phys. Rev. B* 72: 165204.

12 Wang, H.L., Chen, L., and Zhao, J.H. (2013). Enhancement of the Curie temperature of ferromagnetic semiconductor (Ga,Mn)As. *Sci. China-Phys. Mech. Astron.* 56: 99–110.

13 Wang, H.L., Wang, X.L., Xiong, P., and Zhao, J.H. (2016). Control of magnetism in dilute magnetic semiconductor (Ga,Mn)As films by surface decoration of molecules. *Front. Phys.* 4: 00009.

14 Ohno, H., Chiba, D., Matsukura, F., Omiya, T., Abe, E., Dietl, T., Ohno, Y., and Ohtani, K. (2000). Electric-field control of ferromagnetism. *Nature* 408: 944–946.

15 Chiba, D., Yamanouchi, M., Matsukura, F., and Ohno, H. (2003). Electrical manipulation of magnetization reversal in a ferromagnetic semiconductor. *Science* 301: 943–945.

16 Matsukura, F., Tokura, Y., and Ohno, H. (2015). Control of magnetism by electric fields. *Nat. Nanotechnol.* 10: 209–220.

17 Ralph, D.C. and Stiles, M.D. (2008). Spin transfer torques. *J. Magn. Magn. Mater.* 320: 1190–1216.

18 Chernyshov, A., Overby, M., Liu, X.Y., Furdyna, J.K., Lyanda-Geller, Y., and Rokhinson, L.P. (2009). Evidence for reversible control of magnetization in a ferromagnetic material by means of spin-orbit magnetic field. *Nat. Phys.* 5: 656–659.

19 Miron, I.M., Gaudin, G., Auffret, S., Rodmacq, B., Schuhl, A., Pizzini, S., Vogel, J., and Gambardella, P. (2010). Current-driven spin torque induced by the Rashba effect in a ferromagnetic metal layer. *Nat. Mater.* 9: 230–234.

20 Brataas, A., Kent, A.D., and Ohno, H. (2012). Current-induced torques in magnetic materials. *Nat. Mater.* 11: 372–381.

21 Methfessel, S. (1965). Potential applications of magnetic rare earth compounds. *IEEE Trans. Magn.* 1: 144–155.

22 Ascher, E., Rider, H., Schimid, H., and Stossel, H. (1966). Some properties of ferromagnetoelectric nickel-iodine boracite $Ni_3B_7O_{13}I$. *J. Appl. Phys.* 37: 1404–1405.

23 Ohno, H. (2010). A window on the future of spintronics. *Nat. Mater.* 9: 952–954.

24 Dietl, T. (2010). A ten-year perspective on dilute magnetic semiconductors and oxides. *Nat. Mater.* 9: 965–974.

25 Geim, A.K. (2011). Nobel Lecture: random walk to graphene. *Rev. Mod. Phys.* 83: 851–862.

26 Nagaosa, N., Sinova, J., Onoda, S., MacDonal, A.H., and Ong, N.P. (2010). Anomalous Hall effect. *Rev. Mod. Phys.* 82: 1539–1592.

27 Xiao, D., Chang, M.C., and Niu, Q. (2010). Berry phase effects on electronic properties. *Rev. Mod. Phys.* 82: 1959–2007.

28 Chiba, D., Yamanouchi, M., Matsukura, F., Abe, E., Ohno, Y., Ohtani, K., and Ohno, H. (2003). Electric field effect on the magnetic properties of III–V ferromagnetic semiconductor (In, Mn)As and ((Al),Ga,Mn)As. *J. Supercond.* 16179–16182.

29 Chiba, D., Matsukura, F., and Ohno, H. (2006). Electric-field control of ferromagnetism in (Ga,Mn)As. *Appl. Phys. Lett.* 89: 162505.

30 Nazmul, A.M., Kobayashi, S., Sugahara, S., and Tanaka, M. (2004). Electrical and optical control of ferromagnetism in III–V semiconductor heterostructures at high temperature (~100 K). *Jpn. J. Appl. Phys.* 43: L233–L236.

31 Nazmul, A.M., Kobayashi, S., Sugahara, S., and Tanaka, M. (2004). Control of ferromagnetism in Mn delta-doped GaAs-based semiconductor heterostructures. *Phys. E* 21: 937–942.

32 Endo, M., Chiba, D., Shimotani, H., Matsukura, F., Iwasa, Y., and Ohno, H. (2010). Electric double layer transistor with a (Ga,Mn)As channel. *Appl. Phys. Lett.* 96: 022515.

33 Endo, M., Chiba, D., Nishitani, Y., Matsukura, F., and Ohno, H. (2007). Channel thickness dependence of the magnetic properties in (Ga,Mn)As FET structures. *J. Supercond.* 20: 409–411.

34 Chiba, D., Sawicki, M., Nishitani, Y., Nakatani, Y., Matsukura, F., and Ohno, H. (2008). Magnetization vector manipulation by electric fields. *Nature* 455: 515–518.

35 Ma, J.L., Wang, H.L., Yu, Z.F., Wang, X.L., and Zhao, J.H. (2017). Magnetic properties of (Ga,Mn)As (110) epitaxial films. *Europhys. Lett.* 118: 17003.

36 Chiba, D., Nakatani, Y., Matsukura, F., and Ohno, H. (2010). Simulation of magnetization switching by electric-field manipulation of magnetic anisotropy. *Appl. Phys. Lett.* 96: 192506.

37 Balestriere, P., Devolder, T., Wunderlich, J., and Chappert, C. (2010). Electric field induced anisotropy modification in (Ga,Mn)As: a strategy for the precessional switching of the magnetization. *Appl. Phys. Lett.* 96: 142504.

38 Chiba, D., Ono, T., Matsukura, F., and Ohno, H. (2013). Electric field control of thermal stability and magnetization switching in (Ga,Mn) As. *Appl. Phys. Lett.* 103: 142418.

39 Chiba, D., Matsukura, F., and Ohno, H. (2010). Electrically defined ferromagnetic nanodots. *Nano Lett.* 10: 4505–4508.

40 Tian, Y., Ye, L., and Jin, X.F. (2009). Proper scaling of the anomalous Hall effect. *Phys. Rev. Lett.* 103: 087206.

41 Hou, D.Z., Su, G., Tian, Y., Jin, X.F., Yang, S.Y.A., and Niu, Q. (2015). Multivariable scaling for the anomalous Hall effect. *Phys. Rev. Lett.* 114: 217203.

42 Chiba, D., Werpachowska, A., Endo, M., Nishitani, Y., Matsukura, F., Dietl, T., and Ohno, H. (2010). Anomalous Hall effect in field-effect structures of (Ga,Mn)As. *Phys. Rev. Lett.* 104: 106601.

43 Sawicki, M., Chiba, D., Korbecka, A., Nishitani, Y., Majewski, J.A., Matsukura, F., Dietl, T., and Ohno, H. (2010). Experimental probing of the interplay between ferromagnetism and localization in (Ga, Mn)As. *Nat. Phys.* 6: 22–25.

44 Ohya, S., Takata, K., and Tanaka, M. (2011). Nearly non-magnetic valence band of the ferromagnetic semiconductor GaMnAs. *Nat. Phys.* 7: 342–347.

45 Samarth, N. (2012). Battle of the bands. *Nat. Mater.* 11: 360–361.

46 Chen, L., Matsukura, F., and Ohno, H. (2015). Electric-field modulation of damping constant in a ferromagnetic semiconductor (Ga,Mn) As. *Phys. Rev. Lett.* 115: 057204.

47 Owen, M.H.S., Wunderlich, J., Novak, V. et al. (2009). Low-voltage control of ferromagnetism in a semiconductor p–n junction. *New J. Phys.* 11: 023008.

48 Olejnik, K., Owen, M.H.S., Novak, V., Masek, J., Irvine, A.C., Wunderlich, J., and Jungwirth, T. (2008). Enhanced annealing, high Curie temperature, and low-voltage gating in (Ga,Mn)As: a surface oxide control study. *Phys. Rev. B* 78: 054403.

49 Yuan, H.T., Shimotani, H., Tsukazaki, A., Ohtomo, A., Kawasaki, M., and Iwasa, Y. (2009). High-density carrier accumulation in ZnO field-effect transistors gated by electric double layers of ionic liquids. *Adv. Funct. Mater.* 19: 1046–1053.

50 Weisheit, M., Fahler, S., Marty, A., Souche, Y., Poinsignon, C., and Givord, D. (2007). Electric field-induced modification of magnetism in thin-film ferromagnets. *Science* 315: 349–351.

51 Wang, H.L., Ma, J.L., and Zhao, J.H. (2019). Giant modulation of magnetism in (Ga,Mn)As ultrathin films via electric field. *J. Semicond.* 40: 092501.

52 Wang, H.L., and Zhao, J.H., (2018). (unpublished).

53 Yamada, K.T., Koyama, T., Kakizakai, H. et al. (2017). Electrical control of superparamagnetism. *Appl. Phys. Express* 10: 013004.

54 Wang, S.L., Chen, L., Meng, K.K., Xu, P.F., Meng, H.J., Lu, J., Yan, W.S., and Zhao, J.H. (2010). Origin of ferromagnetism in self-assembled Ga1-xMnxAs quantum dots grown on Si. *Appl. Phys. Lett.* 97: 242505.

55 Wang, S.L., Chen, L., Meng, K.K., Xu, P.F., Meng, H.J., Lu, J., and Zhao, J.H. (2011). Fabrication of (Ga, Mn) As magnetic semiconductor quantum dots on Si substrates by droplet epitaxy. *Phys. Status Solidi C* 8: 393–395.

56 Stolichnov, I., Riester, S.W.E., Trodahl, H.J. et al. (2008). Non-volatile ferroelectric control of ferromagnetism in (Ga, Mn)As. *Nat. Mater.* 7: 464–467.

57 Riester, S.W.E., Stolichnov, I., Trodahl, H.J. et al. (2009). Toward a low-voltage multiferroic transistor: magnetic (Ga,Mn)As under ferroelectric control. *Appl. Phys. Lett.* 94: 063504.

58 Stolichnov, I., Riester, S.W.E., Mikheev, E. et al. (2011). Enhanced Curie temperature and nonvolatile switching of ferromagnetism in ultrathin (Ga,Mn)As channels. *Phys. Rev. B* 83: 115203.

59 Stolichnov, I., Riester, S.W.E., Trodahl, H.J. et al. (2011). Ferroelectric polymer gates for non-volatile field effect control of ferromagnetism in (Ga,Mn)As layers. *Nanotechnology* 22: 254004.

60 Mikheev, E., Stolichnov, I., Huang, Z. et al. (2012). Non-volatile ferroelectric gating of magnetotransport anisotropy in (Ga,Mn)(As,P). *Appl. Phys. Lett.* 100: 262906.

61 Misuraca, J., Trbovic, J., Lu, J., Zhao, J.H., Ohno, Y., Ohno, H., Xiong, P., and von Molnar, S. (2010). Band-tail shape and transport near the metal-insulator transition in Si-doped $Al_{0.3}Ga_{0.7}As$. *Phys. Rev. B* 82: 125202.

62 Knoff, W., Swiatek, K., Andrearczyk, T. et al. (2011). Magnetic anisotropy of semiconductor (Ge,Mn)Te microstructures produced by laser and electron beam induced crystallization. *Phys. Status Solidi B* 248: 1605–1608.

63 Haury, A., Wasiela, A., Arnoult, A. et al. (1997). Observation of a ferromagnetic transition induced by two-dimensional hole gas in modulation-doped CdMnTe quantum wells. *Phys. Rev. Lett.* 79: 511–514.

64 Boukari, H., Kossacki, P., Bertolini, M. et al. (2002). Light and electric field control of ferromagnetism in magnetic quantum structures. *Phys. Rev. Lett.* 88: 207204.

65 Koshihara, S., Oiwa, A., Hirasawa, M. et al. (1997). Ferromagnetic order induced by photogenerated carriers in magnetic III–V semiconductor heterostructures of (In,Mn)As/GaSb. *Phys. Rev. Lett.* 78: 4617–4620.

66 Oiwa, A., Slupinski, T., and Munekata, H. (2001). Control of magnetization reversal process by light illumination in ferromagnetic semiconductor heterostructure p-(In,Mn)As/GaSb. *Appl. Phys. Lett.* 78: 518–520.

67 Oiwa, A., Mitsumori, Y., Moriya, R., Slupinski, T., and Munekata, H. (2002). Effect of optical spin injection on ferromagnetically coupled Mn spins in the III–V magnetic alloy semiconductor (Ga, Mn)As. *Phys. Rev. Lett.* 88: 137202.

68 Aoyama, J., Kobayashi, S., and Munekata, H. (2010). All-optical 90-degree switching of magnetization in a ferromagnetic $Ga_{0.98}Mn_{0.02}As$ microbar. *J. Appl. Phys.* 107: 09C301.
69 Kimel, A.V., Astakhov, G.V., Schott, G.M. et al. (2004). Picosecond dynamics of the photoinduced spin polarization in epitaxial (Ga,Mn)As films. *Phys. Rev. Lett.* 92: 237203.
70 Kapetanakis, M.D., Lingos, P.C., Piermarocchi, C., Wang, J., and Perakis, I.E. (2011). All-optical four-state magnetization reversal in (Ga,Mn)As ferromagnetic semiconductors. *Appl. Phys. Lett.* 99: 091111.
71 Qaiumzadeh, A., Bauer, G.E.W., and Brataas, A. (2013). *Phys. Rev. B* 88: 064416.
72 Astakhov, G.V., Kimel, A.V., Schott, G.M. et al. (2005). Magnetization manipulation in (Ga,Mn)As by subpicosecond optical excitation. *Appl. Phys. Lett.* 86: 152506.
73 Munekata, H. (2005). Optical manipulation of coupled spins in magnetic semiconductor systems: a path toward spin optoelectronics. *Phys. E* 29: 475–482.
74 Wang, D.M., Ren, Y.H., Liu, X., Furdyna, J.K., Grimsditch, M., and Merlin, R. (2007). Light-induced magnetic precession in (Ga,Mn)As slabs: hybrid standing-wave Damon-Eshbach modes. *Phys. Rev. B* 75: 233308.
75 Rozkotova, E., Nemec, P., Holodyska, P. et al. (2008). Light-induced magnetization precession in GaMnAs. *Appl. Phys. Lett.* 92: 122507.
76 Hashimoto, Y. and Munekata, H. (2008). Coherent manipulation of magnetization precession in ferromagnetic semiconductor (Ga,Mn)As with successive optical pumping. *Appl. Phys. Lett.* 93: 202506.
77 Chovan, J. and Perakis, I.E. (2008). Femtosecond control of the magnetization in ferromagnetic semiconductors. *Phys. Rev. B* 77: 085321.
78 Zhu, Y.G., Zhang, X.H., Li, T. et al. (2009). Ultrafast dynamics of four-state magnetization reversal in (Ga,Mn)As. *Appl. Phys. Lett.* 95: 052108.
79 Kapetanakis, M.D., Perakis, I.E., Wickey, K.J., Piermarocchi, C., and Wang, J. (2009). Femtosecond coherent control of spins in (Ga,Mn)As ferromagnetic semiconductors using light. *Phys. Rev. Lett.* 103: 047404.
80 Matsuda, T. and Munekata, H. (2016). Mechanism of photoexcited precession of magnetization in (Ga,Mn)As on the basis of time-resolved spectroscopy. *Phys. Rev. B* 93: 075202.
81 Wenisch, J., Gould, C., Ebel, L. et al. (2007). Control of magnetic anisotropy in (Ga,Mn)As by lithography-induced strain relaxation. *Phys. Rev. Lett.* 99: 077201.
82 Edmonds, K.W., van der Laan, G., Farley, N.R.S. et al. (2008). Strain dependence of the Mn anisotropy in ferromagnetic semiconductors observed by x-ray magnetic circular dichroism. *Phys. Rev. B* 77: 113205.
83 Limmer, W., Daeubler, J., Dreher, L. et al. (2008). Advanced resistivity model for arbitrary magnetization orientation applied to a series of compressive- to tensile-strained (Ga,Mn)As layers. *Phys. Rev. B* 77: 205210.
84 Shiogai, J., Schuh, D., Wegscheider, W., Kohda, M., Nitta, J., and Weiss, D. (2011). Magnitude and sign control of lithography-induced uniaxial anisotropy in ultra-thin (Ga,Mn)As wires. *Appl. Phys. Lett.* 98: 083101.
85 De Ranieri, E., Rushforth, A.W., Vyborny, K. et al. (2008). Lithographically and electrically controlled strain effects on anisotropic magnetoresistance in (Ga,Mn)As. *New J. Phys.* 10: 065003.
86 van der Laan, G., Edmonds, K.W., Arenholz, E. et al. (2010). Valence-state model of strain-dependent Mn $L_{2,3}$ x-ray magnetic circular dichroism from ferromagnetic semiconductors. *Phys. Rev. B* 81: 214422.

87 Goennenwein, S.T.B., Althammer, M., Bihler, C. et al. (2008). Piezo-voltage control of magnetization orientation in a ferromagnetic semiconductor. *Phys. Stat. Sol. (RRL)* 2: 96–98.

88 Overby, M., Chernyshov, A., Rokhinson, L.P., Liu, X., and Furdyna, J.K. (2008). GaMnAs-based hybrid multiferroic memory device. *Appl. Phys. Lett.* 92: 192501.

89 Dengel, R.G., Gould, C., Wenisch, J., Brunner, K., Schmidt, G., and Molenkamp, L.W. (2008). Lateral magnetic anisotropy superlattice out of a single (Ga,Mn)As layer. *New J. Phys.* 10: 073001.

90 Rushforth, A.W., De Ranieri, E., Zemen, J. et al. (2008). Voltage control of magnetocrystalline anisotropy in ferromagnetic-semiconductor-piezoelectric hybrid structures. *Phys. Rev. B* 78: 085314.

91 King, C.S., Zemen, J., Olejnik, K. et al. (2011). Strain control of magnetic anisotropy in (Ga,Mn)As microbars. *Phys. Rev. B* 83: 115312.

92 Casiraghi, A., Rushforth, A.W., Zemen, J. et al. (2012). Piezoelectric strain induced variation of the magnetic anisotropy in a high Curie temperature (Ga,Mn)As sample. *Appl. Phys. Lett.* 101: 082406.

93 Bihler, C., Althammer, M., Brandlmaier, A. et al. (2008). $Ga_{1-x}Mn_xAs$/piezoelectric actuator hybrids: a model system for magnetoelastic magnetization manipulation. *Phys. Rev. B* 78: 045203.

94 Thevenard, L., Peronne, E., Gourdon, C. et al. (2010). Effect of picosecond strain pulses on thin layers of the ferromagnetic semiconductor (Ga,Mn)(As,P). *Phys. Rev. B* 82: 104422.

95 Linnik, T.L., Scherbakov, A.V., Yakovlev, D.R., Liu, X., Furdyna, J.K., and Bayer, M. (2011). Theory of magnetization precession induced by a picosecond strain pulse in ferromagnetic semiconductor (Ga,Mn)As. *Phys. Rev. B* 84: 214432.

96 Bombeck, M., Salasyuk, A.S., Glavin, B.A. et al. (2012). Excitation of spin waves in ferromagnetic (Ga,Mn)As layers by picosecond strain pulses. *Phys. Rev. B* 85: 195324.

97 Bombeck, M., Jager, J.V., Salasyuk, A.S. et al. (2013). Picosecond opto-acoustic interferometry and polarimetry in high-index GaAs. *Phys. Rev. B* 87: 060302.

98 Gohler, B., Hamelbeck, V., Markus, T.Z., Kettner, M., Hanne, G.F., Vager, Z., Naaman, R., and Zacharias, H. (2011). Spin selectivity in electron transmission through self-assembled monolayers of double-stranded DNA. *Science* 331: 894–897.

99 Dor, O.B., Yochelis, S., Mathew, S.P., Naaman, R., and Paltiel, Y. (2013). A chiral-based magnetic memory device without a permanent magnet. *Nat. Commun.* 4: 2256–2261.

100 Atodiresei, N., Brede, J., Lazic, P., Caciuc, V., Hoffmann, G., Wiesendanger, R., and Blugel, S. (2010). Design of the local spin polarization at the organic-ferromagnetic interface. *Phys. Rev. Lett.* 105: 066601.

101 Raman, K.V., Kamerbeek, A.M., Mukherjee, A. et al. (2013). Interface-engineered templates for molecular spin memory devices. *Nature* 493: 509–513.

102 Callsen, M., Caciuc, V., Kiselev, N., Atodiresei, N., and Blugel, S. (2013). Magnetic hardening induced by nonmagnetic organic molecules. *Phys. Rev. Lett.* 111: 106805.

103 Wang, X.L., Wang, H.L., Pan, D. et al. (2015). Robust manipulation of magnetism in dilute magnetic semiconductor (Ga,Mn)As by organic molecules. *Adv. Mater.* 27: 8043–8050.

104 Nishitani, Y., Chiba, D., Endo, M., Sawick, M., Matsukura, F., Dietl, T., and Ohno, H. (2010). Curie temperature versus hole concentration in field-effect structures of $Ga_{1-x}Mn_xAs$. *Phys. Rev. B* 81: 045208.

105 Dietl, T., Ohno, H., and Matsukura, F. (2001). Hole-mediated ferromagnetism in tetrahedrally coordinated semiconductors. *Phys. Rev. B* 63: 195205.

106 Wang, X.L., Wang, H.L., Ma, J.L., Zhao, X.P., and Zhao, J.H. (2019). Efficiently rotating the magnetization vector in a magnetic semiconductor via organic molecules. *ACS Appl. Mater. Inter* 11: 6615–6623.

107 Maruyama, T., Shiota, Y., Nozaki, T. et al. (2009). Large voltage-induced magnetic anisotropy change in a few atomic layers of iron. *Nat. Nanotechnol.* 4: 158–161.

108 Endo, M., Kanai, S., Ikeda, S., Matsukura, F., and Ohno, H. (2010). Electric-field effects on thickness dependent magnetic anisotropy of sputtered MgO/$Co_{40}Fe_{40}B_{20}$/Ta structures. *Appl. Phys. Lett.* 96: 212503.

109 Chiba, D., Fukami, S., Shimamura, K., Ishiwata, N., Kobayashi, K., and Ono, T. (2011). Electrical control of the ferromagnetic phase transition in cobalt at room temperature. *Nat. Mater.* 10: 853–856.

110 Shimamura, K., Chiba, D., Ono, S. et al. (2012). Electrical control of Curie temperature in cobalt using an ionic liquid film. *Appl. Phys. Lett.* 100: 122402.

111 Shiota, Y., Bonell, F., Miwa, S., Mizuochi, N., Shinjo, T., and Suzuki, Y. (2013). Opposite signs of voltage-induced perpendicular magnetic anisotropy change in CoFeB vertical bar MgO junctions with different underlayers. *Appl. Phys. Lett.* 103: 082410.

112 Koyama, T., Obinata, A., Hibino, Y., and Chiba, D. (2013). Sign reversal of electric field effect on coercivity in MgO/Co/Pt system. *Appl. Phys. Express* 6: 123001.

113 Bonell, F., Takahashi, Y.T., Lam, D.D. et al. (2013). Reversible change in the oxidation state and magnetic circular dichroism of Fe driven by an electric field at the FeCo/MgO interface. *Appl. Phys. Lett.* 102: 152401.

114 Miwa, S., Matsuda, K., Tanaka, K. et al. (2015). Voltage-controlled magnetic anisotropy in Fe|MgO tunnel junctions studied by x-ray absorption spectroscopy. *Appl. Phys. Lett.* 107: 162402.

115 Chiba, D. and Ono, T. (2013). Control of magnetism in Co by an electric field. *J. Phys. D: Appl. Phys.* 46: 213001.

116 Challener, W.A., Peng, C., Itagi, A.V. et al. (2009). Heat-assisted magnetic recording by a near-field transducer with efficient optical energy transfer. *Nat. Photonics* 3: 220–224.

117 O'Connor, D. and Zayats, A.V. (2010). The third plasmonic revolution. *Nat. Nanotechnol.* 5: 482–483.

118 Shiota, Y., Nozaki, T., Bonell, F., Murakami, S., Shinjo, T., and Suzuki, Y. (2012). Induction of coherent magnetization switching in a few atomic layers of FeCo using voltage pulses. *Nat. Mater.* 11: 39–43.

119 Wang, W.G., Li, M.G., Hageman, S., and Chien, C.L. (2012). Electric-field-assisted switching in magnetic tunnel junctions. *Nat. Mater.* 11: 64–68.

120 Kanai, S., Yamanouchi, M., Ikeda, S., Nakatani, Y., Matsukura, F., and Ohno, H. (2012). Electric field-induced magnetization reversal in a perpendicular-anisotropy CoFeB-MgO magnetic tunnel junction. *Appl. Phys. Lett.* 101: 122403.

121 Kanai, S., Nakatani, Y., Yamanouchi, M., Ikeda, S., Matsukura, F., and Ohno, H. (2013). In-plane magnetic field dependence of electric field-induced magnetization switching. *Appl. Phys. Lett.* 103: 072408.

122 Wang, W.G. and Chien, C.L. (2013). Voltage-induced switching in magnetic tunnel junctions with perpendicular magnetic anisotropy. *J. Phys. D: Appl. Phys.* 46: 074004.

123 Wadley, P., Howells, B., Zelezny, J. et al. (2016). Electrical switching of an antiferromagnet. *Science* 351: 587–590.

124 Jungwirth, T., Marti, X., Wadley, P., and Wunderlich, J. (2016). Antiferromagnetic spintronics. *Nat. Nanotechnol.* 11: 231–241.

125 Baltz, V., Manchon, A., Tsoi, M. et al. (2018). Antiferromagnetic spintronics. *Rev. Mod. Phys.* 90: 015005.

126 Chen, X.Z., Zarzuela, R., Zhang, J. et al. (2018). Antidamping-torque-induced switching in biaxial antiferromagnetic insulators. *Phys. Rev. Lett.* 120: 207204.

127 Jungwirth, T., Sinova, J., Manchon, A., Marti, X., Wunderlich, J., and Felser, C. (2018). The multiple directions of antiferromagnetic spintronics. *Nat. Phys.* 14: 200–203.

128 Gomonay, O., Baltz, V., Brataas, A., and Tserkovnyak, Y. (2018). Antiferromagnetic spin textures and dynamics. *Nat. Phys.* 14: 213–216.

129 Duine, R.A., Lee, K.J., Parkin, S.S.P., and Stiles, M.D. (2018). Synthetic antiferromagnetic spintronics. *Nat. Phys.* 14: 217–219.

130 Zelezny, J., Wadley, P., Olejnik, K., Hoffmann, A., and Ohno, H. (2018). Spin transport and spin torque in antiferromagnetic devices. *Nat. Phys.* 14: 220–228.

131 Nemec, P., Fiebig, M., Kampfrath, T., and Kimel, A.V. (2018). Antiferromagnetic opto-spintronics. *Nat. Phys.* 14: 229–241.

132 Smejkal, L., Mokrousov, Y., Yan, B.H., and MacDonald, A H (2018). Topological antiferromagnetic spintronics. *Nat. Phys.* 14: 242–251.

133 Wadley, P., Reimers, S., Grzybowski, M.J. et al. (2018). Current polarity-dependent manipulation of antiferromagnetic domains. *Nat. Nanotechnol.* 13: 362–365.

134 Barthem, V.M.T.S., Colin, C.V., Mayaffre, H., Julien, M.H., and Givord, D. (2013). Revealing the properties of Mn_2Au for antiferromagnetic spintronics. *Nat. Commun.* 4: 2892–2898.

135 Bodnar, S.Y., Smejkal, L., Turek, I. et al. (2018). Writing and reading antiferromagnetic Mn_2Au by Neel spin-orbit torques and large anisotropic magnetoresistance. *Nat. Commun.* 9: 348–354.

11

Antiferromagnetic Materials and Their Manipulations

Xionghua Liu[1,2] *and Kaiyou Wang*[1,3,4]

[1] *State Key Laboratory of Superlattices and Microstructures, Institute of Semiconductors, Chinese Academy of Sciences, Beijing, China*
[2] *Center of Materials Science and Optoelectronics Engineering, University of Chinese Academy of Sciences, Beijing, China*
[3] *College of Materials Science & Opto-Electronic Technology and Center for Excellence in Topological Quantum Computation, University of Chinese Academy of Science, Beijing, China*
[4] *Beijing Academy of Quantum Information Sciences, Beijing, China*

CHAPTER MENU

11.1 Introduction

During the past three decades, spintronics has been widely investigated worldwide, and its potential applications have drawn a great deal of attention [1]. Conventional spintronic devices rely on the manipulation and detection of spin order in ferromagnetic materials [2]. Recently, antiferromagnetic materials have been generating intense interest in the spintronics community due to their numerous interesting features such as zero stray field, ultrafast spin dynamics, and the generation of large magnetotransport effects [3–5]. The resulting zero net magnetic moment in antiferromagnets (AFMs) means the stored information would be invisible to common magnetic probes, insensitive to disturbing magnetic fields, and importantly, the antiferromagnetic element eliminates crosstalk between neighbors regardless of how densely the elements are arranged in the device [6]. The intrinsic high frequencies of antiferromagnetic dynamics makes spin manipulation in AFMs inherently faster than in ferromagnets (FMs) [3–6]. Accordingly, spintronics based on antiferromagnetic materials (antiferromagnetic spintronics) as an emerging field that makes AFMs useful and spintronics more interesting.

Antiferromagnetic materials represent the overwhelming majority of magnetic materials in nature, and show fascinating physical properties [7]. They are spontaneously long-range ordered but with a zero or vanishingly small net magnetic moment. AFMs are considerably more common than FMs, they are permitted in each magnetic symmetry group, including

Spintronics: Materials, Devices and Applications, First Edition. Edited by Kaiyou Wang, Meiyin Yang, and Jun Luo.

insulators, metals, semimetals, semiconductors or superconductors [4, 8]. Furthermore, AFMs also provide much greater structural flexibility than FMs because of numerous ways to achieve a zero net moments, which show a variety of unique functionalities of either intrinsic or engineered nature. In this chapter, to begin with, I recall the antiferromagnetic materials suitable for antiferromagnetic spintronics and their fundamental properties. I then focus on different effective ways to manipulate the antiferromagnetic states. Finally, the emerging research field that is exploring the links between antiferromagnetic spintronics and topological structures in real and momentum space is introduced.

11.2 Antiferromagnetic Materials

In the early 1930s, Louis Néel was drawn to the issue that some materials with magnetic elements but exhibiting zero remanence at all temperatures did not follow the paramagnetic Curie law, which opens up the prelude for antiferromagnetic research [9]. AFMs have been the subject of intense research for more than 60 years since the discovery of exchange bias in the context of FM/AFM system [10]. Especially, AFM-based devices, such as spin valves and magnetic tunnel junctions, were widely investigated at the beginning of the 21st century because of their technological importance: the AFM acts to pin or harden the FM layer. So, AFM has the role of auxiliary element. The antiferromagnetic spintronics aims to complement or even substitute ferromagnets in the active components of spintronic devices [3–6]. Here I strictly select some antiferromagnetic materials from the perspective of spin transport. Table 11.1 illustrates the AFMs with a bulk Néel temperature above room temperature or playing a key role in the development of antiferromagnetic spintronics, which is divided into three categories relying on their metallic, insulating, or semiconducting or semimetallic nature.

11.2.1 Metallic Antiferromagnets

Metallic AFMs including Mn-based alloy such as PtMn [11, 12], IrMn [13–15], PdMn [16], FeMn [17], or NiMn [16], are mainly produced by magnetron sputtering and are most widely used in industrial applications, such as read heads for magnetic storage or spin valves [18, 19]. Compared with alloys containing light elements (such as Fe, Ni), alloys with heavier elements (such as Pt, Ir, Pd) are more important in exploring spin-transfer effects [20, 21], manipulating magnetic order via spin-transfer torque or spin-orbit torque effects [22–25], enhancing spin pumping near the Néel temperature [26], and utilizing for the deterministic switching of FMs by spin-orbit torque (SOT) [27–29]. Moreover, the Mn metal provides large spontaneous moments, while the noble metal (Pt, Ir or Au) offers large spin-orbit coupling. This combination is ideal in controlling strong magnetic anisotropy phenomena and anisotropic magnetoresistance [30–32]. In particular, the noncollinear spin texture breaks the invariance under the combination of time-reversal symmetry with a crystal symmetry operation leading to a finite anomalous Hall effect [33]. The inverse spin galvanic effects in Mn_2Au has been theoretically predicted [34] and experimentally verified [35, 36]. In addition, the metallic AFMs also contain archetypal materials such as Cr with an intriguing spin density wave configuration [37], which is commonly used to study exchange bias effect in FM/AFM structure [38]. Furthermore, some special metallic metamagnets with an antiferromagnetic to ferromagnetic transition (such as FeRh) give an indirect way to manipulate the antiferromagnetic order through iterative steps consisting of modulating the ferromagnetic

Table 11.1 Néel temperatures of some metallic, insulating and semiconducting or semimetallic antiferromagnets.

Electrical category	Material	$T_{N,bulk}$ (K)	Reference
Metal	$Pt_{50}Mn_{50}$	970	[11]
	$Pt_{25}Mn_{75}$	475	[12]
	$Ir_{20}Mn_{80}$	690	[13]
	$Ir_{25}Mn_{75}$	730	[14]
	$Ir_{50}Mn_{50}$	1145	[15]
	$Pd_{50}Mn_{50}$	780	[16]
	$Fe_{50}Mn_{50}$	490	[17]
	$Ni_{50}Mn_{50}$	1100	[16]
	Mn_2Au	1500	[36]
	Cr	310	[37]
	$Fe_{50}Rh_{50}$	380	[39]
Insulator	NiO	520	[45]
	Cr_2O_3	310	[47]
	CoO	293	[8, 49]
	$BiFeO_3$	653	[51]
	$SrMnO_3$	260	[54]
	$SrTcO_3$	1023	[55]
	MnF_2	68	[59]
	$FeCl_2$	24	[60]
	CrI_3	61	[64]
	$CrBr_3$	37	[65]
	$CrCl_3$	17	[66]
Semiconductor/semimetal	CuMnAs	480	[72]
	Sr_2IrO_4	240	[73]
	MnTe	323	[75]
	$MnSiN_2$	490	[76]
	$CuFeS_2$	825	[77]
	LiMnAs	374	[78]
	Mn_3Sn	420	[83]
	Mn_3Ge	390	[86]

order and undergoing the magnetic phase transition [39, 40]. A small anisotropic magnetoresistance was also found in the antiferromagnetic phase and quasistatic write-read operations were reported in FeRh [41]. Other metallic AFMs, such as rare-earth-based Gd alloys [42], itinerant AFM TiAu [43] or CrB_2 with antiferromagnetism and superconductivity [44] would also be very important in fundamental research of spintronics.

11.2.2 Insulating Antiferromagnets

Insulating AFMs are mostly oxides, sulfides, and halides. The oxides are ideal for investigating magnonic effects such as the spin waves magnon in NiO [45, 46], caloritronic effects such as Seebeck in Cr_2O_3 [47], tunnelling anisotropic magnetoresistance with CoO barrier [48] or the spin current behavior around the phase transition [49]. Moreover, some insulating AFMs exhibit unique multiferroicity and magnetoelectric effects [50], which involves direct coupling between ferroic orders. The typical material is perovskite $BiFeO_3$, where the ferroelectric and antiferromagnetic orders coexist [51]. Recently, the antiferromagnetic magnetoelectric random access memory was demonstrated by utilizing magnetoelectric AFM Cr_2O_3 [52]. In addition, other perovskite AFMs, such as $LaMnO_3$ [53], $SrMnO_3$ [54] and $SrTcO_3$ [55] have be explored due to their interesting properties for antiferromagnetic spintronics. Especially, La_2CuO_4 cuprate show both antiferromagnetism and traces of superconductivity, which opens up fascinating perspectives for superconducting antiferromagnetic spintronics [56]. Furthermore, rare-earth orthoferrites (such as $TmFeO_3$) present a perovskite-like structure and canted antiferromagnetism (weak ferromagnetism), which are known to show a strong temperature-dependent anisotropy. The optical modulation of the antiferromagnetic order of $TmFeO_3$ in the terahertz range was studied [57]. The combination of optics and antiferromagnetic spintronics is considered to induce appealing results in the future [58].

Besides oxides insulating AFMs, the halides such as MnF_2, FeF_2, and $FeCl_2$ have also drawn extensive attention. These materials have Néel temperatures below room temperature and are usually employed to investigate the fundamental issues of exchange bias effect in FM/AFM system [8, 11, 18]. The MnF_2 [59] and $FeCl_2$ [60] are typical AFMs for exhibiting spin-flop and spin-flip transitions, respectively, due to their weak exchange interaction. Mechanisms associated with antiferromagnetic order such as resonant mode degeneracy can be switched on and off due to the transitions [61]. These AFMs were also used to demonstrate the antiferromagnetic spin Seebeck effect [62, 63]. It is worth emphasizing that two-dimensional van der Waals antiferromagnetic materials such as CrI_3 [64], $CrBr_3$ [65] and $CrCl_3$ [66] have recently been attracting scientists' attention worldwide; a great deal of effort has been devoted to this vivid field and numerous novel and exciting phenomena have been reported [67–69]. Moreover, other halides involve complex physical properties such as Bose-Einstein condensation of magnons in antiferromagnetic dimers in $TlCuCl_3$ [70], or fractionalized excitations in the spin liquid state of a kagome-lattice in $ZnCu_3(OD)_6Cl_2$ [71] and so on, which play a vital part in fundamental research.

11.2.3 Semiconducting and Semimetallic Antiferromagnets

Semiconducting and semimetallic AFMs are another class of significant materials in studying spin transport, which have got less attention compared to the metallic and insulating AFMs. Whereas, the recent discovery of electrical switching of semimetallic CuMnAs via SOT shows that AFMs can be electrically manipulated in similar ways to FMs. This work provides a pure AFM memory prototype, with electrical writing and readout, that uses the inverse spin galvanic effect for writing and the planar Hall component of the anisotropic magnetoresistance effect for reading [72]. This work has provoked considerable research into antiferromagnetic spintronics [3–6]. Furthermore, a special oxide semiconducting Sr_2IrO_4 has been reported to display giant anisotropic magnetoresistance and nonvolatile memory [73, 74]. And the multiple-stable anisotropic magnetoresistance memory is also observed in semiconducting MnTe [75] and semimetallic CuMnAs [72]. In addition, there are a large number of Mn- or Fe-based antiferromagnetic semiconductors such as $MnSiN_2$ [76], $CuFeS_2$

[77] or LiMnAs [78], which possess high Néel temperatures and have yet to be developed for antiferromagnetic spintronics. In particular, it has been found that antiferromagnetism can coexist with topologically nontrivial phases of AFMs, which has generated an emerging research field that focuses on the links between AFM and topological structures in real and momentum space. Lots of interesting phenomena, such as Majorana fermions in antiferromagnetic topological superconductor [79], topologically protected antiferromagnetic skyrmions [80, 81], and antiferromagnetic Weyl fermions in Mn_3Sn and Mn_3Ge, have been theoretically predicted and experimentally demonstrated [82–86]. This field is expected to become the focus of essential attention in the near future [79, 87].

11.3 Manipulations of Antiferromagnetic States

The main obstacle to keep AFMs away from application is that they are hard to control and read. Ferromagnetic order can be detected by the magnetic fields it produces, and in turn controlled by an external magnetic field, while AFMs possess zero fringing magnetic fields and are much less sensitive to magnetic fields though they also can be tuned by high enough magnetic fields [59, 60]. If we want to use AFMs for memories like FMs, the difficulty in detecting and controlling the antiferromagnetic order seems a huge barrier. Therefore, how to manipulate the antiferromagnetic order in an effective way is a key issue in the practical application of antiferromagnetic spintronics. To date, much effort has been devoted to trying to solve this problem through magnetic, strain, optical, and electrical manipulations [3–6, 58, 88], see Figure 11.1. Magnetic control contains a strong magnetic field, exchange bias, and field cooling. Strain control includes the magnetic anisotropy effect and metamagnetic transition. Optical control involves thermal and electronic excitation, an inertia-driven mechanism, and terahertz laser control. Electrical control including electric field and electric current is the core issue in antiferromagnetic spintronics. Next, I will introduce the manipulations of antiferromagnetic order utilizing these four methods, respectively.

11.3.1 Magnetic Control of Antiferromagnets

Usually, a magnetic field is an effective way to manipulate the magnetic order of magnetic materials such as ferromagnets or ferrimagnets. However, this is very difficult for AFMs due to their nearly zero net magnetic moment, and hence vanishing applicable Zeeman energy under a normal magnetic field. While the magnetic moments can be controlled via applying a high enough external magnetic field along certain magnetic moments or domains directions, like the spin-flop and spin-flip transitions in MnF_2 [59] and $FeCl_2$ [60]. The critical magnetic field for halides is commonly small (still up to several Tesla) because of the weak exchange interaction. For AFMs with high Néel temperature, this critical magnetic field can reach to several tens Tesla, such as 70 T for Mn_2Au [89].

The more universal way to manipulate the magnetic moments of AFMs is to use the exchange bias effect between FM and AFM layers. This phenomenon was first discovered by Meiklejohn and Bean in oxide-coated Co particle in 1956 [10]. Exchange bias refers to a shift in the hysteresis loop along the magnetic field axis due to the interfacial coupling of FM and AFM when the heterostructures are field-cooled below their Curie and Néel temperatures, respectively [8, 18]. There is a lot of research being conducted on the exchange bias in

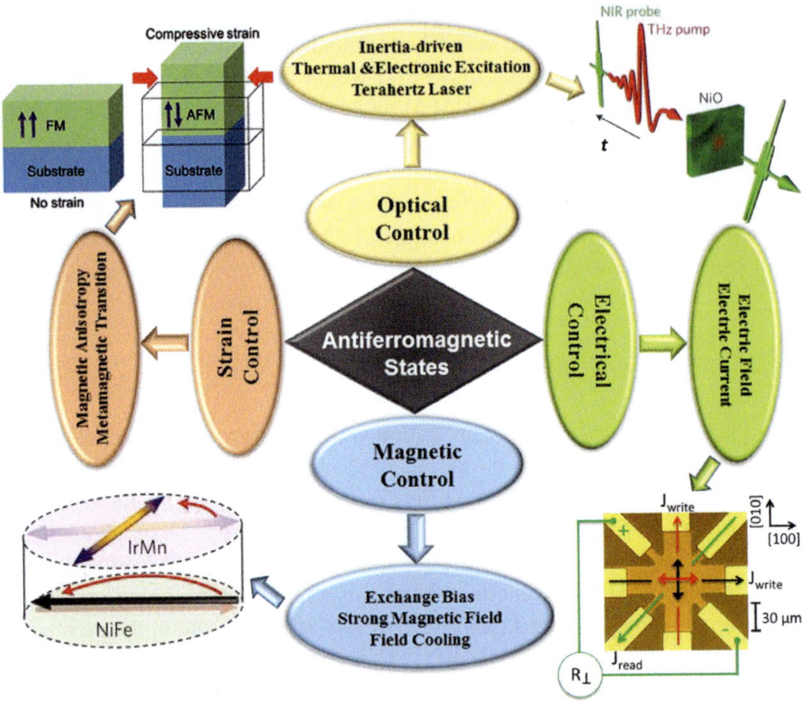

Figure 11.1 Schematic of magnetic, strain, optical, and electrical manipulation methods of antiferromagnetic states. *Source:* Reprinted from [88], Copyright 2018, with permission from IOP Publishing Ltd.

FM/AFM systems such as Co/NiO [90], Co/FeF_2 [91], Fe_3O_4/CoO [19, 92], Fe/NiO [93], or IrMn/[Co/Pt] [94]. It has been found that the exchange bias can be influenced by many factors, such as the temperature, FM magnetization, the film thickness, and the anisotropy of AFM or FM layers [8, 18].

In addition, the exchange spring is an important interfacial exchange coupling that have been experimentally demonstrated in FM/AFM heterostructure. Magnetic control of AFMs utilizing the exchange spring was first realized in the NiFe/IrMn/MgO/Pt system by Park et al [30]. As seen in Figure 11.2(a), the antiferromagnetic configuration is driven by an exchange spring on one side and the tunnel anisotropic magnetoresistance (TAMR) is detected for the other side. However, the large TAMR up to 160% is only obtained at quite a low temperature (4K), which limits the practical application for information storage. Thereafter, Wang et al reported the thermal stability of AFM-based systems with out-of-plane anisotropy to demonstrate room temperature antiferromagnetic TAMR (~0.2%) in [Co/Pt]/IrMn/AlO_x/Pt stacks [95, 96]. Moreover, the antiferromagnetic order can be modulated via field cooling across the magnetic phase transition for some special AFMs like FeRh. A clear TAMR was found in the antiferromagnetic phase and quasistatic write-read operations were demonstrated in this material [39], as illustrated in Figure 11.2(b). Although the magnetic control of AFMs has been studied for more than 60 years, yet more work is needed to further decrease the applied magnetic field and increase the resistance to external perturbations.

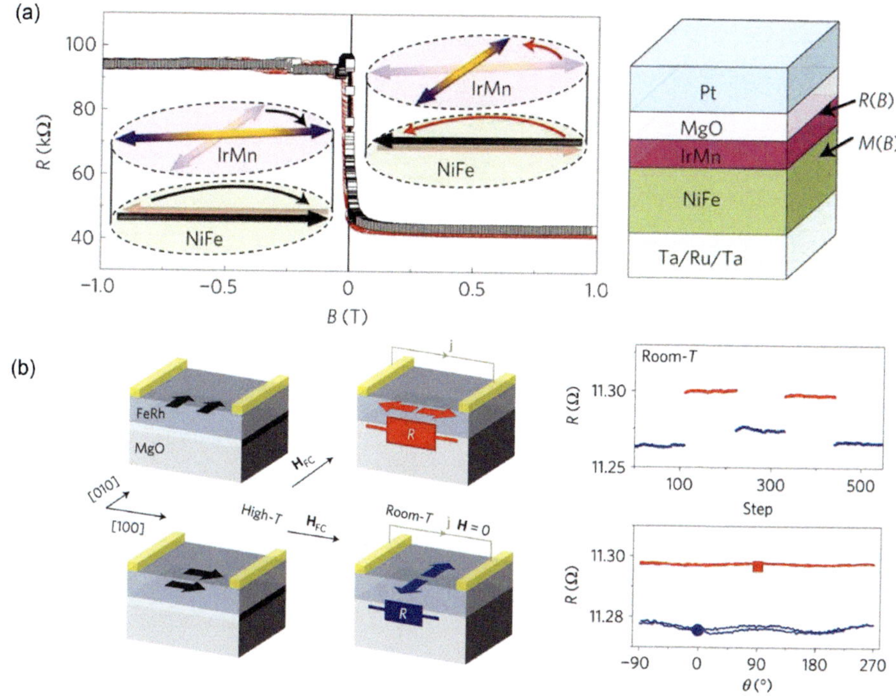

Figure 11.2 (a) Large TAMR signal recorded at 4 K in a NiFe/IrMn(1.5 nm)/MgO/Pt tunnel junction. The insets display rotation of antiferromagnetic moments in IrMn via the exchange-spring effect of the adjacent NiFe ferromagnets. The external magnetic field is sensed by the NiFe ferromagnet, whereas the tunneling transport is determined by the IrMn. *Source:* Reprinted from [30], Nature Publishing Group. (b) Left: Schematic of FeRh memory. For writing, the sample is cooled in a field $\mathbf{H}_{FC}$ from a temperature above the antiferromagnetic-ferromagnetic transition in FeRh. Black arrows indicate the orientation of the magnetic moments in the ferromagnetic phase, while red and blue arrows show two distinct configurations of magnetic moments in the antiferromagnetic phase. **j** is the readout current. Right: Resistance, R, measured at room temperature and zero magnetic field after field-cooling the sample with field parallel (blue) and perpendicular (red) to the current direction (top panel). And the stability of the two memory states after field-cooling (blue dot and red square) at room temperature is observed (bottom panel). *Source:* Adapted from [39], Nature Publishing Group.

11.3.2 Strain Control of Antiferromagnets

Strain is considered as an intriguing way to manipulate antiferromagnetic moments through the magnetic anisotropy effect or metamagnetic transition. Usually, the stain can change the lattice parameters and spin-orbit coupling, which significantly affects the chemical potential, ground-state energy, and density of states of AFMs, leading to the variation of magnetic anisotropy. This method has been theoretically predicted and experimentally demonstrated in metallic AFMs such as Mn_2Au and IrMn [89, 94, 97]. For Mn_2Au and IrMn AFMs, the 5*d* shell of the noble metal (Au, Ir) provides a large spin-orbit coupling and the 3*d* shell of the transition metal offers complementary large spontaneous moments. When a strong strain is exerted on the Mn_2Au (001) crystal structure, a lattice distortion and reduced in-plane symmetry will occur, resulting in a uniaxial anisotropy with preferential Néel vector orientation, compared to that without strain. As a result, the antiferromagnetic moments rotate spontaneously, accompanying with the anisotropic magnetoresistance [97], see Figure 11.3(a).

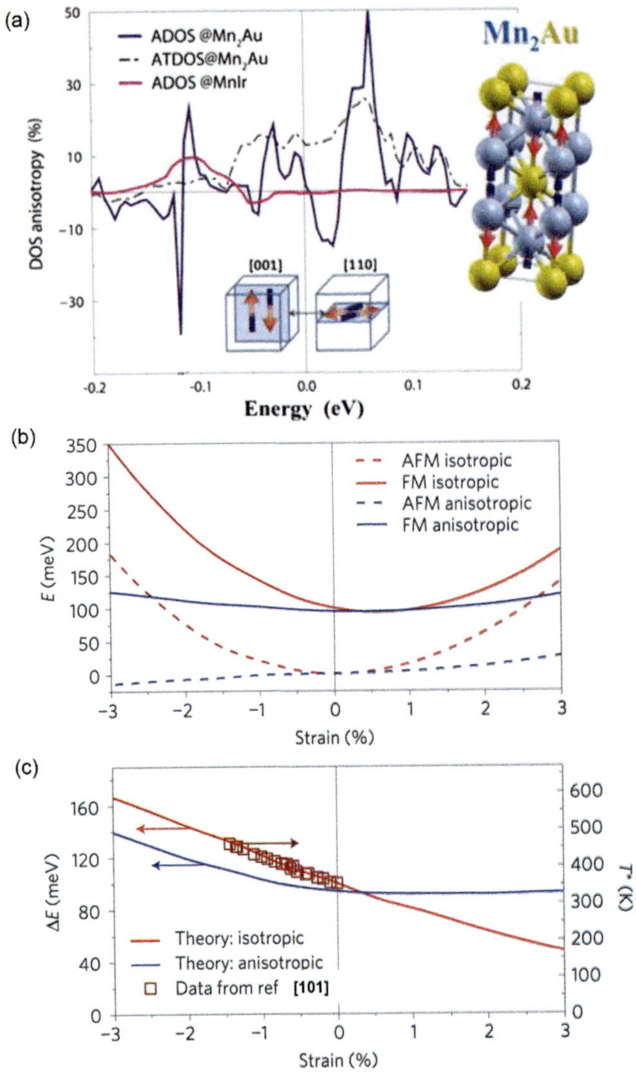

Figure 11.3 (a) Anisotropy of the densities of state between the [001] and [110] axes for Mn_2Au and MnIr films. The inset displays a $MoSi_2$-type body-centered tetragonal structure and the spin arrangement in Mn_2Au. The Fermi level is at 0 eV. *Source:* Adapted from [97], American Physical Society. (b) Energy per FeRh unit formula versus theisotropic and anisotropic strain for antiferromagnetic and ferromagnetic bulk FeRh. (c) Energy difference between the ferromagnetic and antiferromagnetic phases for the isotropic and anisotropic strain. The data for the transition T^* temperature dependence versus isotropic strain are taken from [98]. The vertical scales for both axes are chosen to stress the correspondence between the theory and the data. *Source:* Reprinted from [99], Nature Publishing Group.

The strain control of metamagnetic transition from ferromagnetic to antiferromagnetic phase has been investigated for FeRh thin film [40, 100]. FeRh is antiferromagnetic state at room temperature, and possesses the metamagnetic transition at about 350 K. The phase transition from AFM to FM corresponds to a lattice expansion of 1%; thus, the epitaxial growth process can effectively manipulate the metamagnetic transition via varying the electronic structure under tetragonal lattice distortion, leading to magnetocrystalline anisotropy in thin films [100]. Moreover, the strain controlling of metamagnetic transition in FeRh provides a low-power manipulation method, which is promising for magnetic information storage. For instance, Liu et al realized a large electroresistance (~22%) modulation for FeRh thin film by a small electric field of 2 kV/cm via piezoelectric strain-mediated magnetoelectric coupling and the resulting magnetic phase transition in epitaxial FeRh/$BaTiO_3$ heterostructures [40]. Especially, Cherifi et al reported the increase of metamagnetic transition temperature about 25 K via a low electric field (0.4 kV/cm), which is adequate to isothermally change the FeRh from the AFM to FM [99]. First-principles calculations indicate that the antiferromagnetic state is more stable without strain and with an enhancement of compressive strain [Figure 11.3(b)]. The monotonical increase in ΔE (the energy difference between ferromagnet E_{FM} and antiferromagnets E_{AFM}) means the antiferromagnetic phase becomes progressively stable [98, 99], see Figure 11.3(c). Therefore, using strain or strain-related methods to control magnetic moments of AFMs would expand the usage of antiferromagnetic spintronics.

11.3.3 Optical Control of Antiferromagnets

Optical control is also a practical way to obtain effective manipulation of AFMs. Lights can be utilized to tune the AFM magnetization through optically driven thermal and electronic excitation, as indicated by second-harmonic generation, linear magnetic reflection or other technologies, see Figure 11.4(a). The interaction between light and magnetism gives an invaluable tool for light to detect or tune magnetic materials with an ultrafast timescale, which has drawn a lot of attention in fundamental physics and technological applications [58, 101, 102]. The ultrafast dynamics of antiferromagnetic spins using optical excitation has been demonstrated in many AFMs. For instance ultrafast spin dynamics is found in Cr_2O_3 owing to the spin-lattice interaction and emission of magnetic excitations [103], and the antiferromagnetic order parameter can be reoriented upon optical excitation in NiO [104], or in rare-earth orthoferrites such as $TmFeO_3$ [57], $HoFeO_3$ [105], or $DyFeO_3$ [106].

In particular, the AFMs display an inertial behavior, that the order parameter keeps evolving even after the light pulse is turned off, which drives the antiferromagnetic order reorientation, while this is not observed for FMs. As illustrated in Figure 11.4(b), a strong laser pulse transfers sufficient momentum and kinetic energy to the spin system though the spin orientation hardly varies during the action. When momentum is strong enough the spins overcome the potential barrier and reorient into a new metastable state, resulting in the switching of antiferromagnets [105]. This inertial-driven spin switching in AFMs offers a new insight into ultrafast recording and information storage.

More generally, the study of antiferromagnetic spin-wave modes has attracted a great deal of attention lately with the development of terahertz (THz) research [107, 108]. THz radiation is located between infrared light and microwave radiation with the frequency of about 0.1–10 THz, which offers a universal ultrafast method to dominate previously inaccessible

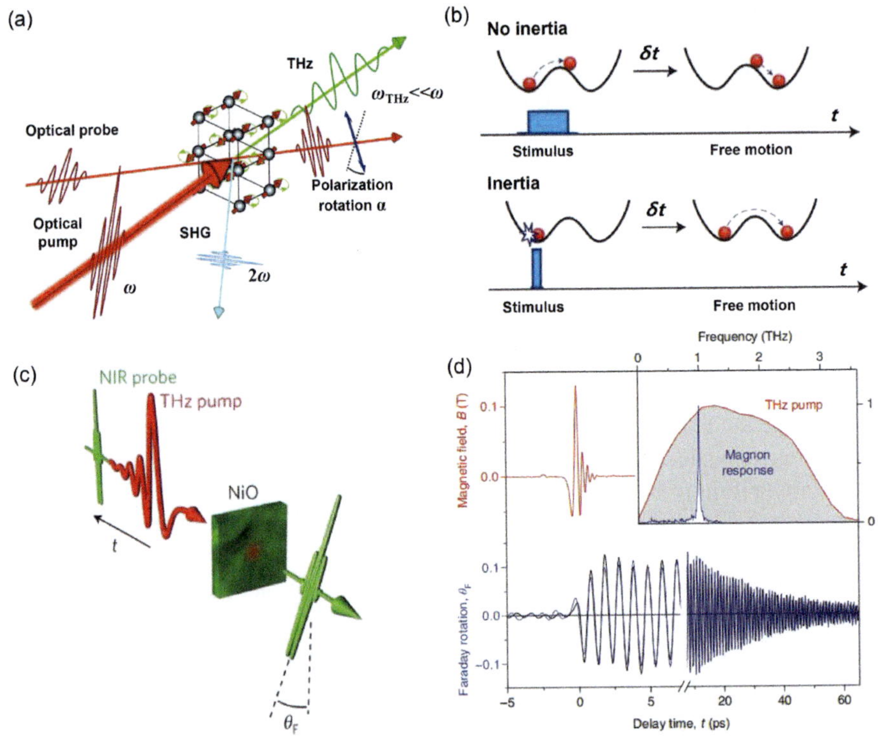

Figure 11.4 (a) An intense optical pump laser pulse, typically in the red or infrared spectral regions, brings the AFM out of equilibrium. Magnetization-assisted second harmonic generation (SHG), or electromagnetic radiation emitted by precessing spins in the THz spectral range, can be used to monitor the dynamic and/or static properties of AFMs. *Source:* Reprinted from [58], Nature Publishing Group. (b) Non-inertial and inertial models to transport a point mass over a potential, typical of ferromagnetic and antiferromagnetic dynamics, respectively. *Source:* Reprinted from [105], Nature Publishing Group. (c) Schematic of femtosecond THz spin resonance. An intense free-space THz transient (red curve) interacts with the electron spins of the NiO thin films and excites magnons. A femtosecond near-infrared (NIR) probe pulse (green curves) samples the induced net magnetization by means of the Faraday effect (Faraday rotation θ_F) after a variable delay time t. (d) Time trance of magnetic field of the incident THz pulse (top panel). Ultrafast Faraday rotation θ_F induced in the NiO sample at room temperature (bottom panel). Harmonic oscillations with a period of 1 ps are due to an antiferromagnetic spin precession. Inset exhibits amplitude spectra of the Faraday transient and driving THz field. *Source:* Adapted from [107], Nature Publishing Group.

magnetic excitations in the electronic ground state. Earlier work showed that low-energy elementary excitations in solids, such as lattice vibrations and plasma oscillations, can be tuned by the terahertz field, which manipulates charge carriers via electric-dipole coupling. Cheng et al proposed THz-frequency generation in a NiO/Pt bilayer through a nonlinear feedback mechanism, which stabilizes the antiferromagnetic precession around the easy axis [109]. Furthermore, Kampfrath et al shown that the magnetic component of intense THz transients causes ultrafast control of the spin degree of freedom [107]. They employed single-cycle THz pulses to turn on and off coherent magnons in NiO at room temperature with a resonance frequency of about 1 THz. An ultrashort optical pulse detects the THz-induced spin dynamics directly in the time domain via time-resolved Faraday rotation, which is the magnetic field generated polarization variation of the probing laser light, see Figures 11.4(c) and 11.4(d). The observation and manipulation of the spin waves and resonance in the time domain through THz technology have been extended to other AFMs such as MnO, $RFeO_3$ or $RMnO_3$ (R is rare earth) [108]. The optical methods employed to manipulate antiferromagnetism reveal ultrafast information recording and processing for purely optical memory devices in the future.

11.3.4 Electrical Control of Antiferromagnets

An ultimate goal of spintronics is to manipulate magnetism via electrical means. Electrical control can be divided into two parts: Electric field and electric current, which are convenient for practical applications. As described in previous section, strain combined with an electric field can tune the antiferromagnetic moments above room temperature (like FeRh). Remarkably, the electric field can control antiferromagnetic order in itself or in combination with other more methods [110, 111]. Usually, the coupling between magnetization and the electric field via magnetoelectric, multiferroic materials can help to realize the tunability of antiferromagnetic states and the design of low power spintronics devices. Early investigations about the controlling of AFM in the electric field was performed utilizing multiferroic (two or more ferroic orders coexist) films. It has been reported that the ferroelectric polarization can be switched by electric field, and different polar states correspond to different signs of chirality of antiferromagnetic domains, indicating the antiferromagnetic domains can be tuned by electric field [112, 113]. The perovskite $BiFeO_3$ is a typical multiferroic material that has been widely studied due to its large ferroelectric polarization and high Néel temperature [51]. The electrical control of antiferromagnetic domains via electric field can happen at room temperature. The switching that originating from the coupling of antiferromagnetic and ferroelectric domains to the underlying ferroelastic domain structure has been experimentally demonstrated with piezoelectric force microscopy and photoemission electron microscopy [114]. Moreover, the magnetotransport properties were found to take place across domain walls via external fields in $BiFeO_3$, which illustrates the control mechanism of antiferromagnetic moments and stimulates the development of multiferroic materials in spintronics [115], see Figure 11.5(a).

In addition, the magnetoelectric Cr_2O_3 is commonly used to study the electric manipulation of antiferromagnetic domains. Because the magnetization reversal and reversible isothermal magnetoelectric switching can be achieved at room temperature. He et al reported that two degenerate 180° antiferromagnetic domains can be tuned by magnetoelectric annealing in (0001) surface of Cr_2O_3 [116]. When the electrically switchable Cr_2O_3 is coupled with FMs, the switchable exchange bias can be implemented [116–118]. Interestingly, the purely antiferromagnetic magnetoelectric random access memory has been demonstrated by utilizing Cr_2O_3 as the antiferromagnetic element [52], as illustrated in Figure 11.5(b), which opens an promising avenue for magnetoelectric AFM research. Moreover, the ionic liquid is employed as the dielectric gate to tune the exchange spring in metallic antiferromagnetic/ferromagnetic heterostructures such as the IrMn/[Co/Pt] system [119, 120]. When the ionic liquid is dropped onto the surface of structures, an electric double layer forms over the channel surface to accumulate electron carriers, which presents an extremely high electric-field effect and penetrates deeper in AFMs relative to the conventional solid gate insulators.

The metallic AFMs do not possess ferroelectric orders or magnetoelectric properties, while they are irreplaceable in traditional spintronics. In order to directly control the metallic

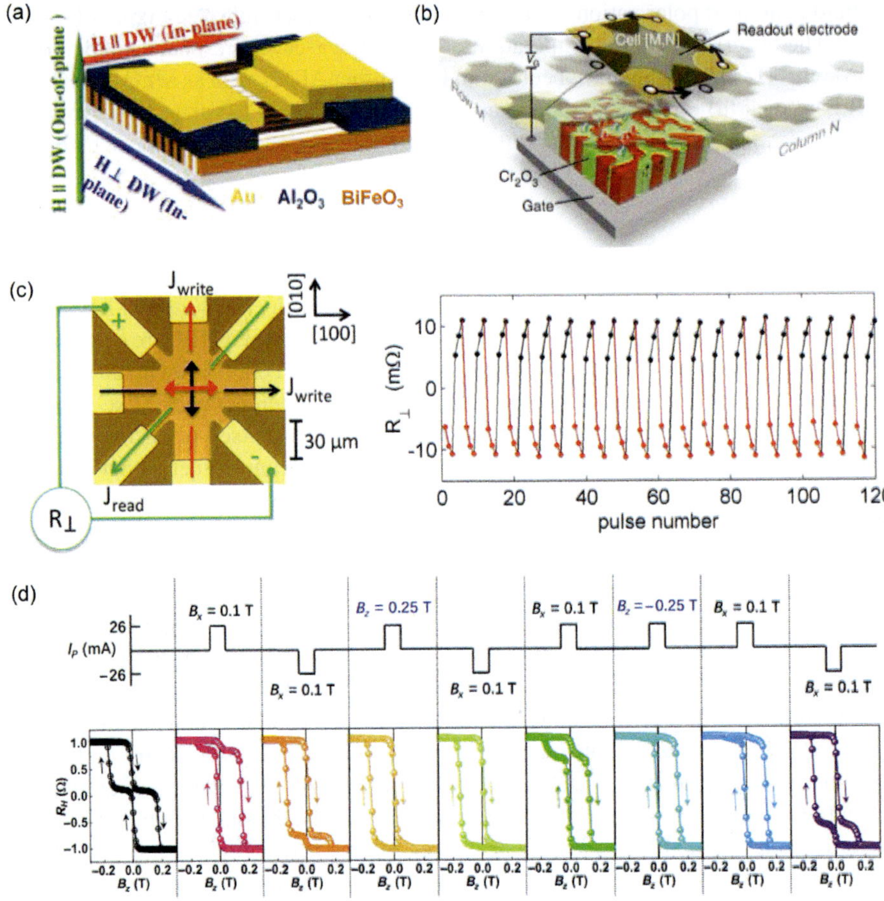

Figure 11.5 (a) Schematic of the device structure, and example of current path parallel to the domain walls. *Source:* Reprinted from [115], American Physical Society. (b) Sketch of one memory cell within a matrix of devices. The arrows indicate the contact permutation to obtain offset free Hall readings. *Source:* Reprinted from [52], Nature Publishing Group. (c) Left: Optical microscopy image of the solid-state memory device made from a CuMnAs antiferromagnet and schematic of the electrical writing and readout geometry. Right: Change in the transverse resistance, $R_{\perp}$, after applying three successive 50 ms writing pulses of amplitude $J_{write} = 4 \times 10^6$ A/cm^2 alternately along the [100] crystal direction of CuMnAs (black arrow in left and black points in right) and the [010] axis (red arrow in left and red points in right). *Source:* Adapted from [72], AAAS. (d) Antiferromagnetic interfacial spins of IrMn tuned by spin-orbit torque. Sequences of current pulses with external magnetic field (B_x or B_z) applied to the sample and anomalous Hall resistance R_H versus B_z curves. The switching between multiple states of the antiferromagnetic interfacial spins can be achieved via SOT combined with external magnetic fields. *Source:* Adapted from [121], American Physical Society.

antiferromagnetic moments without FMs, magnetic fields, or heating, an electric current is the priority in ultrafast spin dynamics. The SOT generated by the current takes an important role in this mechanism to obtain all electrical manipulation of AFMs at room temperature [3, 111]. When flowing electric current across the solid, the relativistic current-induced spin-orbit coupling interaction can produce a magnetic field on every atomic site. A non-zero effect on a particular lattice site can appear if the crystal structure has a broken inversion symmetry, and the reversal of the electric current direction produces different sign of the magnetic field. Therefore, controlling antiferromagnetic moments by electric current needs a special crystal structure [4, 34].

Mn_2Au is the first AFM that was predicted to possess the required crystal structure suitable for switching through electric current. Mn_2Au can be divided into two sublattices, each with a broken inversion symmetry and inducing opposite inverse spin galvanic effects. These two sublattices can form inversion partners, resulting in the observation of Néel SOT [34, 36]. Notably, the first experimental demonstration of Néel SOT switching via electric current is in semimetallic CuMnAs [72], which has a similar crystal structure to Mn_2Au with two symmetric sublattices with respect to Mn sites. The electric current flowing along the basal plane will generate an alternating magnetic field in the whole thin film. The staggered Mn magnetic moments simultaneously rotate following spatially modulated torque, and keep stable along special directions controlled by magnetic anisotropies. To experimentally observe this switching, Wadley et al designed a cross-bar with a ratchet structure, which permits a sequential writing and reading signal. As illustrated in Figure 11.5(c), when current pulses are injected to the device along [100] or [010] directions, the magnetic moments of Mn rotate to be perpendicular to the current. Accordingly, the electrical resistance of the CuMnAs bar varies because of anisotropic magnetoresistance [72]. Soon after, similar works were reported in Mn_2Au [35, 122]. As Mn_2Au is a good conductor with high Néel temperature and has better thermal stability than CuMnAs, which is a very prospective candidate for memory application.

To date, most work has focused on electrical manipulation of the bulk properties of AFMs. Conversely, from the point of view of expanding the functionality and design flexibility in antiferromagnetic spintronic devices, developing methods to tune the interfacial properties of AFMs via SOT is also a vitally significant issue. Although the switching of exchange bias via electric field in FM/AFM heterostructure with magnetoelectric or multiferroic insulators (such as Cr_2O_3 and $BiFeO_3$) has been demonstrated, yet the effective electrical control faces rather challenging for metallic AFMs, which play an irreplaceable role in traditional spintronics. Recently, the manipulation of interfacial spins of AFMs through SOT has been reported [121, 123, 124]. The Pt/Co/IrMn heterostructure is designed to realize the effective tunability of exchange bias. As presented in Figure 11.5(d), the interfacial spins of IrMn can be manipulated by applying current pulses under different external fields, corresponding to variation of negative and positive exchange bias [121]. This electric current-induced switching of exchange bias is more convenient as compared to the conventional field-annealing treatment. Especially, combining field annealing and the current pulse approaches, the smart exchange bias design can be realized within one device. For instance, the magnetic sensors composed of Wheatstone bridges would need different exchange bias directions on different cells, the current pulse offers a convenient approach to tune the exchange bias in different directions [121, 123, 124].

11.4 Topological Antiferromagnetic Spintronics

Topologically protected states of matter are robust because they cannot be destroyed by small variations in system parameters. Materials with topologically protected Dirac or Weyl quasiparticles in their momentum-space bands and those possessing topologically non-trivial real-space spin textures, have both inspired new energy efficient spintronic concepts [79]. Recently, the antiferromagnetism coexisted with topologically nontrivial phases of AFMs has been reported, which creates an emerging research field that focuses on the links between AFM and topological structures in real and momentum space. The roots of topological antiferromagnetic spintronics can be traced to investigations of $SrMnBi_2$, which was found to present quasi-two-dimensional massive Dirac quasiparticles near the Fermi level [125, 126]. Moreover, the control of the Dirac quasiparticle current and the quantum Hall effect by applying strong magnetic field has been demonstrated in antiferromagnetic $EuMnBi_2$ [127], and the two-dimensional massless Dirac cones were observed in the infrared spectral of the antiferromagnetic superconductor $BaFe_2As_2$ [128].

It has been found that topological semimetal states appear when conduction and valence bands touch at discrete points, with lines or planes in a bulk Brillouin zone at energies near the Fermi level. The low energy physics of topological semimetals can be described by effective Dirac or Weyl equations [129, 130]. The 3D Dirac semimetal state is not possible in FMs but can exist in an AFM such as the orthorhombic phase of CuMnAs [131]. Many interesting effects have been proposed in topological Dirac semimetal AFMs, such as the coexistence of Dirac Fermions and SOTs in CuMnAs. This indicates a new phase transition mechanism called the topological metal-insulator transition [87]. When time reversal or spatial inversion symmetry, or both, is broken and the double band degeneracy is lifted, the touching points of two non-degenerate bands can form a 3D Weyl semimetal [129]. Weyl points act as monopole source of Berry curvature flux and produce a topological charge, which always come in pairs with opposite topological charges and usually do not depend on any special symmetry protection [132]. Recently, the Weyl fermions were predicted [82] and experimental verified [133] in Mn_3Sn, a non-collinear AFM from the Heusler family that is potentially more relevant for metallic spintronics because of its high Néel temperature (420K) [83]. Many emergent phenomena such as the large anomalous Hall effect, anomalous Nernst effect, chiral anomaly, and optical gyrotropy can be observed in this material and also in antiferromagnet Mn_3Ge [82–86, 129, 132].

In addition, a topological spin Hall effect was predicted for skyrmions [137], where the spin Hall response happens even in the absence of spin-orbit coupling. Antiferromagnetic skyrmion crystals were proposed to show non-zero topological spin Hall effect but without the topological Hall effect [81]. Antiferromagnetic skyrmions can be considered as two interpenetrating ferromagnetic skyrmions, where the index (+1, −1) labels the two antiferromagnetic sublattices, as displayed in Figure 11.6(a). Microscopically, skyrmionic magnetization manipulation is induced by the Dzyaloshinskii-Moriya interaction (DMI) of non-centrosymmetric crystals or because of the inversion asymmetry at the interfaces [138]. Importantly, the DMI in the bulk is more abundant in AFMs than FMs [80]. The manipulation of antiferromagnetic skyrmions by an electric current was theoretically predicted [80, 134, 139] but still no experimental report, though the room-temperature stabilization of antiferromagnetic skyrmions in synthetic antiferromagnets has been observed [140]. Micromagnetic simulations present that antiferromagnetic skyrmions move faster than ferromagnetic skyrmions, which can be driven with lower current densities and move in straight lines, see Figure 11.6(b). The skyrmion would represent the smallest micromagnetic object that can store information. For example, in the skyrmion racetrack memory illustrated in Figure 11.6(c), the magnetic information can be stored in skyrmions instead of magnetic

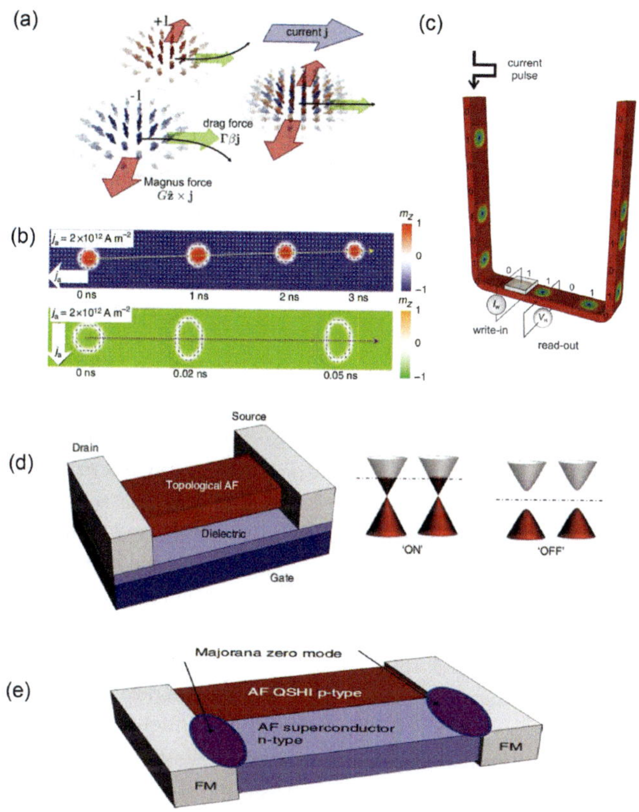

Figure 11.6 (a) The Magnus force associated with skyrmion motion (relative to either a static background or an electronic or magnonic flow) is proportional to the topological charge in the ferromagnetic case. This transverse force is absent in antiferromagnets as the topological charge is odd under reversal of the local spin orientation and thus cancels between the two tightly exchange-bound antiferromagnetic sublattices. *Source:* Adapted from [80], American Physical Society. (b) Micromagnetic simulation of ferromagnetic (upper panel) and antiferromagnetic (lower panel) skyrmion motion driven by a spin-orbit torque. The ferromagnetic skyrmion is deflected by the Magnus force, while the antiferromagnetic skyrmion can move in a straight line because of mutual compensation between Magnus forces from the two magnetic sublattices. *Source:* Adapted from [134], American Institute of Physics. (c) The skyrmion racetrack memory concept. Magnetic information is stored in a topological change Q instead of the magnetization of the magnetic domain. "1" encodes the skyrmion with Q = 1, while "0" corresponds to a uniform domain with Q = 0. Spin-transfer torque shifts the register. Bits are read by the topological Hall effect and written by nucleating or deleting the skyrmion. *Source:* Reprinted from [135], Nature Publishing Group. (d) A concept for a topological antiferromagnetic transistor, one can obtain a highly mobile bulk Dirac quasiparticle current (display in the "ON" state) that is tunable by gating or by ultrafast spin-orbit torque. *Source:* Adapted from [136], Nature Publishing Group. (e) A Majorana qubit in an antiferromagnetic FeSe monolayer. QSHI, quantum spin Hall insulator. *Source:* Reprinted from [79], Nature Publishing Group.

domains [135, 141]. Moreover, one can foresee the possibility of a topological transistor that operates at high frequencies and low current densities, as shown in Figure 11.6(d), where one can achieve a highly mobile bulk Dirac quasiparticle current (show in the "ON" state) that is tunable by gating or by ultrafast spin-orbit torque [136]. Another scenario is forming a p–n junction in a single layer of FeSe by gating to get a superconducting region and a region with a quantum spin Hall effect [142]. Coupling this system to one ferromagnetic electrode from each side would localize Majorana modes at the interfaces, which offers an alternative realization of Fu-Kane Majorana fermion proposal [143], as illustrated in Figure 11.6(e).

11.5 Summaries and Prospects

Antiferromagnetic materials are considered to represent the future of spintronic applications. The recent experimental achievements undertake the promises for future outstanding developments. The wide diversity under which antiferromagnetism exists in nature gives an appealing playground for physicists, materials scientists, and engineers. Whereas, tremendous challenges such as all-electrical manipulation of antiferromagnetic magnetic tunnel junctions and large antiferromagnetic tunnel magnetoresistance at room temperature, etc are still needed to be overcome before AFMs can become active elements of real spintronic devices. Remarkably, there has difficulty in collecting systematic and reproducible data because of significant sample-to-sample variability and uncontrolled size effects. The antiferromagnetic texture is sensitive to the layer thickness, temperature, or growth conditions, and the antiferromagnetic order on the spin transport properties are still needed to be accurately understood. To provoke continued advancement of this emerging field, more suitable and functional materials should be studied along with novel methods that are efficient in energy and operation. Accordingly, the intrinsic mechanism of coupling between antiferromagnetic spins and different methods also need more attention, which will offer guidance for experimental research. Finally, emerging materials with strong spin-orbit coupling, such as antiferromagnetic topological insulators or Weyl semimetals, would also be an inspiring research direction in the near future.

References

1 Žutić, I., Fabian, J., and Das Sarma, S. (2004). Spintronics: fundamentals and applications. *Rev. Mod. Phys.* 76: 323.

2 Bhatti, S., Sbiaa, R., Hirohata, A., Ohno, H., Fukami, S., and Piramanayagam, S.N. (2017). Spintronics based random access memory: a review. *Materials Today* 20: 530.

3 Manchon, A., Železný, J., Miron, I.M., Jungwirth, T., Sinova, J., Thiaville, A., Garello, K., and Gambardella, P. (2019). Current-induced spin-orbit torques in ferromagnetic and antiferromagnetic systems. *Rev. Mod. Phys.* 91: 035004.

4 Baltz, V., Manchon, A., Tsoi, M., Moriyama, T., Ono, T., and Tserkovnyak, Y. (2018). Antiferromagnetic spintronics. *Rev. Mod. Phys.* 90: 015005.

5 Železný, J., Wadley, P., Olejník, K., Hoffmann, A., and Ohno, H. (2018). Spin transport and spin torque in antiferromagnetic devices. *Nat. Phys.* 14: 220.

6 Jungwirth, T., Marti, X., Wadley, P., and Wunderlich, J. (2016). Antiferromagnetic spintronics. *Nat. Nanotechnol.* 11: 231.

7 Dong, S., Liu, J.M., Cheong, S.W., and Ren, Z. (2015). Multiferroic materials and magnetoelectric physics: symmetry, entanglement, excitation, and topology. *Adv. Phys.* 64: 519–626.

8 Nogués, J. and Schuller, I.K. (1999). Exchange bias. *J. Magn. Magn. Mater.* 192: 203–232.

9 Néel, L. (1971). Magnetism and Local Molecular Field. *Science* 174: 985–992.

10 Meiklejohn, W.H. and Bean, C.P. (1956). New magnetic anisotropy. *Phys. Rev.* 102: 1413.

11 Umetsu, R.Y., Fukamichi, K., and Sakuma, A. (2006). Electrical and magnetic properties, and electronic structures of pseudo-gap-type antiferromagnetic $L1_0$-type MnPt alloys. *Mater. Transactions* 47: 2.

12 Krén, E., Kádár, G., Pál, L., and Szabó, P. (1967). Investigation of the first-order magnetic transformation in Mn_3Pt. *J. Appl. Phys.* 38: 1265.

13 Lacour, D., Jaffrès, H., Nguyen Van Dau, F., Petroff, F., Vaurès, A., and Humbert, J. (2002). Field sensing using the magnetoresistance of IrMn exchange-biased tunnel junctions. *J. Appl. Phys.* 91: 4655.

14 Yamaoka, T. (1974). Antiferromagnetism in γ-phase Mn-Ir alloys. *J. Phys. Soc. Japan* 36: 445.

15 Umetsu, R.Y., Miyakawa, M., Fukamichi, K., and Sakuma, A. (2004). Pseudogap in the density of states and the highest Néel temperature of the $L1_0$-type MnIr alloy system. *Phys. Rev. B* 69: 104411.

16 Pál, L., Krén, E., Kádár, G., Szabó, P., and Tarnóczi, T. (1968). Magnetic structures and phase transformations in Mn-based CuAu-I type alloys. *J. Appl. Phys.* 39: 538.

17 Marrows, C. H. (2003). Three-dimensional exchange bias in $\{Co/Pd\}_N$/FeMn. *Phys. Rev. B* 68: 012405.

18 Nogués, J., Sort, J., Langlais, V., Skumryev, V., Suriñach, S., Muñoz, J.S., and Baró, M.D. (2005). Exchange bias in nanostructures. *Physics Reports* 422: 65–117.

19 Liu, X.H., Liu, W., Zhang, Z.D., and Chang, C.F. (2018). Mediating exchange bias by Verwey transition in CoO/Fe_3O_4 thin film. *J. Appl. Phys.* 123: 083903.

20 Acharyya, R., Nguyen, H.Y.T., Pratt, W.P., and Bass, J. (2010). Spin-flipping associated with the antiferromagnet IrMn. *IEEE Trans. Magn.* 46: 1454.

21 Merodio, P., Ghosh, A., Lemonias, C., Gautier, E., Ebels, U., Chshiev, M., Béa, H., Baltz, V., and Bailey, W.E. (2014). Penetration depth and absorption mechanisms of spin currents in $Ir_{20}Mn_{80}$ and $Fe_{50}Mn_{50}$ polycrystalline films by ferromagnetic resonance and spin pumping. *Appl. Phys. Lett.* 104: 032406.

22 Wei, Z., Sharma, A., Nunez, A., Haney, P.M., Duine, R.A., Bass, J., MacDonald, A.H., and Tsoi, M. (2007). Changing exchange bias in spin valves with an electric current. *Phys. Rev. Lett.* 98: 116603.

23 Mendes, J.B.S., Cunha, R.O., Alves Santos, O., Ribeiro, P.R.T., Machado, F.L.A., Rodríguez-Suárez, R.L., Azevedo, A., and Rezende, S.M. (2014). Large inverse spin Hall effect in the antiferromagnetic metal $Ir_{20}Mn_{80}$. *Phys. Rev. B* 89: 140406.

24 Zhang, W., Jungfleisch, M.B., Jiang, W., Pearson, J.E., Hoffmann, A., Freimuth, F., and Mokrousov, Y. (2014). Spin Hall effects in metallic antiferromagnets. *Phys. Rev. Lett.* 113: 196602.

25 Itoh, R., Takeuchi, Y., DuttaGupta, S., Fukami, S., and Ohno, H. (2019). Stack structure and temperature dependence of spin-orbit torques in heterostructures with antiferromagnetic PtMn. *Appl. Phys. Lett.* 115: 242404.

26 Frangou, L., Auffret, S., Vila, L., Gambarelli, S., and Baltz, V. (2016). Enhanced spin pumping efficiency in antiferromagnetic IrMn thin films around the magnetic phase transition. *Phys. Rev. Lett.* 116: 077203.

27 Brink, A.V.D., Vermijs, G., Solignac, A., Koo, J., Kohlhepp, J.T., Swagten, H.J.M., and Koopmans, B. (2016). Field-free magnetization reversal by spin-Hall effect and exchange bias. *Nat. Commun.* 7: 10854.

28 Fukami, S., Zhang, C., DuttaGupta, S., Kurenkov, A., and Ohno, H. (2016). Magnetization switching by spin-orbit torque in an antiferromagnet/ferromagnet bilayer system. *Nat. Mater.* 15: 535.

29 Lau, Y.C., Betto, D., Rode, K., Coey, J.M.D., and Stamenov, P. (2016). Spin-orbit torque switching without an external field using interlayer exchange coupling. *Nat. Nanotechnol.* 11: 758.

30 Park, B.G., Wunderlich, J., Martí, X., Holý, V., Kurosaki, Y. et al. (2011). A spin-valve-like magnetoresistance of an antiferromagnet-based tunnel junction. *Nat. Mater.* 10: 347.

31 Wang, Y.Y., Song, C., Cui, B., Wang, G.Y., Zeng, F., and Pan, F. (2012). Room-temperature perpendicular exchange coupling and tunneling anisotropic magnetoresistance in an antiferromagnet-based tunnel junction. *Phys. Rev. Lett.* 109: 137201.

32 Galceran, R., Fina, I., Fernández, J.C., Bozzo, B., Frontera, C. et al. (2016). Isothermal anisotropic magnetoresistance in antiferromagnetic metallic IrMn. *Sci. Rep.* 6: 35471.

33 Chen, H., Niu, Q., and MacDonald, A.H. (2014). Anomalous Hall effect arising from noncollinear antiferromagnetism. *Phys. Rev. Lett.* 112: 017205.

34 Železný, J., Gao, H., Výborný, K., Zemen, J., Mašek, J. et al. (2014). Relativistic Néel-order fields induced by electrical current in antiferromagnets. *Phys. Rev. Lett.* 113: 157201.

35 Meinert, M., Graulich, D., and Matalla-Wagner, T. (2018). Electrical switching of antiferromagnetic Mn_2Au and the role of thermal activation. *Phys. Rev. Appl.* 9: 064040.

36 Barthem, V.M.T.S., Colin, C.V., Mayaffre, H., Julien, M.H., and Givord, D. (2013). Revealing the properties of Mn_2Au for antiferromagnetic spintronics. *Nat. Commun.* 4: 2892.

37 Fawcett, E. (1988). Spin-density-wave antiferromagnetism in chromium. *Rev. Mod. Phys.* 60: 209.

38 Kim, K.Y., Hwang, Y.S., Park, J.G., Torikai, N., Takeda, M., Han, S.W., and Shin, S.C. (2007). Exchange bias and uncompensated spins in a Fe/Cr(100) bilayer. *Phys. Stat. Sol. (b)* 244: 4499.

39 Marti, X., Fina, I., Frontera, C., Liu, J., Wadley, P. et al. (2014). Room-temperature antiferromagnetic memory resistor. *Nat. Mater.* 13: 367.

40 Liu, Z.Q., Li, L., Gai, Z., Clarkson, J.D., Hsu, S.L. et al. (2016). Full electroresistance modulation in a mixed-phase metallic alloy. *Phys. Rev. Lett.* 116: 097203.

41 Chen, X.Z., Feng, J.F., Wang, Z.C., Zhang, J., Zhong, X.Y. et al. (2017). Tunneling anisotropic magnetoresistance driven by magnetic phase transition. *Nat. Commun.* 8: 449.

42 Tung, L.D., Buschow, K.H.J., Franse, J.J.M., and Thuy, N.P. (1996). Magnetic coupling between rare earth moments in some antiferromagnetic Gd compounds. *J. Magn. Magn. Mater.* 154: 96.

43 Svanidze, E., Wang, J.K., Besara, T., Liu, L., Huang, Q. et al. (2015). An itinerant antiferromagnetic metal without magnetic constituents. *Nat. Commun.* 6: 7701.

44 Brasse, M., Chioncel, L., Kuneš, J., Bauer, A., Regnat, A. et al. (2013). de Haas-van Alphen effect and Fermi surface properties of single crystal CrB_2. *Phys. Rev. B* 88: 155138.

45 Wang, H., Du, C., Hammel, P.C., and Yang, F. (2014). Antiferromagnetic spin transport from $Y_3Fe_5O_{12}$ into NiO. *Phys. Rev. Lett.* 113: 097202.

46 Wang, Y., Zhu, D., Yang, Y., Lee, K., Mishra, R. et al. (2019). Magnetization switching by magnon-mediated spin torque through an antiferromagnetic insulator. *Science* 366: 1125.

47 Seki, S., Ideue, T., Kubota, M., Kozuka, Y., Takagi, R. et al. (2015). Thermal generation of spin current in an antiferromagnet. *Phys. Rev. Lett.* 115: 266601.

48 Wang, K., Sanderink, J.G.M., Bolhuis, T., van der Wiel, W.G., and de Jong, M.P. (2015). Tunnelling anisotropic magnetoresistance due to antiferromagnetic CoO tunnel barriers. *Sci. Rep.* 5: 15498.

49 Qiu, Z., Li, J., Hou, D., Arenholz, E., N'Diaye, A.T. et al. (2016). Spin-current probe for phase transition in an insulator. *Nat. Commun.* 7: 12670.

50 Martin, L.W., Crane, S.P., Chu, Y.H., Holcomb, M.B., Gajek, M. et al. (2008). Multiferroics and magnetoelectrics: thin films and nanostructures. *J. Phys.: Condens. Matter* 20: 434220.

51 Sando, D., Barthélémy, A., and Bibes, M. (2014). $BiFeO_3$ epitaxial thin films and devices: past, present and future. *J. Phys.: Condens. Matter* 26: 473201.

52 Kosub, T., Kopte, M., Hühne, R., Appel, P., Shields, B. et al. (2017). Purely antiferromagnetic magnetoelectric random access memory. *Nat. Commun.* 8: 13985.

53 Wu, L., Li, C., Chen, M., Zhang, Y., Han, K. et al. (2017). Interface-induced enhancement of ferromagnetism in insulating $LaMnO_3$ ultrathin films. *ACS Appl. Mater. Interfaces* 9: 44931.

54 Takeda, T. and Ohara, S. (1974). Magnetic structure of the cubic perovskite type $SrMnO_3$. *J. Phys. Soc. Jpn.* 37: 275.

55 Rodriguez, E.E., Poineau, F., Llobet, A., Kennedy, B.J., Avdeev, M. et al. (2011). High temperature magnetic ordering in the 4*d* perovskite $SrTcO_3$. *Phys. Rev. Lett.* 106: 067201.

56 Coldea, R., Hayden, S.M., Aeppli, G., Perring, T.G., Frost, C.D. et al. (2001). Spin waves and electronic interactions in La_2CuO_4. *Phys. Rev. Lett.* 86: 5377.

57 Kimel, A.V., Kirilyuk, A., Tsvetkov, A., Pisarev, R.V., and Rasing, Th. (2004). Laser-induced ultrafast spin reorientation in the antiferromagnet $TmFeO_3$. *Nature* 429: 850.

58 Němec, P., Fiebig, M., Kampfrath, T., and Kimel, A.V. (2018). Antiferromagnetic opto-spintronics. *Nat. Phys.* 14: 229.

59 Jacobs, I.S. (1961). Spin-flopping in MnF_2 by high magnetic fields. *J. Appl. Phys.* 32: S61.

60 Jacobs, I.S. and Lawrence, P.E. (1967). Metamagnetic phase transitions and hysteresis in $FeCl_2$. *Phys. Rev.* 164: 866.

61 Hagiwara, M., Katsumata, K., Yamaguchi, H., Tokunaga, M., Yamada, I. et al. (1999). A complete frequency-field diagram for the antiferromagnetic resonance in MnF_2. *Int. J. Infrared Millimeter Waves.* 20: 617.

62 Wu, S.M., Zhang, W., Amit, KC, Borisov, P., Pearson, J.E. et al. (2016). Antiferromagnetic spin Seebeck effect. *Phys. Rev. Lett.* 116: 097204.

63 Rezende, S.M., Rodríguez-Suárez, R.L., and Azevedo, A. (2016). Theory of the spin Seebeck effect in antiferromagnets. *Phys. Rev. B* 93: 014425.

64 McGuire, M.A., Dixit, H., Cooper, V.R., and Sales, B.C. (2015). Coupling of crystal structure and magnetism in the layered, ferromagnetic insulator CrI_3. *Chem. Mater.* 27: 612.

65 Tsubokawa, I. (1960). On the magnetic properties of a $CrBr_3$ single crystal. *J. Phys. Soc. Jpn.* 15: 1664.

66 MacGuire, M.A., Clark, G., Santosh, KC, Chance, W.M., Jellison, G.E. et al. (2017). Magnetic behavior and spin-lattice coupling in cleavable van der Waals layered $CrCl_3$ crystals. *Phys. Rev. Mater.* 1: 014001.

67 Huang, B., Clark, G., Moratalla, E.N., Klein, D.R., Cheng, R. et al. (2017). Layer-dependent ferromagnetism in a van der Waals crystal down to the monolayer limit. *Nature* 546: 270.

68 Chen, W., Sun, Z., Wang, Z., Gu, L., Xu, X., Wu, S., and Gao, C. (2019). Direct observation of van der Waals stacking-dependent interlayer magnetism. *Science* 366: 983.

69 Kim, H.H., Yang, B., Li, S., Jiang, S., Jin, C. et al. (2019). Evolution of interlayer and intralayer magnetism in three atomically thin chromium trihalides. *Proc. Natl. Acad. Sci. USA* 116: 11131.

70 Giamarchi, T., Rüegg, C., and Tchernyshyov, O. (2008). Bose-Einstein condensation in magnetic insulators. *Nat. Phys.* 4: 198.

71 Han, T.H., Helton, J.S., Chu, S., Nocera, D.G., Rivera, J.A.R. et al. (2012). Fractionalized excitations in the spin-liquid state of a kogame-lattice antiferromagnet. *Nature* 492: 406.

72 Wadley, P., Howells, B., Železný, J., Andrews, C., Hills, V. et al. (2016). Electrical switching of an antiferromagnet. *Science* 351: 587.

73 Fina, I., Marti, X., Yi, D., Liu, J., Chu, J.H. et al. (2014). Anisotropic magnetoresistance in an antiferromagnetic semiconductor. *Nat. Commun.* 5: 4671.

74 Wang, H., Lu, C., Chen, J., Liu, Y., Yuan, S.L. et al. (2019). Giant anisotropic magnetoresistance and nonvolatile memory in canted antiferromagnet Sr_2IrO_4. *Nat. Commun.* 10: 2280.

75 Kriegner, D., Výborný, K., Olejník, K., Reichlová, H., Novák, V. et al. (2016). Multiple-stable anisotropic magnetoresistance memory in antiferromagnetic MnTe. *Nat. Commun.* 7: 11623.

76 Esmaeilzadeh, S., Halenius, U., and Valldor, M. (2006). Crystal growth, magnetic, and optical properties of the ternary nitride $MnSiN_2$. *Chem. Mater.* 18: 2713.

77 Nagaev, E.L. (1975). Ferromagnetic and antiferromagnetic semiconductors. *Soviet Physics Uspekhi* 18: 863.

78 Beleanu, A., Kiss, J., Kreiner, G., Köhler, C., Müchler, L. et al. (2013). Large resistivity change and phase transition in the antiferromagnetic semiconductors LiMnAs and LaOMnAs. *Phys. Rev. B* 88: 184429.

79 Šmejkal, L., Mokrousov, Y., Yan, B., and MacDonald, A.H. (2018). Topological antiferromagnetic spintronics. *Nat. Phys.* 14: 242.

80 Barker, J. and Tretiakov, O.A. (2016). Static and dynamical properties of antiferromagnetic skyrmions in the presence of applied current and temperature. *Phys. Rev. Lett.* 116: 147203.

81 Göbel, B., Mook, A., Henk, J., and Mertig, I. (2017). Antiferromagnetic skyrmion crystals: generation, topological Hall, and topological spin Hall effect. *Phys. Rev. B* 96: 060406.

82 Yang, H., Sun, Y., Zhang, Y., Shi, W., Parkin, S.S.P., and Yan, B. (2017). Topological Weyl semimetals in the chiral antiferromagnetic materials Mn_3Ge and Mn_3Sn. *New J. Phys.* 19: 015008.

83 Nakatsuji, S., Kiyohara, N., and Higo, T. (2015). Large anomalous Hall effect in a non-collinear antiferromagnet at room temperature. *Nature* 527: 212.

84 Tsai, H., Higo, T., Kondou, K., Nomoto, T., Sakai, A. et al. (2020). Electrical manipulation of a topological antiferromagnetic state. *Nature* 580: 608.

85 Nayak, A.K., Fischer, J.E., Sun, Y., Yan, B., Karel, J. et al. (2016). Large anomalous Hall effect driven by a nonvanishing Berry curvature in the noncollinear antiferromagnet Mn_3Ge. *Sci. Adv.* 2: e1501870.

86 Ogasawara, T., Kim, J., Ando, Y., and Hirohata, A. (2019). Structural and antiferromagnetic characterization of noncollinear $D0_{19}$ Mn_3Ge polycrystalline film. *J. Magn. Magn. Mater.* 473: 7.

87 Šmejkal, L., Železný, J., Sinova, J., and Jungwirth, T. (2017). Electric control of Dirac quasiparticles by spin-orbit torque in an antiferromagnet. *Phys. Rev. Lett.* 118: 106402.

88 Song, C., You, Y., Chen, X., Zhou, X., Wang, Y., and Pan, F. (2018). How to manipulate magnetic states of antiferromagnets. *Nanotechnology* 29: 112001.

89 Sapozhnik, A.A., Abrudan, R., Skourski, Y., Jourdan, M., Zabel, H. et al. (2017). Manipulation of antiferromagnetic domain distribution in Mn_2Au by ultrahigh magnetic fields and by strain. *Phys. Status Solidi* 11: 1600438.

90 Ohldag, H., Scholl, A., Nolting, F., Anders, S., Hillebrecht, F.U., and Stöhr, J. (2001). Spin reorientation at the antiferromagnetic NiO(001) surface in response to an adjacent ferromagnet. *Phys. Rev. Lett.* 86: 2878.

91 Roy, S., Fitzsimmons, M.R., Park, S., Dorn, M., Petracic, O. et al. (2005). Depth profile of uncompensated spins in an exchange bias system. *Phys. Rev. Lett.* 95: 047201.

92 Ijiri, Y., Schulthess, T.C., Borchers, J.A., van der Zaag, P.J., and Erwin, R.W. (2007). Link between perpendicular coupling and exchange biasing in Fe_3O_4/CoO multilayers. *Phys. Rev. Lett.* 99: 147201.

93 Kim, W., Jin, E., Wu, J., Park, J., Arenholz, E., Scholl, A., Hwang, C., and Qiu, Z.Q. (2010). Effect of NiO spin orientation on the magnetic anisotropy of the Fe film in epitaxially grown Fe/NiO/Ag(001) and Fe/NiO/MgO(001). *Phys. Rev. B* 81: 174416.

94 Wang, Y.Y., Song, C., Zhang, J.Y., and Pan, F. (2017). Role of an ultrathin platinum seed layer in antiferromagnet-based perpendicular exchange coupling and its electrical manipulation. *J. Magn. Magn. Mater.* 428: 431.

95 Wang, Y.Y., Song, C., Wang, G., Miao, J., Zeng, F., and Pan, F. (2014). Anti-ferromagnet controlled tunneling magnetoresistance. *Adv. Funct. Mater.* 24: 6806.

96 Wang, Y.Y., Song, C., Wang, G.Y., Zeng, F., and Pan, F. (2014). Evidence for asymmetric rotation of spins in antiferromagnetic exchange-spring. *New J. Phys.* 16: 123032.

97 Shick, A.B., Khmelevskyi, S., Mryasov, O.N., Wunderlich, J., and Jungwirth, T. (2010). Spin-orbit coupling induced anisotropy effects in bimetallic antiferromagnets: a route towards antiferromagnetic spintronics. *Phys. Rev. B* 81: 212409.

98 Zakharov, A.I., Kadomtseva, A.M., Levitin, R.Z., and Ponyatovskii, E.G. (1964). Magnetic and magnetoelastic properties of a metamagnetic iron-rhodium alloy. *Sov. Phys. JETP* 19: 1348.

99 Cherifi, R.O., Ivanovskaya, V., Phillips, L.C., Zobelli, A., Infante, I.C. et al. (2014). Electric-field control of magnetic order above room temperature. *Nat. Mater.* 13: 345.

100 Fina, I. and Fontcuberta, J. (2020). Strain and voltage control of magnetic and electric properties of FeRh films. *J. Phys. D: Appl. Phys.* 53: 023002.

101 Manz, S., Matsubara, M., Lottermoser, T., Büchi, J., Iyama, A. et al. (2016). Reversible optical switching of antiferromagnetism in $TbMnO_3$. *Nat. Photonics* 10: 653.

102 Fiebig, M., Phuc, N., Satoh, T., Van Aken, B.B., Miyano, K. et al. (2008). Ultrafast magnetization dynamics of antiferromagnetic compounds. *J. Phys. D: Appl. Phys.* 41: 164005.

103 Satoh, T., Van Aken, B.B., Duong, N.P., Lottermoser, T., and Fiebig, M. (2007). Ultrafast spin and lattice dynamics in antiferromagnetic Cr_2O_3. *Phys. Rev. B* 75: 155406.

104 Duong, N.P., Satoh, T., and Fiebig, M. (2004). Ultrafast manipulation of antiferromagnetism of NiO. *Phys. Rev. Lett.* 93: 117402.

105 Kimel, A.V., Ivanov, B.A., Pisarev, R.V., Usachev, P.A., Kirilyuk, A., and Rasing, Th. (2009). Inertial-driven spin switching in antiferromagnets. *Nat. Phys.* 5: 727.

106 Afanasiev, D., Ivanov, B.A., Kirilyuk, A., Rasing, T., Pisarev, R.V., and Kimel, A.V. (2016). Control of the ultrafast photo induced magnetization across the morin transition in $DyFeO_3$. *Phys. Rev. Lett.* 116: 097401.

107 Kampfrath, T., Sell, A., Klatt, G., Pashkin, A., Mährlein, S. et al. (2011). Coherent terahertz control of antiferromagnetic spin waves. *Nat. Photonics* 5: 31.

108 Kirilyuk, A., Kimel, A.V., and Rasing, T. (2010). Ultrafast optical manipulation of magnetic order. *Rev. Mod. Phys.* 82: 2731.

109 Cheng, R., Xiao, D., and Brataas, A. (2016). Terahertz antiferromagnetic spin Hall nano-oscillator. *Phys. Rev. Lett.* 116: 207603.

110 Song, C., Cui, B., Peng, J.J., Mao, H.J., and Pan, F. (2016). Electrical control of magnetism in oxides. *Chin. Phys. B* 25: 067502.

111 Fina, I., and Marti, X. (2017). Electric control of antiferromagnets. *IEEE Trans. Magn.* 53: 2500107.

112 Fina, I., Fàbrega, L., Marti, X., Sánchez, F., and Fontcuberta, J. (2011). Chiral domains in cycloidal multiferroic thin films: switching and memory effects. *Phys. Rev. Lett.* 107: 257601.

113 Fina, I., Skumryev, V., Flynn, D.O., Balakrishnan, G., and Fontcuberta, J. (2013). Phase coexistence and magnetically tuneable polarization in cycloidal multiferroics. *Phys. Rev. B* 88: 100403.

114 Zhao, T., Scholl, A., Zavaliche, F., Lee, K., Barry, M. et al. (2006). Electrical control of antiferromagnetic domains in multiferroic $BiFeO_3$ films at room temperature. *Nat. Mater.* 5: 823.

115 He, Q., Yeh, C.H., Yang, J.C., Singh-Bhalla, G., Liang, C.W. et al. (2012). Magnetotransport at domain walls in $BiFeO_3$. *Phys. Rev. Lett.* 108: 067203.

116 He, X., Wang, Y., Wu, N., Caruso, A.N., Vescovo, E. et al. (2010). Robust isothermal electric control of exchange bias at room temperature. *Nat. Mater.* 9: 579.

117 Echtenkamp, W., Street, M., Mahmood, A., and Binek, C. (2017). Tuning the effective anisotropy in a voltage-susceptible exchange-bias heterosystem. *Phys. Rev. Appl.* 7: 034015.

118 Borisov, P., Hochstrat, A., Chen, X., Kleemann, W., and Binek, C. (2005). Magnetoelectric switching of exchange bias. *Phys. Rev. Lett.* 94: 117203.

119 Wang, Y.Y., Zhou, X., Song, C., Yan, Y., Zhou, S., Wang, G., Chen, C., Zeng, F., and Pan, F. (2015). Electrical control of the exchange spring in antiferromagnetic metals. *Adv. Mater.* 27: 3196.

120 Diez, L.H., Mantel, A.B., Vila, L., Warty, P., Ono, S., Givord, D., and Ranno, L. (2014). Electric-field assisted depinning and nucleation of magnetic domain walls in $FePt/Al_2O_3$/liquid gate structures. *Appl. Phys. Lett.* 104: 082413.

121 Liu, X.H., Edmonds, K.W., Zhou, Z.P., and Wang, K.Y. (2020). Tuning interfacial spins in antiferromagnetic-ferromagnetic-heavy-metal heterostructures via spin-orbit torque. *Phys. Rev. Appl.* 13: 014059.

122 Bodnar, S.Y., Šmejkal, L., Turek, I., Jungwirth, T., Gomonay, O. et al. (2018). Writing and reading antiferromagnetic Mn_2Au by Néel spin-orbit torques and large anisotropic magnetoresistance. *Nat. Commun.* 9: 348.

123 Lin, P.H., Yang, B.Y., Tsai, M.H., Chen, P.C., Huang, K.F., Lin, H.H., and Lai, C.H. (2019). Manipulating exchange bias by spin-orbit torque. *Nat. Mater.* 18: 335.

124 Zhou, Z.P., Liu, X.H., and Wang, K.Y. (2020). Controlling vertical magnetization shift by spin-orbit torque in ferromagnetic/antiferromagnetic/ferromagnetic heterostructure. *App. Phys. Lett.* 116: 062403.

125 Park, J., Lee, G., Wolff-Fabris, F., Koh, Y.Y., Eom, M.J. et al. (2011). Anisotropic Dirac fermions in a Bi square net of $SrMnBi_2$. *Phys. Rev. Lett.* 107: 126402.

126 Wang, K., Graf, D., Lei, H., Tozer, S.W., and Petrovic, C. (2011). Quantum transport of two-dimensional Dirac fermions in $SrMnBi_2$. *Phys. Rev. B* 84: 220401.

127 Masuda, H., Sakai, H., Tokunaga, M., Yamasaki, Y., Miyake, A. et al. (2016). Quantum Hall effect in a bulk antiferromagnet $EuMnBi_2$ with magnetically confined two-dimensional Dirac fermions. *Sci. Adv.* 2: e1501117.

128 Richard, P., Nakayama, K., Sato, T., Neupane, M., Xu, Y.M. et al. (2010). Observation of Dirac cone electronic dispersion in $BaFe_2As_2$. *Phys. Rev. Lett.* 104: 137001.

129 Burkov, A.A. (2016). Topological semimetals. *Nat. Mater.* 15: 1145.

130 Armitage, N.P., Mele, E.J., and Vishwanath, A. (2018). Weyl and Dirac semimetals in three-dimensional solids. *Rev. Mod. Phys.* 90: 015001.

131 Tang, P., Zhou, Q., Xu, G., and Zhang, S.C. (2016). Dirac fermions in an antiferromagnetic semimetal. *Nat. Phys.* 12: 1100.

132 Yan, B., and Felser, C. (2017). Topological materials: Weyl semimetals. *Annu. Rev. Condens. Matter Phys.* 8: 337.

133 Kuroda, K., Tomita, T., Suzuki, M.T., Bareille, C., Nugroho, A.A. et al. (2017). Evidence for magnetic Weyl fermions in a correlated metal. *Nat. Mater.* 16: 1090.

134 Jin, C., Song, C., Wang, J., and Liu, Q. (2016). Dynamics of antiferromagnetic skyrmion driven by the spin Hall effect. *Appl. Phys. Lett.* 109: 182404.

135 Zhang, S., Baker, A.A., Komineas, S., and Hesjedal, T. (2015). Topological computation based on direct magnetic logic communication. *Sci. Rep.* 5: 15773.

136 Xue, Q.K. (2011). Nanoelectronics: a topological twist for transistors. *Nat. Nanotechnol.* 6: 197.

137 Yin, G., Liu, Y., Barlas, Y., Zang, J., and Lake, R.K. (2015). Topological spin Hall effect resulting from magnetic skyrmions. *Phys. Rev. B* 92: 024411.

138 Soumyanarayanan, A., Reyren, N., Fert, A., and Panagopoulos, C. (2016). Emergent phenomena induced by sipn-orbit coupling at surfaces and interfaces. *Nature* 539: 509.

139 Zhang, X., Zhou, Y., and Ezawa, M. (2016). Magnetic bilayer-skyrmions without skyrmion Hall effect. *Nat. Commun.* 7: 10293.

140 Legrand, W., Maccariello, D., Ajejas, F., Collin, S., Vecchiola, A. et al. (2020). Room-temperature stabilization of antiferromagnetic skyrmions in synthetic antiferromagnets. *Nat. Mater.* 19: 34.

141 Fert, A., Cros, V., and Sampaio, J. (2013). Skyrmions on the track. *Nat. Nanotechnol.* 8: 152.

142 Tsai, W.F. and Lin, H. (2016). Topological insulators and superconductivity: the integrity of two sides. *Nat. Mater.* 15: 927.

143 Fu, L. and Kane, C.L. (2008). Superconducting proximity effect and Majorana fermions at the surface of a topological insulator. *Phys. Rev. Lett.* 100: 096407.

12

Prospects

Meiyin Yang[1] *and Kaiyou Wang*[2,3,4]

[1]*Key Laboratory of Microelectronic Devices and Integrated Technology, Institute of Microelectronics, Chinese Academy of Sciences, Beijing, China*
[2]*State Key Laboratory of Superlattices and Microstructures, Institute of Semiconductors, Chinese Academy of Sciences, Beijing, China*
[3]*College of Materials Science & Opto-Electronic Technology and Center for Excellence in Topological Quantum Computation, University of Chinese Academy of Science, Beijing, China*
[4]*Beijing Academy of Quantum Information Sciences, Beijing, China*

Spintronics deals with spin and interactions between spin and others, such as electron charge, photon, orbit, magnetic field, and electric current etc. As a result, a new degree of freedom can be utilized compared with conventional charged-based electronics. Since 1960s, spintronics study involves a broad range of materials,including ferromagnetic and antiferromagnetic metals, semimetals, magnetic semiconductors, insulators, topological insulators, multiferroic materials in bulk, thin filmand one/two dimensional systems etc.. Besides these magnetic materials, non-magnetic materials, such as heavy metals, hetero-structure and van der Waals layered materials,which exhibit large spin and charge conversion ability, are also regarded as the cutting-edge research. Spin injection, detection and manipulation have been demonstrated in these materials as well as in various devices such as anisotropy magnetoresistance devices, spin valves, magnetic tunneling junctions (MTJ). Many interesting physics and phenomenon have been discovered, such as (quantized) spin Hall effect, giant magnetoresistance, and tunneling magnetoresistance etc. Some spintronic devices have been commercialized, which significantly increased the density and speed of storage and profoundly impacted the information industry. It will promisingly encompass a variety of application in future due to the fast, energy efficient and robust characteristics of spin-based materials and devices. As an emerging research field, we expect the field of spintronics will experience a rapid development due to the increasing interesting both from fundamental research as well industrial applications.

Driven by the practical demand for future intelligent society, fast and energy efficient spintronics materials, devices and relative new physics will attract more attention and become important than ever before. Pico-second electrical pulses and femtosecond light pulses have been successfully to control the spin of magnetic materials. Electrical way to control the spins are more suitable for large scale circuit manufacture. Thus, exploring new materials that is manipulated by electrical means for terahertz range are fascinating for this field. Antiferromagnetic materials which exhibits terahertz precession frequency are promising candidates. Spin precession takes minimum times, which limits the spin switching speed. Alternatively, other methods, including controlling the ferromagnetic and antiferromagnetic phase transition, could be solutions to push the speed out of the limits confined by spin precession.

Pursuing the faster spin manipulation often accompanies the dramatically increasing energy consumption, which is another crucial topic for spintronics. Looking for new

Spintronics: Materials, Devices and Applications, First Edition. Edited by Kaiyou Wang, Meiyin Yang, and Jun Luo.

materials aimed to enhance the energy efficiency are key solutions. Materials with high charge to spin conversion efficiency, including topological insulator, transition metal chalcogenide (TMD)/FM, Dirac semimetal, and Weyl semimetal, are under extensive research. Some options are very promising. Obtaining high quality with large scale manufacture compatibility is still aquestion for these new materials. Furthermore, new magnetic materials and materials engineering may lead important scientific discoveries and fruitful physics.

The applications of the spintronics materials are examined in various devices, which can be fallen into two main categories, sensors and memories, according to their different functions. The spintronic sensors, which transfer magnetic field to electric signals by the effect of magnetoresistance, are very popular and demanded in the market. They are broadly used in position, angle detection and geomagnetic navigation. Recently, researchers have demonstrated the biomedical spintronic sensors, but the success application of biomedical spintronics devices also relies on the scientific and technological development of biography and medical science. In future, the spintronic sensors with much higher sensitivity are still under development to detect more weaker magnetic fields from brain and neural cells, which are expected to significantly improve the human heathy for early disease diagnose, assistant treatment and brain computer interface.

Another important spintronics devices are for information storage, which often adopt MTJ as a unit for magnetic random access memory (MRAM). The devices must be fit and ready for another surge in technological innovation, such as the rapid development of virtual reality and artificial intelligences etc., which will affect every corner of the society. To support this, we need find solutions to efficiently process, store, and transport massive data. Thus high speed, low energy consumption and high density memories devices are highly demanded, where spintronic devices can make a difference. Nowadays, STT-MRAM has already been commercialized, which is used in aerospace vehicles and smart devices. To replace SRAM and DRAM, the industry requires MRAM to process low writing current, fast writing speed, high endurance and large capacity to fit for more advanced CMOS technology node. However, due to the limitation of the endurance and capacity for the commercialized MRAM, it only occupies a very small portion of memory markets currently. To broaden the use of MRAM in different scenarios, efforts have been put for both STT-MRAM and SOT-MRAM for low energy consumption and high endurance, and we still have large room to improve. Methods, including adopting voltage-controlled magnetic anisotropy effect, looking for high spin polarization and low damping materials and designing novel device structures have been investigated, and large progresses are on going.

SOT-MRAM has faster and higher endurance than STT-MRAM at the cost of larger volume per bit. It is a promising technique not only in the cache memory for the replacement of SRAM, but also in future logic-in-memory, where data transfer is avoided. CMOS compatible methods to solve the free-field switching in high density SOT-MRAM are still under investigation. In addition, concatenability is another issue for SOT-MRAM based logic-in-memory, as the results of the one calculation cannot participate the next calculation, which means the data still require to be transferred. However, for the memory in artificial intelligence network, where calculations only relies on simple multiplication and accumulation, SOT-MRAM has its advantageous due to low power consumption and high speed.

In addition, finding novel applications for spintronic devices is another direction. Other than to replace the memories in the market, e.g., SRAM and DRAM, spintrioncs devices are candidates for future non-conventional calculation architecture hardware. Utilizing its stochastic nature, which is a disadvantage for memory and logic for conventional calculation, spintronics device can be a basis unit for probabilistic computing, such as quantum and neuromorphic computing. It also can be used in the memory which does not require high write error rate, e.g. block chain technology. In these new scenarios, spintronics show supreme properties in terms of speed, energy consumption and endurance.

Other topics for spintronics materials and devices, including magnetic semiconductors, skymions, magneto-electric coupling, spin nano-oscillator, and antiferromagnets, are most in the area of fundamental research rather than industry development. The application of such areas will be expected to take a longer time. For example, magnetic semiconductors still requires to rise its Curie temperature. Skyrmions are difficult in methods to steady generation, detection and control of. Spin nano-oscillator needs to improve in terms of the high power output, low bandwidth and low energy consumptions. However, the physics of these areas including spin and charger conversion, spin dynamics in particular systems, and spin textures, are very fascinating and attracts intensive attention. These areas of research have broadened our vision in terms of spin-related phenomena, and they will potentially make a real change to human life in future.We believe that new opportunities for spintronic materials and devices are drawing great interests from both scientific and industrial perspectives, paving the way for more ambitious and futuristic applications.

Index

Note: Page numbers followed by "*f*", "*t*" and "*b*" indicate figures, tables and boxes, respectively.

Spintronics: Materials, Devices and Applications, First Edition. Edited by Kaiyou Wang, Meiyin Yang, and Jun Luo.

n

Printed and bound by CPI Group (UK) Ltd, Croydon, CR0 4YY
17/09/2026
14964269-0003